Springer Proceedings in Mathematics & Statistics

Volume 400

This book series features volumes composed of selected contributions from workshops and conferences in all areas of current research in mathematics and statistics, including data science, operations research and optimization. In addition to an overall evaluation of the interest, scientific quality, and timeliness of each proposal at the hands of the publisher, individual contributions are all refereed to the high quality standards of leading journals in the field. Thus, this series provides the research community with well-edited, authoritative reports on developments in the most exciting areas of mathematical and statistical research today.

Victor Manuel López Sánchez •
Francisco Gaudêncio Mendonça Freires •
João Carlos Gonçalves dos Reis •
Joana Maria Costa Martins das Dores
Editors

Industrial Engineering and Operations Management

XXVIII IJCIEOM, Mexico City, Mexico, July 17–20, 2022

Springer

Editors

Victor Manuel López Sánchez
Engineering
Universidad Anáhuac
Anáhuac Huixquilucan, Mexico

Francisco Gaudêncio Mendonça Freires
Mechanical Engineering
Federal University of Bahia
Salvador, Bahia, Brazil

João Carlos Gonçalves dos Reis
Industrial Engineering and Management
Universidade Lusófona
Lisbon, Portugal

Joana Maria Costa Martins das Dores
DEGEIT
University of Aveiro
Aveiro, Portugal

ISSN 2194-1009 ISSN 2194-1017 (electronic)
Springer Proceedings in Mathematics & Statistics
ISBN 978-3-031-14762-3 ISBN 978-3-031-14763-0 (eBook)
https://doi.org/10.1007/978-3-031-14763-0

Mathematics Subject Classification: 90-XX, 90-06, 90-10, 90Bxx

This Springer imprint is published by the registered company Springer Nature Switzerland AG
The registered company address is: Gewerbestrasse 11, 6330 Cham, Switzerland

Preface

Introduction

This year, the 28th International Joint Conference on Industrial Engineering and Operations Management (IJCIEOM) focuses on the contributions to business models design and innovation. The IJCIEOM 2022 took place at Universidad Anáhuac Mexico, between July 17 and 20, in Mexico City, Mexico. The IJCIEOM is a result of an agreement between five industrial engineering Associations, namely ABEPRO (Brazilian Association of Production Engineering), ADINGOR (Asociación para el Desarrollo de la Ingeniería de Organización), IISE (Institute of Industrial and Systems Engineers), AIM (European Academy for Industrial Management), and ASEM (American Society for Engineering Management). Over the years, these organizations have done an outstanding work in bringing together industrial engineering researchers and perspectives. This joint conference has a long tradition of strengthening ties between students in industrial engineering and operations management, researchers, professionals, and academics from around the world. Since its early days, the goal has been to stimulate the networking and the sharing of new and exciting knowledge. Thus, following the conference series in Industrial Engineering and Operations Management (e.g., [1–3]), part of the book series of Springer Proceedings in Mathematics & Statistics (PROMS), we look forward to continuing to support the entire academic community with insightful publications.

Book Overview

Business models design and innovation has been chosen due its centrality to industrial engineering and operations management in designing organizational architecture and value creation. As modern societies are learning to reinvent themselves based on sustainable and digital practices, industrial engineering and operations management is making a remarkable contribution to the advancement of manufacturing and service production. Due to COVID-19 and regional wars that affect the entire globe, organizations are forced to operate in volatile, uncertain, complex, and ambiguous (VUCA) environments. To meet this need, industrial engineering in specific areas, such as operations management, has been called upon to face the challenges of modern production systems that must align the management of limited resources with the needs of society. In light of the above, industrial engineers are primarily responsible for incorporating innovative concepts and disruptive technologies into product-service operations. In the forthcoming years, demographic drivers, including urbanization and the growth in world population, will put the next generation of industrial engineers under great pressure concerning the management of resources and the modernization of sustainable processes. It is for this reason that there is a need to prepare new industrial engineers to face a quite uncertain and ambiguous future.

Final Remarks

Looking ahead, this conference aims to combine theory and practice of social, economic, and environmental aspects through the lens of new business models. To achieve that purpose, the 28th IJCIEOM brought together scientists from all over the world, as it received around 116 submissions, from 112 institutions and 12 different countries. Of the total number of articles submitted, only a third of the papers were selected for publication in PROMS. This edition focused on 5 most relevant areas, within 25 possible ones, namely (1) operations and process management, (2) sustainability, (3) healthcare operations management, (4) strategy and organizational engineering, and (5) logistics and supply chain management.

Acknowledgments

The editors of this book of proceedings are grateful for the support of the entire industrial engineering and operations management community. Special thanks also to the anonymous reviewers, who worked hard to review all manuscripts assigned to them in a timely manner. Last but not least, our publishers who have been tireless in managing the process of publishing the best IJCIEOM papers, bringing knowledge to the entire scientific community.

References

1. Reis, J., Pinelas, S., Melão, N. (eds.): Industrial engineering and operations management I. In: IJCIEOM 2018: Springer Proceedings in Mathematics & Statistics, vol. 280. Springer, Cham (2019). https://doi.org/10.1007/978-3-030-14969-7
2. Thomé, A., Barbastefano, R., Scavarda, L., Reis, J., Amorim, M. (eds.): Industrial engineering and operations management. In: IJCIEOM 2020: Springer Proceedings in Mathematics & Statistics, vol. 337. Springer, Cham (2020). https://doi.org/10.1007/978-3-030-56920-4
3. Thomé, A., Barbastefano, R., Scavarda, L., Reis, J., Amorim, M. (eds.): Industrial engineering and operations management. In: IJCIEOM 2021: Springer Proceedings in Mathematics & Statistics, vol. 367. Springer, Cham (2021). https://doi.org/10.1007/978-3-030-78570-3

Contents

Eco-efficiency Analysis of the Brazilian Public Transport Based on Data Envelopment Analysis

Wagner Rui Correa Olaya, Victor Welder Santiago Araújo, Nathália Jucá Monteiro, Andreia Zanella, and Renata Oliveira

Abstract This study reports the eco-efficiency quantification of urban public transportation in Brazilian states. The modal focused on is urban bus. A composite indicator was specified to meet the premises of Sustainable Consumption and Production (SC&P) in the urban planning domain. This paper can be considered applied research that considers the Benefit of the doubt (BoD) approach in Data Envelopment Analysis (DEA) models. DEA is non-parametric a technique recommended by the Organization for Economic Co-operation and Development (OECD) for performance assessments using composite indicators (CI). The scope of this research encompasses 26 capitals of Brazilian states and the capital of the Federal District. This paper also reports the framework of Key Performance Indicators (KPIs) reflecting urban transportation system in Brazil. The results using data of 2019 showed that only the Brazilian Federal District was considered eco-efficient whilst more economically developed and populous states showed undesirable levels of CO_2 emissions (e.g. Sao Paulo and Rio de Janeiro).

Keywords Data Envelopment Analysis · Eco-efficiency · Composite Indicator · Urban Transportation

W. R. C. Olaya · V. W. S. Araújo
Universidade do Estado do Para (UEPA), Belém, PA, Brazil
e-mail: renata.oliveira@uepa.br

N. J. Monteiro
Universidade do Estado do Para (UEPA), Belém, PA, Brazil

Pontifical Catholic University of Parana (PUC-PR), Curitiba, PR, Brazil

A. Zanella
Universidade Federal de Santa Catarina (UFSC), Florianópolis, SC, Brazil

R. Oliveira (✉)
Universidade do Estado do Para (UEPA), Belém, PA, Brazil

Conselho Regional de Agronomia e Engenharia do Para (CREA-Pa), Belem, PA, Brazil
e-mail: renata.oliveira@uepa.br

V. M. López Sánchez et al. (eds.), *Industrial Engineering and Operations Management*, Springer Proceedings in Mathematics & Statistics 400,
https://doi.org/10.1007/978-3-031-14763-0_1

1 Introduction

Eco-friendly urban development and Eco-efficient public transportation solutions are considered a major challenge in the context of Agenda 2030 [1]. On the one hand, the challenge to reduce air emissions rates is particularly relevant to fighting climate change. On the other hand, there is a policy toward individual fuel-based vehicles instead of collective urban transportation in Latin America.

A few Latin American countries reported particularly high rates of air emissions in recent years. For instance, Brazil occupied the sixth position in the ranking of countries with the highest emission of carbon dioxide in 2016. This performance is strongly related to the emissions generated by urban road transportation in this country. Road transportation is responsible for approximately 94% of the Brazilian energy consumption [2] and also for 90% of CO_2 emissions in cities. It is worth noting that 70% of CO_2 emissions come from private cars, 13% come from collective road transport, and 7% come from motorcycles [3].

The quantification of the eco-efficiency of urban transportation from a quantitative perspective has become more relevant to the assessment of public policies in large cities. For instance, decisions taken according to the current Brazilian transportation policy resulted in the increase of the private fuels-based vehicles fleet and the consequent increase in air emissions. For instance, the Institute of Applied Economic Research (IPEA) [2] registered an increase of 50% in the road network of the cities that was installed between the years 1997 and 2011. In the opposite direction, a range of countries in Europe managed to reduce their CO_2 emissions during the same period. This is due to their formulation of policies oriented toward the use of public transportation. And these nations also prioritized policies for the purchase of vehicles powered by clean energy [4].

Considering the context presented, monitoring the eco-efficiency of urban transportation using key performance indicators (KPIs) recommended in the literature can support the decision-making of public policies toward a more sustainable future. Therefore, the main objective of this study is to specify a composite indicator (CI) to quantify the eco-efficiency of the bus urban transportation system in 27 Brazilian states. The selected KPIs reflect inputs and outputs of criteria such as air emissions; the number of service users and the consumption of fuels. The composite indicator developed is estimated using an optimization model based on Data Envelopment Analysis (DEA) [5] and it follows the Benefit of the Doubt (BoD) approach [6]. The use of the BoD approach is recommended in the literature by The Organization of Economic Country Development (OECD) [7] and there is a body of research using DEA to quantify the eco-efficiency of a range of sectors [8–12] (e.g., public transportation, mining, cities, countries).

This article is structured as follows. Section 2 reports a literature review on Data Envelopment Analysis. Section 3 contains the methodology of this study and describes the procedures to select and use the data. Section 4 discusses the results found and the insights extracted. Finally, Sect. 5 presents the conclusions attained and it foresees opportunities for subsequent research.

2 Data Envelopment Analysis

The literature in the field of performance assessment is dense and considered multidisciplinary. This paper focuses on the evaluation of relative efficiency using the non-parametric frontier method named Data Envelopment Analysis (DEA).

DEA enables the estimation of relative efficiency using multiple inputs and outputs. It is based on a comparison among a homogeneous set of Decision Making Units (DMUs). This method was introduced by Charnes et al. [5]. These three authors operationalized the propositions of Farrell [13] using linear programming to solve deterministic problems of performance assessment.

The contribution of Charnes et al. [5] considers a performance assessment of n DMUs ($j = 1, \ldots, n$), each consuming inputs x_{ij} to produce outputs y_{rj}. The DMU under assessment k (k = 1, .., n) consumes inputs x_{ik} ($i = 1, \ldots, m$) to produce outputs y_{rk} ($r = 1, \ldots, r$). The relative efficiency scores (θ_k) ranges between zero and one. When $\theta_k = 1$, the DMU is considered efficient, and when $\theta_k < 1$, the DMU k is considered inefficient.

Formulation (1) reports an input-oriented DEA formulation assuming Constant Returns to Scale (CRS) [5].

$$\begin{aligned} \theta_k : \quad & \max \sum_{r=1}^{s} u_r y_{rk} \\ & \text{s.t.} \\ & \sum_{r=1}^{s} u_r y_{rj} - \sum_{i=1}^{s} v_i x_{ij} \leq 0 \qquad j = 1, \ldots, n \\ & \sum_{i=1}^{m} v_i x_{ik} = 1 \qquad j = 1, \ldots, n \\ & u_r \geq \varepsilon, \qquad r = 1, \ldots, s \\ & v_i \geq \varepsilon, \qquad i = 1, \ldots, m \end{aligned} \tag{1}$$

For each DMU under assessment, an individual set of weights v_i and u_r is chosen to evaluate its efficiency in the best possible light. These weights are the decision variables of the problem. ε is a mathematical infinitesimal ensuring that the weights are strictly positive so that all inputs and outputs are taken into account in the evaluation.

A by-product of the optimization is the quantification of target values based on the convex combination of the inputs and output levels of the benchmark DMUs. These targets can guide inefficient DMUs to reach efficient levels of inputs or outputs. Therefore, assessments based on DEA allow discrimination between efficient and inefficient entities while promoting feasible improvements for efficient operational performance.

Performance assessments based on DEA became popular in a range of applications. The popularity of this technique can be explained by some features. They are discussed in the next paragraph.

The efficiency scores are calculated from individual real observations. DEA can simultaneously account for multiple outputs and multiple inputs without requiring unit transformation or normalization procedures. Since the weights are the decision variables of the optimization, DEA models do not require prior specification of weights for inputs or outputs. Also, the DEA-based models work under assumptions of the input and output relationship giving the returns to scale. Finally, the use of DEA can accommodate value judgments of the decision when required.

There is a body of research dedicated to performance assessment using DEA. Both methodological refinements and applications in many fields can be traced in the literature since the 1970s. For instance, Tavares [14], Gattoufi et al. [15], Emrouznejad and Yang [16], and Zhou et al. [17] reported bibliometric analysis covering the period between 1978 and 2018.

In this context, the specification of composite indicators (CIs) based on DEA has become popular since the 2000s in a variety of performance evaluations (e.g., Environmental Performance Index, Human Development Index). A CI is an aggregation of a set of sub-indicators into a single performance measure.

The Organisation for Economic Co-operation and Development (OECD) recommends a range of techniques for the choice of adequate weights for aggregating the CI. One of the OECD's recommendations is the use of DEA for reducing subjectivity in the choice of weights [7].

The estimation of the CI of assessing the eco-efficiency of public transportation using DEA is presented in the methodology section of this paper.

3 Methodology

3.1 Eco-efficiency Composite Indicator

The DEA-based Composite Indicator (CI) to quantify the Eco-efficiency of urban transportation was specified as follows The CI aggregates a range of KPIs representing multiple dimensions of the eco-efficiency. The bus lines are the homogeneous set of Decision Making-Units (DMUs).

The idea of a DEA-based Composite Indicator was introduced by Cook et al. [18]. Years later, the work of Cherchye et al. [6] popularized this concept. These authors introduced a DEA-based model of the relatively easy application from a mathematical point of view. This approach was called the "Benefit-of-the-doubt (BoD) composite indicator". The BoD means that the weights of each KPI are assigned in the optimization endogenously for each DMU under evaluation.

Therefore, every DMU is evaluated with an individual set of weights that puts it in the best light. Formulation (2) was specified according to the BoD principles [6, 14]. Model (2) is equivalent to the DEA model introduced by Charnes et al. [19]. Thus, formulation (2) is input-oriented and assumes constant returns to scale.

$$
\begin{aligned}
CI_k:\ \max \sum_{r=1}^{s} u_r Y_{rk} & \\
\text{s.t.} \sum_{r=1}^{s} u_r Y_{rk} \leq 1 \qquad & j = 1, \ \dots \ , \ n \\
u_r \geq \varepsilon, \qquad & r = 1, \ \dots \ , \ s
\end{aligned}
\tag{2}
$$

Model (2) is equivalent to the input-oriented model under CRS, with a unitary input assigned to the DMUs under assessment. Y_r are the output KPIs be expanded. $j\ (j = 1, \dots, n)$ is the homogeneous set of DMUs. The weights u_r (r = 1, ..., s) are the decision variables that give the objective function eco-efficiency score (CI_k^*) for each DMU k under assessment. The eco-efficiency scores CI_k^* range from 0 (bottom performance) to 1 (top performance). ε is a mathematical infinitesimal ensuring that the weights u_r are strictly positive so that all KPIs are taken into account in the evaluation.

Formulation (2) follows the proposition of Cook [20]. This author proposed that CI_k^* is the optimized aggregation of outputs assuming all DMUs are similar in terms of inputs. Hence, the inputs of all DMUs assessed in (2) are a dummy variable equal to one. Since all inputs are unitary, this dummy variable can be interpreted as a "helmsman" attempting to steer the DMUs toward the maximization of outputs.

This type of DEA-based CI is unit invariant, which makes the normalization constraint redundant [21]. The applications of DEA based on this approach dedicate special attention to the selection of meaningful KPIs.

4 Results

4.1 KPIs Framework Specification

Three key performance indicators (KPIs) were specified to quantify the eco-efficiency of urban public transportation. KPIs are performance indicators based on the proportional relationship between two measures, giving performance scores that are unit-free [22]. Although KPIs are frequently formulated by comparing outputs and inputs, it is possible to compare only outputs or only inputs.

This eco-efficiency framework for public transportation selected KPIs to represent eco-efficiency criteria such as climate change, fleet efficiency, and public policies.

Table 1 reports the KPIs developed for this study. The framework presented in Table 1 gives scores ranging between 0 and 1 in the dataset. The adoption of ratio indicators has the additional advantage of facilitating the performance comparison between states with high population numbers and small populations with low population numbers.

Table 2 reports the KPI values for each federation unit. The calculation of the KPIs was effectuated after data collection procedures. The base year was 2019.

Table 1 Framework

Nature	Description	KPI
Undesirable output	Environmental burden	Fleet air emissions / 100,000 Inhab (Y_1)
Desirable output	The social benefit of mobility	Fleet size / 100,000 Inhas (Y_2)
Desirable output	The social benefit of responsible driving	Professional drivers licensed /100,000 Inhab (Y_3)

Table 2 Dataset of KPIs 2019

DMU	State	Y_1	Y_2	Y_3
1	Acre	−0.039	0.139	0.271
2	Alagoas	−0.037	0.250	0.167
3	Amapa	−0.041	0.147	0.189
4	Amazonas	−0.039	0.224	0.157
5	Bahia	−0.074	0.280	0.187
6	Ceara	−0.039	0.201	0.229
7	Federal District	−0.044	0.442	0.535
8	Espírito Santo	−0.088	0.380	0.364
9	Goiás	−0.139	0.339	0.364
10	Maranhao	−0.051	0.137	0.121
11	Mato Grosso	−0.290	0.361	0.363
12	Mato Grosso do Sul	−0.172	0.360	0.389
13	Minas Gerais	−0.111	0.382	0.339
14	Para	−0.091	0.223	0.158
15	Paraíba	−0.038	0.192	0.204
16	Parana	−0.171	0,400	0.440
17	Pernambuco	−0.051	0.216	0.215
18	Piaui	−0.058	0.253	0.174
19	Rio de Janeiro	−0.046	0.258	0.330
20	Rio Grande do Norte	−0.045	0.216	0.228
21	Rio Grande do Sul	−0.114	0.371	0.437
22	Rondoônia	−0.160	0.358	0.378
23	Roraima	−0.051	0.201	0.243
24	Santa Catarina	−0.130	0.285	0.534
25	Sao Paulo	−0.091	0.358	0.488
26	Sergipe	−0.049	0.325	0.225
27	Tocantins	−0.235	0.360	0.295

Data of public domain were collected from official Federal institutions of statistics and climate change monitoring. The observations collected for this research were certified by independent third parties. Estimates of fleets' air emissions (CO_2 eq) were collected from SEEG [23]. Data on fleet sizes of urban buses and the population of professional licensed drivers were acquired from the National Traffic

Table 3 Descriptive statistics of the dataset

Descriptive statistics	Y_1	Y_2	Y_3
Average	−0.092	0.284	0.297
Standard deviation	0.066	0.088	0.121
Minimum	−0.290	0.137	0.121
Maximum	−0.037	0.442	0.535

Secretary (SENATRAN) [24]. Finally, the population of the states in millions of inhabitants was collected from the Brazilian Institute of Geography and Statistics (IBGE) [25].

Note that in the dataset reported, air emissions (Y_2) were assigned to negative signals to reflect the undesirable output nature of this indicator. This procedure is recommended by the character of Air emissions. This procedure was recommended by Seiford and Zhu [26] to accommodate undesirable outputs in DEA-based models.

In Table 3, the main measures of descriptive statistics of the dataset of KPIs are reported.

4.2 *Eco-efficiency Analysis*

The results of the ecoefficiency analysis of the Brazilian public transportation (modal bus) are reported in Table 4. The CI (θ_k) is the eco-efficiency score of DMU k estimated using formulation (2). The variable intensity (λ) expresses the relationship between the one benchmark state in the set (Federal District, DMU 7) and the other states in the set. λ is estimated using the by the dual of formulation (2), assuming Constant Returns to Scale (CRS).

Model (2) also enables estimating radial improvement targets for each KPI of DMUs assessed as inefficient. To illustrate this feature of the CI DEA-based model, the following paragraphs discuss improvement opportunities for the state of Roraima (DMU 23).

Roraima is the northernmost state in Brazil. Located in the Amazonian region, this state is has a territory with low demographic density according to IBGE (2.01 inhab/km^2) [25]. Despite this peculiarity, Roraima occupies the 24th position in eco-efficiency in urban transportation in the Brazilian ranking. On the other hand, Roraima reports the smallest fleet in the set.

Therefore, a benchmarking exercise between Roraina (DMU 23) and its peer Federal District (DMU 7) could provide DMU 23 with some enlightenment to reach higher eco-efficiency standards. This exercise involves the estimation of target values reflecting more ecoefficient levels of operations for Roraima. For instance, the radial target values for desirable outputs KPIs are given by the expression ($Y_{rk}' = \frac{1}{\theta_k} \times Y_{rk}$). Roraima ($\theta_{23} = 45.49\ \%$) should increase urban mobility by enlarging the state bus fleet and by increasing the number of licensed drivers. Therefore, the target value for Y_2 is given by the expression $Y_2' = \frac{1}{0.4549} \times 0,201 = 0.4418$.

Table 4 Results of the Eco-efficiency analysis

DMU	State	IC(θ_k)	Rank	Benchmark DMU (λ)
1	Acre	50.69%	19	7 (0.5069)
2	Alagoas	56.65%	17	7 (0.5664)
3	Amapa	35.29%	26	7 (0.3529)
4	Amazonas	50.77%	18	7 (0.5077)
5	Bahia	63.35%	14	7 (0.6335)
6	Ceara	45.49%	23	7 (0.4549)
7	Federal District	100.00%	1	–
8	Espirito Santo	86.05%	6	7 (0.8605)
9	Goias	76.72%	12	7 (0.7672)
10	Maranhao	31.13%	27	7 (0.3113)
11	Mato Grosso	81.76%	8	7 (0.8176)
12	Mato Grosso do Sul	81.58%	9	7 (0.8158)
13	Minas Gerais	86.61%	5	7 (0.8661)
14	Para	50.44%	20	7 (0.5044)
15	Paraiba	43.52%	25	7 (0.4352)
16	Parana	90.66%	4	7 (0.9066)
17	Pernambuco	48.99%	21	7 (0.4899)
18	Piau	57.40%	16	7 (0.5740)
19	Rio de Janeiro	61.70%	15	7 (0.6170)
20	Rio Grande do Norte	48.90%	22	7 (0.4891)
21	Rio Grande do Sul	84.08%	7	7 (0.8408)
22	Rondonia	80.96%	11	7 (0.8096)
23	Roraima	45.49%	24	7 (0.4549)
24	Santa Catarina	99.67%	2	7 (0.9967)
25	Sao Paulo	91.23%	3	7 (0.9123)
26	Sergipe	73.52%	13	7 (0.7352)
27	Tocantins	81.50%	10	7 (0.8150)

5 Conclusions

This work presented a composite indicator formulation based on DEA to quantify the eco-efficiency of urban transportation in the Brazilian states. The approach adopted was “Benefit-of-the-doubt” (BoD).

The framework reported is based on three KPIs following recommendations of the literature. The specification of unit-invariant KPIs enabled the comparison of states of different sizes and populations. This is a particularly interesting feature that the BoD model can offer.

Although the individual set of weights given by DEA models has the advantage of permitting one to quantify the eco-efficiency while putting each DMU in the best light, it also has the potential disadvantage of enabling the use of weights virtually similar to zero. Therefore, if the performance of a DMU in a particular KPI is unflattering, virtually zero weights can camouflage the phenomenon.

Finally, as a research opportunity the development of refinements in the model to incorporate virtual weight restrictions is highly desirable to prevent overlooking KPIs in the assessment. Another foreseen opportunity is the estimation of a temporal progressive analysis of the public transportation eco-efficiency evolution in a post COVID-19 pandemic era.

References

1. UN: Transforming our world: the 2030 agenda for sustainable development. In: General Assembley 70 Session, vol. 16301, no. October. pp. 1–35 (2015). United Nations. Homepage, https://www.un.org/en/development/desa/population/migration/generalassembly/docs/globalcompact/A_RES_70_1_E.pdf. Last accessed 29 Nov 2021
2. de Carvalho, C.H.R.: TD2198 – Textos para discussão: Desafios da mobilidade urbana no Brasil. IPEA, Rio de Janeiro (2016)
3. Albuquerque, I., et al.: Análise das Emissões Brasileiras de Gases de Efeito Estufa e suas implicações para as metas de clima do Brasil, São Paulo (2020)
4. Rupprecht Consult – Forschung & Beratung GmbH: SUMP Planning for More Resilient and Robust Urban Mobility. Amsterdam (2019)
5. Charnes, A., Cooper, W.W., Rhodes, E.: Measuring the efficiency of decision making units. Eur. J. Oper. Res. **2**(6), 429–444 (1978)
6. Cherchye, L., Moesen, W., Rogge, N., Puyenbroeck, T.V.: An introduction to 'benefit of the doubt' composite indicators. Soc. Indic. Res. **82**(1), 111–145 (2007)
7. Nardo, M., Saisana, M., Saltelli, A., Tarantola, S., Hoffman, A., Giovannini, E.: Handbook on Constructing Composite Indicators, 1st edn. OECD Publications, Paris (2008)
8. Yoshino, D., Fujiwara, A., Zhang, J.: Environmental efficiency model based on data envelopment analysis and its application to environmentally sustainable transport policies. Transp. Res. Rec. J. Transp. Res. Board. **2163**(1), 112–123 (2010)
9. Oliveira, R., Camanho, A.S., Zanella, A.: Expanded eco-efficiency assessment of large mining firms. J. Clean. Prod. **142**, 2364–2373 (2017)
10. Oliveira, R., Camanho, A., Zanella, A.: Eco-efficiency assessment at firm level: an application to the mining sector. In: Operations Research and Big Data, IO2015-XVII Congress of Portuguese Association of Operational Research (APDIO), vol. 15, pp. 149–157 (2015)
11. Milanović, T., Savić, G., Martić, M., Milanović, M., Petrović, N.: Development of the waste management composite index using DEA method as circular economy indicator: the case of European Union countries. Polish J. Environ. Stud. **31**(1), 771–784 (2022)
12. Huang, J., Xia, J., Yu, Y., Zhang, N.: Composite eco-efficiency indicators for China based on data envelopment analysis. Ecol. Indic. **85**, 674–697 (2018)
13. Farrell, M.J.: The measurement of productive efficiency. J. R. Stat. Soc. **120**(3), 253–290 (1957)
14. Tavares, G.: A Bibliography of Data Envelopment Analysis (1978–2001). New Jersey (2002)
15. Gattoufi, S., Oral, M., Reisman, A.: Data envelopment analysis literature: a bibliography update (1951–2001). Socioecon. Plann. Sci. **38**(2–3), 159–229 (2004)
16. Emrouznejad, A., Yang, G.: A survey and analysis of the first 40 years of scholarly literature in DEA: 1978–2016. Socioecon. Plann. Sci. **61**, 4–8 (2018)
17. Zhou, H., Yang, Y., Chen, Y., Zhu, J.: Data envelopment analysis application in sustainability: the origins, development and future directions. *Eur. J. Oper. Res.* **264**(1), 1–16 (2018)

18. Cook, W.D., Kazakov, A., Seiford, L.M.: A data envelopment approach to measuring efficiency: case analysis of highway maintenace patrols. J. Socio. Econ. **20**(1), 83–103 (1991)
19. Blancard, S., Hoarau, J.-F.: A new sustainable human development indicator for small island developing states: a reappraisal from data envelopment analysis. Econ. Model. **30**(1), 623–635 (2013)
20. Cook, W.D., Kress, M.: A data envelopment model for aggregating prefence rankings. Manage. Sci. **36**(11), 1302–1310 (1990)
21. Cherchye, L., et al.: Creating composite indicators with DEA and robustness analysis: the case of the Technology Achievement Index. J. Oper. Res. Soc. **59**(2), 239–251 (2008)
22. Charnes, A., Cooper, W.W.W., Huang, Z.M.M., Sun, D.B.B.: Relations between half-space and finitely generated cones in polyhedral cone-ratio DEA models. Int. J. Syst. Sci. **22**(11), 2057–2077 (1991)
23. SEEG, "SEEG MUNIC´iPIOS," *Base de dados*, Aug-2019. Online. Available: https://plataforma.seeg.eco.br/map. Accessed: 26 Oct 2021
24. SENATRAN, "Secretaria Nacional de Trânsito," *Portal de Serviços*, Jun-2021. [Online]. Available: https://portalservicos.senatran.serpro.gov.br. Accessed: 26 Oct 2021
25. IBGE, "Instituto Brasileiro de Geografia e Estatística," *Estados*, Aug-. [Online]. Available: https://www.ibge.gov.br. Accessed: 26 Oct 2021
26. Seiford, L.M., Zhu, J.: Modeling undesirable factors in efficiency evaluation. Eur. J. Oper. Res. **142**(1), 16–20 (2002)

Success Factors in the Application of Lean and Six-Sigma Methodologies to Healthcare: A Literature Review

Ana Rita Queijo, Sara Reis, Luís Coelho, Luís Pinto Ferreira, and F. J. G. Silva

Abstract To provide a safe and fair-value health service that ensures quality, hospitals must provide efficient processes, trained and committed personnel, appropriate technology and a strategic platform which integrates these aspects effectively. At present, a broad set of tools and methodologies are available, associated to the reconfiguration of processes for enhancing efficiency and enabling excellence and sustainability. Of these, the most noteworthy are Lean and Six-Sigma methodologies. A literature review was performed covering the implementation of these methodologies in health services over the last 5 years. The aim was to determine the current approach in this sector and propose guidelines aligned with the future challenges and the needs of healthcare managers. The influence of team management strategies in the final project outcomes has also been addressed representing a novelty.

Keywords Lean · Six Sigma · Healthcare · Review · Improvement

A. R. Queijo
ISEP – School of Engineering, Polytechnic of Porto, Porto, Portugal

S. Reis · L. Coelho
ISEP – School of Engineering, Polytechnic of Porto, Porto, Portugal

CIETI – Center for Innovation in Engineering and Industrial Technology, Porto, Portugal

L. P. Ferreira (✉) · F. J. G. Silva
ISEP – School of Engineering, Polytechnic of Porto, Porto, Portugal

INEGI – Instituto de Ciência e Inovação em Engenharia Mecânica e Engenharia Industrial, Porto, Portugal
e-mail: lpf@isep.ipp.pt

V. M. López Sánchez et al. (eds.), *Industrial Engineering and Operations Management*, Springer Proceedings in Mathematics & Statistics 400,
https://doi.org/10.1007/978-3-031-14763-0_2

1 Introduction

Innovation is generally seen as a driver of economic and social development. To ensure the development, sustainability, and survival of organizations, particularly in the context of an increasingly competitive market, investments must be made in innovative models, strategies, and processes. Thus, it is each organization's responsibility to address new challenges, whether economic, technological, or social, by finding new practices, models, and tools, which will actually lead to greater competitiveness. Only in this way will they be able to meet the demands and changes resulting from a constantly developing world [1]. The health sector currently occupies a rather conspicuous area of our society, both in economic and social terms. In fact, the analysis of the health sector, as well as the search for mechanisms to improve its performance and thus best meet the population's needs, can also be viewed from an economic perspective. In fact, there has been a steady growth in the role of management as an instrument of understanding, and even intervention, in the field of health and its organization [2]. In recent years, there has been a great demand for health care, as well as a growing concern in responding more efficiently to the population's needs. In addition, the demand to improve the quality of the services provided is further triggered by constant technological innovation. All these issues have culminated in a sharp rise in health expenditure in many countries [3–5]. Thus, health systems across the globe are facing a new set of challenges: expenditure is growing constantly, and budgets are becoming increasingly tight; at the same time, the services requested and the quality expected have continued to rise [6]. Depending on the perspective and context considered, there are various definitions of 'quality' in the health sector. Indeed, there is no single universally accepted definition; the quality of health services is even more complex to define and measure than in other sectors. The distinct characteristics inherent to this area, such as its intangibility, heterogeneity, and concurrency, have contributed to making this definition and measurement difficult [7]. Identified by Health Quality Ontario, the nine attributes of a high-quality health system are: accessibility, efficacy, safety, patient-centered health, equity, efficiency, and a system possessing adequate resources, which is integrated in and focused on the population's health [8]. The quality of the assistance provided is, therefore, not only proportional to financial expenditure; of equal importance is the manner in which the system is organized, particularly with regard to its efficiency and, more specifically, the effectiveness of the scant resources allocated to achieve the best desired result [9]. In recent decades, there has been an intense debate concerning the quality of health care. In many countries, there is compelling evidence pointing to the fact that health care is often not provided according to scientifically established professional standards. This ultimately jeopardizes thousands of people every year and leads to the waste of the limited resources available [10]. Indeed, medical errors, variations in practice, the competence of health professionals, the scarcity and, even, lack of resources, have all provided reasons for the discussion of the quality of care [11]. International literature indicates that the inefficient management of health institutions has been one of the main causes of increased health expenditure [4, 10]. Management inefficiency, inappropriate use of means, enormous waste and, in addition to high expenses, further accumulation of great amounts of debt are some of the known problems. This has, therefore, pointed to the need to make the health system much more efficient, without jeopardizing the quality of services [3]. To overcome these challenges in a sustainable manner, it is imperative that health systems improve their services with regard to costs, response times, quality of service and use of resources, as well as by adopting new methods of hospital management [1, 6]. As a result of this pressure, there have been a number of initiatives in several European countries, which aim to ensure and improve the quality of health care. The application of quality strategies/methods in the hospital environment constitutes one of these actions [12]. Since the 1930s, the industry has demonstrated that systematic, reproducible and evidence-based methodologies are of great benefit when aiming to achieve improvements in quality. The advantages in these strategies are intensified when applied to the complex health environment. Firstly, they objectively establish the processes that must be followed to understand the problem, and guide decision-making by identifying potential aspects for improvement. Secondly, these processes are flexible and can be adjusted to focus on improving a wide range of quality measures. A framework is subsequently provided to analyze the improvement - or not - of the interventions performed; at the same time, other equally problematic areas can be identified. Finally, continuous evaluation and reevaluation are promoted, thus creating a cycle of improvements in a system whose needs and resources may vary over time [9]. In the past two decades, Six-Sigma (SS) and Lean have become successful Quality Improvement (IQ) or Quality Improvement strategies for organizations aiming to achieve operational excellence, high quality standards, and a reduction in costs to become more efficient. Both SS and Lean have evolved in an integrated manner since, and despite different approaches, they are aligned with the same objective, thus supplementing each other. The integration of LSS ultimately benefits from the advantages presented in the two continuous improvement methodologies. Thus, Lean reinforces the philosophy of structure and provides a strategic direction towards improvement, guiding the overall dynamics of the system and providing information as to the current state of operations. Six Sigma, in turn, identifies and focuses on improvement, leading the system to the desired state [13]. This combination allows for the implementation of the concepts and practices of both methodologies, complementing and applying their various tools in accordance with the need and level of quality aimed for [14]. Although Lean and SS methodologies initially focused on the contexts of product manufacture, many health organizations have begun to implement them as strategies to reduce operational costs and process inefficiencies, as well as to enhance service levels, make administrative processes more efficient, and minimize medical errors [15]. The general objectives of this article focus on two central questions: "What is the impact of the application of Lean and Six-Sigma methodologies on the optimization of processes in health services?" and "What are the success factors in Lean and Six-Sigma implementation in the healthcare area?". Accordingly, it was proposed to develop a literature review about these methodologies in health services in the last 5 years, thus determining current approaches, as well as gaining an insight into their impact and proposing future guidelines, which are consistent with the requirements of professionals.

2 Methodology and Results

For this review, the main studies published in the last 5 years pertaining to the practice of Lean, Six-Sigma and Lean Six-Sigma methodologies in the healthcare area were considered. The time span was based on the great amount of normalized number of publications within the mentioned timeframe. Six major scientific repositories (PubMed, Web of Knowledge, Google Scholar, Scopus, IEEExplore and Medlib), widely accepted in scientific research, were queried using a combination of the keywords 'Lean', 'Six Sigma' and 'Lean Six Sigma', which were combined and complemented (AND/OR) with terms related to health and quality processes, for instance: 'healthcare', 'hospital', 'sustainability', 'quality improvement', 'quality management', 'quality systems' or 'process optimization'. Following a PRISMA inspired methodology, from the initial 115 articles found, 32 have been considered for this review and the remaining were excluded due to the criteria: (a) not in English; (b) out of scope; (c) not practical cases; (d) adjacent topic.

2.1 *Integration of Lean and Six-Sigma*

Lean and Six-Sigma are complementary methodologies that can be used alone or integrated. Thus, to evaluate their effectiveness it is important to know the distribution frequency for each case in the selected studies. The combination of Lean and Six-Sigma is prevalent, representing 75% of the total. Both methodologies are present with nearly the same frequency. The discrepancies are related to the different types of problems with which the three methodologies deal separately. Therefore, each one applied the methodology according to the needs. Since Lean/Six-Sigma represents the integration of the two methodologies, it is common to be the most used strategy. Six-Sigma provided an average improvement of 39% while Lean allowed to achieve 44%. Their combination had an average improvement of 48%, leading to the highest score. It must not be forgotten that the source values for computing the statistics are from different project with distinct objectives and they must be interpreted with flexibility.

2.2 Targeted Healthcare Services

The selected cases were mapped according to the implementation department. This was the chosen way to identify trends in the application of the methodologies under analysis. In general, Lean, Six-Sigma, and Lean/Six-Sigma were implemented to improve patient care and safety. The main aim was to reduce costs associated with the provided services. Figure 1 shows a distribution of services that benefited from improvement interventions. Looking at the figure, it can be noted that the service with the highest number of reports on the implementation of the methodologies is surgery (16%), followed by oncology and pharmacy (both with 13%), and radiology (9%). The fact that these four services stand out suggests that these areas are at the focus of quality improvement initiatives in hospitals when it comes to Lean, SS and LSS methodologies. The factors that led to the choice of these services were also collected during the review of the articles. Regarding the surgical departments, there is a particular attention for the high number of resources that its operation requires and for its great contribution in the hospital budgets. On the other hand, the complexity of the surgical process, based on flows with high variability and with a high probability of failure, can also be pointed out as justification. Still important is the reduction of risks resulting from complications in the operating room (infections, delays, among others). The analyzed articles addressed aspects related to the operationalization of health services. In general, the proposed methodologies aim to improve the efficiency of processes, the quality of

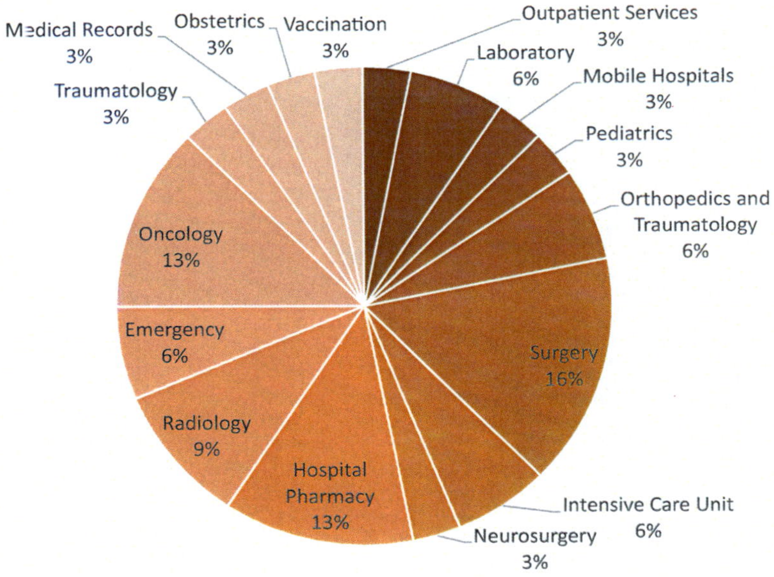

Fig. 1 Percentage of services where improvement projects were implemented

medical care and the management. These directly benefits the services' quality/cost ratio, from the provider's perspective, while, from the user's perspective, the user-experience is improved. Concerning the context, three main objectives can be identified in the analyzed articles: (1) time reduction, (2) cost reduction and (3) error reduction.

In Table 1 it can be observed a summary of the main improvement actions described in the literature. The table is ordered according to the improvement intention so that the reader can easily find the studies best suited to their objectives. Regarding the areas of interventions, approaches that focused on activities for supporting hospital routine (sometimes transversal) and others that focused on clinical services can be identified. In the former intervention area, there are cases of optimization of various processes, including the dispatch of medical reports [16], optimization of the medical registration process [17], optimization of the patient registration process [17], reduction of patient waiting time [15], reduction of medication errors [18], improvement in the discharge process [19], elimination of activities that do not add value in vaccination processes [20] (allowing both reducing the nursing staff as well as the required working hours), reducing the rate of hospital falls [21] and improving the information systems for the sake of process efficiency [16, 22]. In the clinical area the performed analysis have focused on a large spectrum of services: (a) in the case of the surgical departments the reviewed approaches allowed to improve aspects such as the reduction of the process time, the improvement of specialized professionals routine tasks, in the reduction of patients' infections, in the reduction of delays in the start of surgeries, in programming interventions in order to reduce team overtime and in the improvement of procedures [23]; (b) in emergency services, optimization cases involving staff and patients awareness about urgent and non-urgent differentiation and, patients flow improvement are reported [24]; (c) Laboratory services were also targeted for increasing workflow efficiency, improving protocols, making standardizations and reducing the number of exam requests [23]. (d) Improvements aimed at the treatment process for cancer patients, such as reduced chemotherapy preparation time or reduced error in oral prescriptions have also been observed [22, 25]; (e) The main objectives of the implementations reported for Radiology services were to reduce costs and, mainly, waiting periods for exams and results, since these, when long, limit the ability to offer a quality service [26].

2.3 *Success Factors in Project Management*

The use of improvement methodologies by itself does not always lead to optimal results and in some cases the final achievements can be minimal. Many factors can contribute to successful project. Hence, in addition to the improvements resulting from LSS implementation, other aspects of the interventions reported in the selected case studies were collected and analyzed to provide a detailed view on the process of developing quality improvement projects in the hospital environment.

Table 1 Improvement action according to improvement intentions

Improvement intention	Service/Process	Actions/Improvements
Time reduction	Hospital stay	Decrease patients' hospital stay [19, 25, 28–32]
		Minimize infections caused by long stays [30]
	Transversal processes	Improve contact and collaboration between doctors, specialists, technicians and staff in general [15, 22, 26, 33]
		Improve patient flow in processes [23, 26, 28, 34–37]
	Laboratory	Improve prescriptions, reduce analysis time and time to deliver results [16, 26, 35, 38]
	Surgery	Improve rooms' usage [33]
		Decrease processes' duration [27, 29, 39]
	Emergency	Improve the differentiation between urgent and non-urgent cases [24, 40, 41]
		Improve coordination between different areas of the hospital or departments during transfers [41]
Cost reduction	Laboratory	Improve samples' management [38]
		Decrease the number of requisitions [23]
	Surgery	Improve the programming of interventions to reduce team overtime [27, 31, 33]
		Rearrange processes in operating rooms [27, 39, 42]
		Prevent incidents or infections [23, 30, 39]
		Improve coordination between different areas of the hospital or departments during transfers [42]
	Transversal processes	Reduce overtime [20, 43, 44]
		Reduce requests [33]
		Improve resource usage [24, 25, 32, 33, 45]
		Reduce inpatients' falls [21]
Error reduction	Laboratory	Improve protocol to reduce handling errors outputs [38]
		Improve workflow to reduce errors [38]
	Surgery	Reduce complications from anesthesia administration or other types of interventions [31]
	Medication administration	Improve medical prescriptions [22, 25, 45]
		Reduce errors on medication administration [18, 22]
	Support systems	Improving the information system efficiency [16, 17, 22]

These complementary aspects are identified as three qualitative guidelines that can assist in future cases of implementation: (1) the implementation time, (2) project team composition and (3) the auxiliary methods and tools used for goal achievement. Information regarding the duration of the projects (including analysis, implementation and evaluation) was clearly indicated in 27 of the 32 cases analyzed. Among these 27 study cases, 3 (12%) had an intervention time of less than 6 months, 12 (44%) had a duration of 6 months to 1 year and 12 (44%) took more than 1 year. The type of process to be intervened, the objectives to be achieved and the

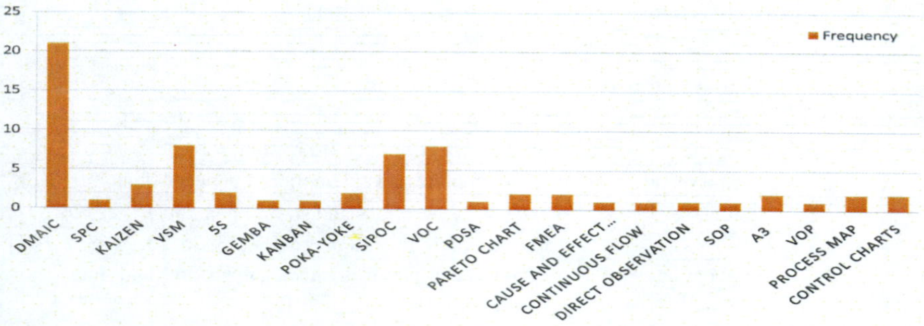

Fig. 2 Usage of complementary analysis and management tools (absolute values)

resources allocated are variables that directly affect the duration of the project but, considering the cases analyzed, it can be predicted a duration of about 1 year to achieve the expected results. It is also repeatedly emphasized that approaches based on LSS require, in addition to the commitment of administration and management and the involvement and commitment of the team, a high theoretical and technical knowledge of the professionals involved, and the use of auxiliary tools allows that meaningful and systematic improvements are achieved. Of the 32 selected articles, in 15 case studies (47%) a multidisciplinary team was formed to carry out the project. In most cases, specialist doctors, nurses, hospital quality directors and external support teams were added, ultimately contributing significantly to the benefits of the project. For example, the team in Tagge et al. [27], was composed of a nursing manager, an administrator, a financial analyst, surgeons, anesthetists and the director of the surgical center in question. Agarwal et al. [23] counted with the participation of managers and directors, doctors, nurses and a specialist in external quality improvement projects. Before project start, both teams received training in LSS, proved to be fundamental for the intervention success. In addition to the greater team commitment, a common language showed the way forward. In all cases, it is noticeable that the presence of specialist doctors responsible in the multidisciplinary team is crucial for project success, since they have practical knowledge of all department procedures. Lean and Six-Sigma methodologies are applied in conjunction with several analysis and management tools that are frequently mentioned in the reviewed articles. In Fig. 2, a list of the support tools for project development and the number of times each have been used can be observed. The use of some of these tools stands out clearly, with a greater preference for a small subset. DMAIC (Define, Measure, Analyze, Improve and Control), VSM (value stream mapping), SIPOC (suppliers, inputs, process, outputs, and customers) and VOC (voice of the customer) were the most recurring methods and tools for performing process optimization interventions in a hospital environment.

3 Discussion

Lean and six sigma are well established methodologies in the industrial area but only in recent years they have started to be applied to the healthcare sectors. The USA leads the presence of LSS applications in the revised literature which may be compatible with the strong industrial tradition of the country. The existence of a big private healthcare sector, in a dynamic and liberal economy, can also increase the motivation to implement continuous improvement methodologies and promote the development of very effective processes. Healthcare systems are evolving worldwide, always having in mind quality and efficiency. Regarding the use of each method independently or the option for combining the two, the results were clear. LSS, as an integrated methodology, provided the highest average improvement. This is an important result is two perspectives: (a) Lean and Six Sigma target slightly different perspectives but both help to seek similar objectives, so the combined use of both can lead to a nearly optimal solution with long lasting results; (b) The improvement process requires resources (people, time and money) allocated to the task and the implementation of changes requires an adaptation period and can imply training. If the improvement step can be higher in a single iteration, using a combination of methodologies, then the efforts in change management can be minimized. The Lean and Six-Sigma methodologies, which focus almost entirely on operational improvements and on reducing process variability, which ends up directly influencing cycle time, quality and process reliability indicators, appear to be very suitable for the optimization of healthcare processes. In fact, considering the number of references presented in Table 1, regarding the reduction of time, it can be concluded that Lean, SS and LSS can have a high potential for impacting this requirement. Thus, it can be observed that, in all cases, improvements have been reported. Hence, obtaining positive results, through the correct implementation of Lean Six-Sigma, is very likely. The application of the described methodologies can also have a significant financial impact. In the case of approaches aimed at cost reduction, the financial impact is direct. In cases designed to optimize processes or reduce errors, the financial benefits are also relevant, and may be direct or indirect. With more optimized processes and shorter cycle times, it becomes possible to increase the number of procedures performed, opening the possibility of a greater economic return. Many errors result from faulty systems and ineffective processes, not from improper practices or useless professionals. Therefore, the early detection of problems and the use of monitoring and improvement strategies is essential for the quality of services. The reduction of waste and clinical error is also important in the financial viewpoint. Finally, to achieve better results, the project should be carefully planned considering several factors: (a) the project team should be multidisciplinary including senior LSS experts, management and onsite healthcare experts daily involved with the processes; (b) raining must be provided to those unaware of LSS methodologies and purposes; (c) project should target specific services and aim to clear objectives; (d) project duration should be around 1 year;

(e) a combination of methodologies and a wise selection of support analytical tools must be done.

4 Conclusions

The processes and their operationalization must be subject to a continuous improvement effort, as this is the only way to guarantee the adequacy to the technical evolution and the survival of the organization. Health systems are no exception and persistence in traditional practices is no solution to guarantee modern, competitive, and quality health systems. The Lean methodology is a management strategy that focuses on creating value for the customer with a scarce use of resources. Thus, health organizations around the world have been promoting its use since the beginning of the twenty-first century. The decision to implement Lean in the healthcare field is clear in terms of quality, cost, waiting times, patient safety and satisfaction of the hospital team, as it allows hospitals to improve the quality of patient care, eliminate barriers and support patients, employees and health professionals for better interaction within health processes. The main objective of this literature review was to explore the use and utility of the Lean Six-Sigma methodology in the optimization of processes in the health sector and, after careful analysis of all selected case studies, the benefits of implementing these strategies, in hospital environment, were transversal, ranging from increased satisfaction of customers and employees, to improvements in processes, to positive financial impacts, to the elimination of waste and the reduction of time. The presented review summarizes the most frequent improvements, organized by improvement intention by hospital service. Practical aspects related to difficulties and success factors are also mentioned. References are made about the importance of creating well-trained multidisciplinary teams in the implementation of strategies and about the need to choose appropriate complementary support tools. Finally, the average duration of implementation for the cases studied is presented. With this study, a snapshot of the current state of the art that can be used as a guide for future implementation of six-sigma methodologies is provided, serving as a base reference for future developments of the technique as well. It is hoped that all aspects mentioned in this review can contribute to the continuous improvement of the performance of health organizations and to provide better healthcare.

References

1. Rego, G., Nunes, R., Costa, J.: The challenge of corporatisation: the experience of Portuguese public hospitals. Eur. J. Health Econ. **11**, 367–381 (2010)
2. Sacristán, J.A.: How to assess the value of low-value care. BMC Health Serv. Res. **20**, 1000 (2020). https://doi.org/10.1186/s12913-020-05825-y

3. Morais Nunes, A.: Fourteen years of business administration practices in health management: the Portuguese experience. Rev. Ciênc. Saúde. **6**, 154–162 (2017). https://doi.org/10.5585/rgss.v6i2.283
4. Folland, S., Goodman, A.C., Stano, M.: The Economics of Health and Health Care. Pearson, Upper Saddle River (2013)
5. OECD: Health at a Glance 2019: OECD Indicators. OECD (2019). https://doi.org/10.1787/4dd50c09-en
6. Arcidiacono, G., Matt, D.T., Rauch, E.: Axiomatic design of a framework for the comprehensive optimization of patient flows in hospitals. J. Healthc. Eng. **2017**, 2309265 (2017). https://doi.org/10.1155/2017/2309265
7. Allen-Duck, A., Robinson, J.C., Stewart, M.W.: Healthcare quality: a concept analysis. Nurs. Forum. **52**, 377–386 (2017). https://doi.org/10.1111/nuf.12207
8. Mannion, R., Davies, H.: Understanding organisational culture for healthcare quality improvement. BMJ. **363**, k4907 (2018). https://doi.org/10.1136/bmj.k4907
9. Mason, S.E., Nicolay, C.R., Darzi, A.: The use of Lean and Six Sigma methodologies in surgery: a systematic review. Surgeon. **13**, 91–100 (2015). https://doi.org/10.1016/j.surge.2014.08.002
10. Jacobs, R., Smith, P.C., Street, A.: Measuring Efficiency in Health Care: Analytic Techniques and Health Policy. Cambridge University Press (2006). https://doi.org/10.1017/CBO9780511617492
11. OECD: Improving Value in Health Care, https://www.oecd-ilibrary.org/content/publication/9789264094819-en. (2010)
12. Alzoubi, M.M., Hayati, K.S., Rosliza, A.M., Ahmad, A.A., Al-Hamdan, Z.M.: Total quality management in the health-care context: integrating the literature and directing future research. Risk Manag. Healthc. Policy. **12**, 167–177 (2019). https://doi.org/10.2147/RMHP.S197038
13. Venanzi, D., Laporta, B.P.: Lean Six Sigma. South Am. Dev. Soc. J. **1**, 66–84 (2017)
14. Salah, S., Rahim, A.: The Integration of Six Sigma and Lean: Combining Lean Six Sigma with Process Improvement. Presented at the (2019). https://doi.org/10.1007/978-3-319-99034-7_4
15. Ortiz Barrios, M., Felizzola Jiménez, H.: Reduction of average lead time in outpatient service of obstetrics through Six Sigma methodology. In: Bravo, J., Hervás, R., Villarreal, V. (eds.) Ambient Intelligence for Health, pp. 293–302. Springer International Publishing, Cham (2015)
16. Basta, Y.L., Zwetsloot, I.M., Klinkenbijl, J.H.G., Rohof, T., Monster, M.M.C., Fockens, P., Tytgat, K.M.A.J.: Decreasing the dispatch time of medical reports sent from hospital to primary care with Lean Six Sigma. J. Eval. Clin. Pract. **22**, 690–698 (2016). https://doi.org/10.1111/jep.12518
17. Bhat, S., Ev, G., Jnanesh, N.: Productivity and performance improvement in the medical records department of a hospital: an application of Lean Six Sigma. Int. J. Product. Perform. Manag. **65** (2015). https://doi.org/10.1108/IJPPM-04-2014-0063
18. Trakulsunti, Y., Antony, J., Dempsey, M., Brennan, A.: Reducing medication errors using lean six sigma methodology in a Thai hospital: an action research study. Int. J. Qual. Reliab. Manage. (2020). https://doi.org/10.1108/IJQRM-10-2019-0334. Ahead-of-print
19. Improta, G., Balato, G., Romano, M., Carpentieri, F., Bifulco, P., Alessandro Russo, M., Rosa, D., Triassi, M., Cesarelli, M.: Lean Six Sigma: a new approach to the management of patients undergoing prosthetic hip replacement surgery. J. Eval. Clin. Pract. **21**, 662–672 (2015). https://doi.org/10.1111/jep.12361
20. Ha, C., Taylor, C., Modi, J.R.: Mass vaccinations at the United States Naval Academy. Health Secur. **14**, 382–388 (2016). https://doi.org/10.1089/hs.2016.0030
21. Bargmann, A.L., Brundrett, S.M.: Implementation of a Multicomponent Fall Prevention Program: contracting with patients for fall safety. Mil. Med. **185**, 28–34 (2020). https://doi.org/10.1093/milmed/usz411
22. Lamm, M.H., Eckel, S., Daniels, R., Amerine, L.B.: Using lean principles to improve outpatient adult infusion clinic chemotherapy preparation turnaround times. Am. J. Health Syst. Pharm. **72**, 1138–1146 (2015). https://doi.org/10.2146/ajhp140453

23. Agarwal, S., Gallo, J.J., Parashar, A., Agarwal, K.K., Ellis, S.G., Khot, U.N., Spooner, R., Murat Tuzcu, E., Kapadia, S.R.: Impact of Lean Six Sigma process improvement methodology on cardiac catheterization laboratory efficiency. Cardiovasc. Revasc. Med. **17**, 95–101 (2016). https://doi.org/10.1016/j.carrev.2015.12.011
24. Alowad, A., Samaranayake, P., Ahsan, K., Alidrisi, H., Karim, A.: Enhancing patient flow in emergency department (ED) using lean strategies–an integrated voice of customer and voice of process perspective. Bus. Process. Manag. J. (2020). https://doi.org/10.1108/BPMJ-11-2019-0457. Ahead-of-print
25. Downen, J., Jaeger, C.: Quality improvement of intravenous to oral medication conversion using Lean Six Sigma methodologies. BMJ Open Qual. **9** (2020). https://doi.org/10.1136/bmjoq-2019-000804
26. Cheung, Y.Y., Goodman, E.M., Osunkoya, T.O.: No more waits and delays: streamlining workflow to decrease patient time of stay for image-guided musculoskeletal procedures. Radiographics. **36**, 856–871 (2016). https://doi.org/10.1148/rg.2016150174
27. Tagge, E.P., Thirumoorthi, A.S., Lenart, J., Garberoglio, C., Mitchell, K.W.: Improving operating room efficiency in academic children's hospital using Lean Six Sigma methodology. J. Pediatr. Surg. **52**, 1040–1044 (2017). https://doi.org/10.1016/j.jpedsurg.2017.03.035
28. Beck, M.J., Gosik, K.: Redesigning an inpatient pediatric service using Lean to improve throughput efficiency. J. Hosp. Med. **10**, 220–227 (2015). https://doi.org/10.1002/jhm.2300
29. Hultman, C.S., Kim, S., Lee, C.N., Wu, C., Dodge, B., Hultman, C.E., Roach, S.T., Halvorson, E.G.: Implementation and analysis of a Lean Six Sigma program in microsurgery to improve operative throughput in perforator flap breast reconstruction. Ann. Plast. Surg. **76**(Suppl 4), S352–S356 (2016). https://doi.org/10.1097/SAP.0000000000000786
30. Montella, E., Di Cicco, M.V., Ferraro, A., Centobelli, P., Raiola, E., Triassi, M., Improta, G.: The application of Lean Six Sigma methodology to reduce the risk of healthcare-associated infections in surgery departments. J. Eval. Clin. Pract. **23**, 530–539 (2017). https://doi.org/10.1111/jep.12662
31. Sayeed, Z., Anoushiravani, A., El-Othmani, M., Barinaga, G., Sayeed, Y., Cagle, P., Saleh, K.J.: Implementation of a hip fracture care pathway using Lean Six Sigma methodology in a Level I Trauma Center. J. Am. Acad. Orthop. Surg. **26**, 881–893 (2018). https://doi.org/10.5435/JAAOS-D-16-00947
32. Trzeciak, S., Mercincavage, M., Angelini, C., Cogliano, W., Damuth, E., Roberts, B.W., Zanotti, S., Mazzarelli, A.J.: Lean Six Sigma to reduce intensive care unit length of stay and costs in prolonged mechanical ventilation. J. Healthc. Qual. **40**, 36–43 (2018). https://doi.org/10.1097/JHQ.0000000000000075
33. Brown, R., Grehan, P., Brennan, M., Carter, D., Brady, A., Moore, E., Teeling, S.P., Ward, M., Eaton, D.: Using Lean Six Sigma to improve rates of day of surgery admission in a national thoracic surgery department. Int. J. Qual. Health Care. **31**, 14–21 (2019). https://doi.org/10.1093/intqhc/mzz083
34. Al Hroub, A., Obaid, A., Yaseen, R., El-Aqoul, A., Zghool, N., Abu-Khudair, H., Al Kakani, D., Alloubani, A.: Improving the workflow efficiency of an outpatient pain clinic at a specialized oncology center by implementing Lean principles. Asia Pac. J. Oncol. Nurs. **6**, 381–388 (2019). https://doi.org/10.4103/apjon.apjon_21_19
35. Godley, M., Jenkins, J.B.: Decreasing wait times and increasing patient satisfaction: a Lean Six Sigma approach. J. Nurs. Care Qual. **34**, 61–65 (2019). https://doi.org/10.1097/NCQ.0000000000000332
36. Vijaya Sunder, M., Mahalingam, S., Sai Nikhil Krishna, M.: Improving patients' satisfaction in a mobile hospital using Lean Six Sigma – a design-thinking intervention. Prod. Plan. Control. **31**, 512–526 (2020). https://doi.org/10.1080/09537287.2019.1654628
37. Wang, H.-M.S., Wang, S.-P., Lee, W.: A case study for reducing client waiting time in a health evaluation center using design for Six Sigma. Eng. Manag. J. **26**, 62–73 (2015). https://doi.org/10.1080/10429247.2014.11432011

38. Inal, T.C., Ozturk, O.G., Kibar, F., Cetiner, S., Matyar, S., Daglioglu, G., Yaman, A.: Lean Six Sigma methodologies improve clinical laboratory efficiency and reduce turnaround times. J. Clin. Lab. Anal. **32**, e22180 (2018). https://doi.org/10.1002/jcla.22180
39. Polanski, W.H., Klingelhoefer, L., Zolal, A., Guenther, S., Klein, J., Schackert, G., Reichmann, H., Sobottka, S.B.: Clinical validation of quality improvements using the Six Sigma concept: a case study for deep brain stimulation in Parkinson's disease. Stereotact. Funct. Neurosurg. **97**, 195–201 (2019). https://doi.org/10.1159/000502654
40. Birmingham, L.E., Sedorovich, A., Mann, N., George, R.L.: Using Lean Six Sigma to improve delayed intracranial hemorrhage screening in a geriatric trauma population. Qual. Manag. Health Care. **27**, 199–203 (2018). https://doi.org/10.1097/QMH.0000000000000186
41. Vashi, A.A., Sheikhi, F.H., Nashton, L.A., Ellman, J., Rajagopal, P., Asch, S.M.: Applying Lean principles to reduce wait times in a VA Emergency Department. Mil. Med. **184**, e169–e178 (2019). https://doi.org/10.1093/milmed/usy165
42. Gleich, S.J., Nemergut, M.E., Stans, A.A., Haile, D.T., Feigal, S.A., Heinrich, A.L., Bosley, C.L., Tripathi, S.: Improvement in patient transfer process from the operating room to the PICU using a Lean and Six Sigma-based quality improvement project. Hosp. Pediatr. **6**, 483–489 (2016). https://doi.org/10.1542/hpeds.2015-0232
43. Al-Balushi, M.M., Al-Mandhari, Z.: Implementing Lean management techniques at a Radiation Oncology Department. Sultan Qaboos Univ. Med. J. **18**, e362–e366 (2018). https://doi.org/10.18295/squmj.2018.18.03.016
44. Hynes, J.P., Murray, A.S., Murray, O.M., Eustace, S.K., Gilchrist, S., Dolan, A., Lawler, L.P.: Use of Lean Six Sigma methodology shows reduction of inpatient waiting time for peripherally inserted central catheter placement. Clin. Radiol. **74**(733), e5–733.e9 (2019). https://doi.org/10.1016/j.crad.2019.04.022
45. Alkuwaiti, A.: Application of Six Sigma methodology to reduce medication errors in the outpatient pharmacy unit: a case study from the King Fahd University Hospital, Saudi Arabia. Int. J. Qual. Res. **10**, 267–278 (2016). https://doi.org/10.18421/IJQR10.02-03

Optimizing the Chemical Collection Used in Cork Composites Manufacture

I. Ferreira, F. J. G. Silva, A. Barreiras, Luís Pinto Ferreira, and M. T. Pereira

Abstract With global development, companies need to seek opportunities for improvement and act accordingly to ensure the optimization of industrial production processes. Companies also try to ensure the quality of the final product, reducing process costs to increase the company's sustainability. Many industries are going through restructuring processes and new challenges are emerging.

The present investigation respects to the operations of an industrial unit, which produces rubberized cork agglomerates. Cork industry invests in search for new products, increasing its competitiveness in the market, expanding its range of products and promoting a better flow of this limited natural resource. This work identifies the causes behind some problems in terms of the final product quality, allowing the creation of corrective actions and opportunities for improvement in terms of product quality, working conditions and product line optimization. It is known that linear programming can make a strong contribution to the optimization of processes within companies. This work presents the application of linear programming as a tool through which it was possible to optimize the discharge process of chemicals from the production line, reorganizing the locker arrangement, and obtaining a reduction in the time spent in production operations of certain articles.

Keywords Linear programming · Results optimization · Production · Rubberized cork

I. Ferreira · A. Barreiras
ISEP – School of Engineering, Polytechnic of Porto, Porto, Portugal

F. J. G. Silva · L. P. Ferreira (✉) · M. T. Pereira
ISEP – School of Engineering, Polytechnic of Porto, Porto, Portugal

INEGI – Instituto de Ciência e Inovação em Engenharia Mecânica e Engenharia Industrial, Porto, Portugal
e-mail: lpf@isep.ipp.pt

V. M. López Sánchez et al. (eds.), *Industrial Engineering and Operations Management*, Springer Proceedings in Mathematics & Statistics 400,
https://doi.org/10.1007/978-3-031-14763-0_3

1 Introduction

The search for sustainability in industrial production, with a particular emphasis on the reduction of consumption of natural resources and the reuse of materials, has deserved a special attention in the policies of industrial companies. Thus, management has been cared more successfully [1]. Cork is a natural material with vast applicability. However, cork is limited by the natural growth cycle that characterizes the cork oak [2, 3]. With the aim of providing the market with products that meet customer needs and meet the required requirements. Moreover, cork companies also need to have a focus on sustainability. Thus, cork processing industries have sought to take advantage of this material as much as possible, starting by applying it in noble products (stopper, among others [4]), but also in cork granules that are applied together with other materials, forming sheets of composite material with countless applications [5]. The characteristics of the final product will depend on the materials added to the cork. Cork-rubber composites have been increasingly accepted by the market for numerous applications, with emphasis on vibration isolation [6]. The agglomeration of this composite with the incorporation of cork granules allows practically full use of all the cork extracted from the cork oak. However, rubber, used as a binder, being a synthetic product, follows an elaborate recipe of properly dosed chemical products, which allow the properties required of the composite material to be effectively achieved. This work aims at optimizing the collection of these chemical products, which is carried out in specific equipment, furnished with reservoirs for each of the chemical products, with a container that moves between some of these reservoirs, depending on the recipe, to collect the correct amount of each one, with a view to later promote the homogenization of the mixture.

2 Literature Review

2.1 Linear Programming: Assignment Problem

In order to ensure the optimisation of industrial production processes and the quality of the final product, companies need to look for improvement and act accordingly [7]. A basic management task is to make decisions. Optimization problems are usually solved based on mathematical models. When it is intended to improve a system, it is important to be aware that, reducing time on a resource that is not a bottleneck, has no impact on reducing the complete cycle time [8]. Otherwise, any time reduction in a bottleneck, represents the same time reduction in the cycle time [9]. Thus, it is necessary to start with the optimisation of a bottleneck and try to remove it or, at least, to guarantee its maximum use [10, 11].

One of the most used techniques in solving optimisation problems, is linear programming. Linear programming consists of minimizing or maximizing a linear

Table 1 Concepts of Linear Programming [10, 11]

Concepts	Description
Objective function	Considering the decisions variables, the function will be maximizing or minimizing. Mathematical function that determines the target-value to be reached.
Decision variables	Unknown decisions that will be determined by the model solution
Restrictions	Limits the system and directly influences the values of the decision variables

objective function of several decision variables, subject to restrictions, represented by linear equations and inequations, which redefine its domain, i.e., it can be considered a technique that allows the optimisation of linear functions, subject to equally linear restrictions [12]. Table 1 shows a brief description of these concepts [10, 11].

2.2 *Assignment Problem: Solver*

The assignment problem is a specific case of the Linear Programming problem which has been used to optimize several solutions in many scientific fields [13–15]. The assignment problem consists of allocating a great number, *i*, of individuals to a great number, *j*, of tasks, with the goal of minimizing the total cost involved in this process, so that each individual performs one and only one task, and each task is performed by one and only one individual. This problem can be understood as a combining optimisation problem, considering *x!* the total number of possible allocations. The assignment problem can also be presented as follows:

$$X_{ij} = \begin{cases} 1, & if\ the\ individual\ “i”\ is\ allocated\ to\ task\ “j” \\ 0, & if\ the\ individual\ “i”\ is\ not\ allocated\ to\ task\ “j” \end{cases}$$

The mathematical formalisation of the assignment problem can be performed with the goal of maximizing or minimizing. For example: (a) to maximize the productivity of a line; (b) to minimize storage costs. The different meanings that can be given to the assignment problem, are related to the coefficients of the decision variables in the objective function. This technique is based on mathematical models. When more complex problems are considered, it is really important to use software to obtain the solution. As it is a logical tool that considers existing variables and restrictions, Solver is a good resource, as it is an MS Excel® tool, that allows the achievement of optimal solutions for linear programming problems. This also enables simulations to be carried out from parameter changes in mathematical modelling [10, 11].

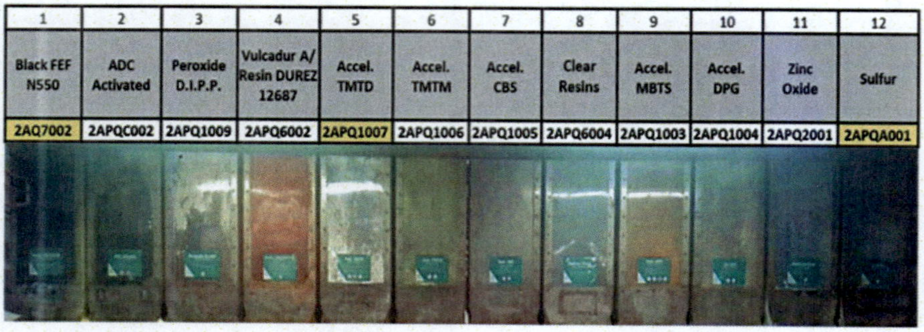

Fig. 1 Current layout of chemical product lockers

3 Problem Description and Modeling Approach

3.1 *Optimization of the Chemical Dosing Process*

During the products production, it was found that the dosing of the chemicals is a time-consuming task, as the trolley with the chemical collection container travels a long distance unnecessarily. Sulfur, for example, which is a chemical used in practically all products produced by the company, is placed in the last cupboard. Figure 1 shows the current arrangement of chemical product lockers. For the analysis of this procedure, the production cycle of product 1521 was considered, as it is the class A product with the greatest impact on the production level of the CR1 line. This observation raised questions about the principle of operation and organization of this locker structure and, consequently, gave rise to a study.

3.2 *Problem Explanation*

An analysis of the chemical product locker operation showed that there was a considerable number of unnecessary trips, and a lot of time could be saved in preparing the mixture of chemicals needed for each recipe. An example of operation for the product with reference 1521 is given below:

- It is produced with chemicals Black FEF N505, Accelerator TMTM and Sulphur (highlighted in yellow in Fig. 1);
- The cycle time, which is the time the trolley takes from leaving the origin until it returns, is 60 seconds;
- Considering the trajectory and the fact that there are 12 lockers, it is concluded by the mathematical expression: $60 \div 2 \div 12$, that the trolley takes 2.5 seconds to move from one locker to another;

- Taking the locker position into account, the current travel time for a container of chemicals was determined using the following calculation:

$$(2.5) + (2.5 \times (5 - 1)) + (2.5 \times (12 - 5)) + 30 = 60 \text{ s}$$

- To produce a block of product *XPTO*, three containers of chemicals are required;
- Thus, through the calculation presented below, it was concluded that the travelling time spent over the year 2020 in the production of article *XPTO* was approximately 13.37 days.

$$\begin{aligned} Travelling\ Time &= (6417 \text{ blocks} \times 3 \text{ containers}) \times 60 \text{ s} \\ &= 1155060 \text{ s} \approx 13.37 \text{ days} \end{aligned}$$

3.3 *Suggestion for Improvement*

In order to overcome this problem, a study was conducted to conclude what would be the correct disposal of the chemicals in the 12 lockers available in the CR1 line. To achieve the 17 chemicals used in the production of CR1 line and all class A products were selected and the matrix of Table 2 was constructed, in which $P_{ij} = 1$ means that the chemical j (j = 1,...,17) enters the composition of the product i.

From Table 2, it was concluded that in Class A products, only 10 of the 12 chemicals that would correspond to the 12 lockers available in the line were used. The 10 chemicals are highlighted in Table 2, within a red rectangle. For this

Table 2 Selection of chemicals used in Class A articles

		Chemical products CLASS A													
Code	Article *i* / Chemical prod. *j*	1521	2549	NL76	CDM01	S112	VC95	1049	1237	H70L	NC733	1028	1235	Qnt. of Article A	Lockers
2APQA001	Sulfur	1	1	1	1	1	1	1	1	1	1	1	1	12	1
2APQ6004	Clear Resins	0	0	0	0	1	1	1	0	1	1	1	0	6	2
2APQ1007	Accel. TMTD	1	1	0	0	0	0	0	1	0	0	1	1	5	3
2APQ1006	Accel. TMTM	0	0	1	0	1	0	0	0	0	0	0	0	2	4
2APQ2001	Zinc Oxide	0	0	1	0	1	0	0	0	0	0	0	0	2	5
2APQ6002	V.A/Resin DUREZ 12687	0	0	0	0	0	1	0	0	0	1	0	0	2	6
2APQ7002	Black FEF N550	1	0	0	0	0	0	0	1	0	0	0	0	2	7
2APQC002	ADC Activated	0	0	1	1	0	0	0	0	0	0	0	0	2	8
2APQ1003	Accel. DPG	0	0	0	0	0	1	0	0	0	0	0	0	1	9
2APQ1004	Accel. MBTS	0	0	0	0	0	1	0	0	0	0	0	0	1	10
2APQ5001	Iron(III) oxide	0	0	0	0	0	0	0	0	0	0	0	0	0	11
2APQ1005	Accel. CBS	0	0	0	0	0	0	0	0	0	0	0	0	0	12
2APQ1009	Peroxide D.I.P.P	0	0	0	0	0	0	0	0	0	0	0	0	0	13
2APQ2002	Stearic acid	0	0	0	0	0	0	0	0	0	0	0	0	0	14
2APQ2004	Sodium Stearate	0	0	0	0	0	0	0	0	0	0	0	0	0	15
2APQ5012	Oppasin Yellow	0	0	0	0	0	0	0	0	0	0	0	0	0	16
2APQA002	Magnesium oxide	0	0	0	0	0	0	0	0	0	0	0	0	0	17
	Production %	0,183	0,160	0,100	0,061	0,058	0,055	0,038	0,031	0,026	0,026	0,024	0,022		

reason, Class B products were also considered for all calculations. To solve this mathematical problem, linear programming was used. Of all the approaches to linear programming, the assignment problem method was chosen. The positioning of the chemicals in the lockers can be interpreted as an assignment problem since each chemical is placed in only one of the lockers and in each of the lockers only one chemical can be placed. In order to model the problem in question as an assignment problem, it is fundamental to define the function coefficients. For this purpose, a study was elaborated based on the class A and B products produced in the year 2020, which were considered for the optimization method calculations. Thus, the index "i" was used to represent the n class A and class B products analysed. It is represented by q_i, ∀i = 1, ..., n the product i weighting factor, where q_i was determined in the calculation of the production frequency when performing the ABC analysis. The q_i is obtained by dividing the production value of an article in 2020 by the total production value in that year.

$$i - Product; (i = 1, \ldots, n)$$

$$Qi - Quantity\ of\ product\ i\ produced\ in\ the\ year\ 2020$$

$$qi = \frac{Qi}{\sum_{i=1}^{n} Qi} \tag{1}$$

Table 3 summarizes what was previously mentioned. Thus, this table can be interpreted as a matrix with n lines, which correspond to the products of classes A and B, whose production was analysed and m columns, corresponding to the 17 chemicals that must be positioned in the 12 lockers. In this way, a $n \times m$ matrix was built in which n represents the line of products and m represents the column of chemicals. This matrix was entitled by P_{ij}, where j corresponds to the chemical product with $(j = 1, ..., m)$. As can be seen in Table 3, the last line corresponds to the weighting factor qi, corresponding to each product i.

Thus, if $P_{ij} = 1$ it means that the chemical product j integrates the composition of the product i. Taking into account that to formulate the assignment problem it is necessary to have the same number of tasks and individuals, a square matrix was built. In this way, for a number of chemicals the same number of lockers has to be assigned. As the 17 chemicals used in the CR1 line and the 12 lockers available in the line were being considered, it was necessary to add five fictitious lockers to the matrix. Thus:

- k, is the position index in the locker: $(k = 1, \ldots, m)$
- t_k, is the time it takes for the trolley to travel, considering the layout of the lockers on the line.
- C_{jk} is the cost of putting the chemical j in position in locker k, and is calculated from the following mathematical expression:

Table 3 P_{ij} Matrix – selection of chemical products used in Class A and B products and calculation of q_i

Article *i* / Chemical *j*	Chemical products CLASS A												Chemical products CLASS B			
	1521	2549	NL76	CDM01	S112	VC95	1049	1237	H70L	NC733	1028	1235	1523	WI02	N709	GT56
Accel. DPG	0	0	0	0	0	1	0	0	0	0	0	0	0	0	0	0
Accel. MBTS	0	0	0	0	0	1	0	0	0	0	0	0	0	0	0	0
Accel. CBS	0	0	0	0	0	0	0	0	0	0	0	0	0	0	0	0
Accel. TMTM	0	0	1	0	1	0	0	0	0	0	0	0	0	0	0	0
Accel. TMTD	1	1	0	0	0	0	0	1	0	0	1	1	1	0	0	0
Peroxide D.I.P.P	0	0	0	0	0	0	0	0	0	0	0	0	0	0	0	0
Zinc Oxide	0	0	1	0	1	0	0	0	0	0	0	0	0	0	0	0
Stearic acid	0	0	0	0	0	0	0	0	0	0	0	0	0	0	0	0
Stearic sodium	0	0	0	0	0	0	0	0	0	0	0	0	0	0	0	0
Iron(III) oxide	0	0	0	0	0	0	0	0	0	0	0	0	1	0	0	0
Oppasin Yellow	0	0	0	0	0	0	0	0	0	0	0	0	0	0	0	0
Vulcadur A/Resin DUREZ 12687	0	0	0	0	0	1	0	0	0	1	0	0	0	1	0	0
Clear Resins	0	0	0	0	1	1	1	0	1	1	1	0	0	1	0	0
Black FEF N550	1	0	0	0	0	0	0	1	0	0	0	0	0	0	0	0
Sulfur	1	1	1	1	1	1	1	1	1	1	1	1	1	1	1	1
Magnesium oxide	0	0	0	0	0	0	0	0	0	0	0	0	0	0	0	0
ADC Activated	0	0	1	1	0	0	0	0	0	0	0	0	0	0	0	0
qi (Production %)	18,299	16,029	10,035	6,074	5,846	5,544	3,776	3,094	2,589	2,584	2,418	2,170	1,808	1,506	1,397	1,009

Table 4 C_{jk} matrix example – Cost of placing chemical *j* from the locker in position *k*

	Chemical prod. j / Position k	1	2	3	4	5
1	Accel. DPG	13,86	27,72	41,58	55,44	69,29
2	Accel. MBTS	13,86	27,72	41,58	55,44	69,29
3	Accel. CBS	0	0	0	0	0
4	Accel. TMTM	43,48	86,76	130,14	173,52	216,9
5	Accel. TMTD	114,73	229,47	344,2	458,94	573,67
6	Peroxide D.I.P.P	1,44	2,88	4,32	5,76	7,2
7	Zinc Oxide	44,94	89,88	134,82	179,77	224,71
8	Stearic acid	0	0	0	0	0

$$C_{jk} = \sum_{i=1}^{n} Pij \times qi \times t_k \tag{2}$$

In this case, the "cost" refers to the time needed for the cart to travel to the lockers of the chemicals that constitute the product being fed to the cork composite composition. An example of the results obtained is shown in Table 4.

It is important to mention that the amount of chemical used in each article is not a relevant data, since we are only evaluating the movements that the trolley needs to accomplish. Finally, the X_{jk} matrix was performed in order to affect one chemical to each available locker to minimize the unloading time and, consequently, to optimize the production line. In order to formulate this assignment problem, the following decision variables were established:

In this way, with support of a MS Excel® Solver tool, the results presented in Table 5 were obtained, which allows to obtain the most convenient position of the lockers into the chemical feeding equipment, as depicted in Fig. 2.

Decision variables:	$X_{jk} = \begin{cases} 1, & if\ the\ chemical\ j\ is\ allocated\ in\ position\ k \\ 0 & otherwise \end{cases}$	$X_{ij} \in \{0, 1\}$
Objective function:	Min $z = \sum_{j=1}^{m} \sum_{k=1}^{m} C_{jk} \times X_{jk}$	(3)
Restrictions 1:	In each locker can only be placed one chemical: $\sum_{j=1}^{m} X_{jk} = 1 \qquad \forall k = 1, \ldots, m$	(4)
Restrictions 2:	Each chemical can only be assignment to one locker. $\sum_{k=1}^{m} X_{jk} = 1 \qquad \forall j = 1, \ldots, m$	(5)

Table 5 Assignment of the chemical to the respective locker

Order	Chemical prod. j / Locker position k	1	2	3	4	5	6	7	8	9	10	11	12	13	14	15	16	17	Restriction 2
1	Sulfur	1	0	0	0	0	0	0	0	0	0	0	0	0	0	0	0	0	1
2	Accel. TMTD	0	1	0	0	0	0	0	0	0	0	0	0	0	0	0	0	0	1
3	Clear Resins	0	0	1	0	0	0	0	0	0	0	0	0	0	0	0	0	0	1
4	Black FEF N550	0	0	0	1	0	0	0	0	0	0	0	0	0	0	0	0	0	1
5	ADC Activated	0	0	0	0	1	0	0	0	0	0	0	0	0	0	0	0	0	1
6	Zinc Oxide	0	0	0	0	0	1	0	0	0	0	0	0	0	0	0	0	0	1
7	Accel. TMTM	0	0	0	0	0	0	1	0	0	0	0	0	0	0	0	0	0	1
8	Vulcadur A/Resin DUREZ 12687	0	0	0	0	0	0	0	1	0	0	0	0	0	0	0	0	0	1
9	Accel. MBTS	0	0	0	0	0	0	0	0	1	0	0	0	0	0	0	0	0	1
10	Accel. DPG	0	0	0	0	0	0	0	0	0	1	0	0	0	0	0	0	0	1
11	Iron(III) oxide	0	0	0	0	0	0	0	0	0	0	1	0	0	0	0	0	0	1
12	Peroxide D.I.P.P	0	0	0	0	0	0	0	0	0	0	0	1	0	0	0	0	0	1
13	Oppasin Yellow	0	0	0	0	0	0	0	0	0	0	0	0	1	0	0	0	0	1
14	Sodium stearate	0	0	0	0	0	0	0	0	0	0	0	0	0	1	0	0	0	1
15	Stearic acid	0	0	0	0	0	0	0	0	0	0	0	0	0	0	1	0	0	1
16	Accel. CBS	0	0	0	0	0	0	0	0	0	0	0	0	0	0	0	1	0	1
17	Magnesium oxide	0	0	0	0	0	0	0	0	0	0	0	0	0	0	0	0	1	1
	Restriction 1	1	1	1	1	1	1	1	1	1	1	1	1	1	1	1	1	1	

1	2	3	4	5	6	7	8	9	10	11	12
Sulfur	Accel. TMTD	Clear Resins	Black FEF N550	ADC Activated	Zinc Oxide	Accel. TMTM	Accel. MBTS	Accel. DPG	Vulcadur A/ Resin DUREZ 12687	Red Iron Oxide	Peroxide D.I.P.P.
2APQA001	2APQ1007	2APQ6004	2AQ7002	2APQC002	2APQ2001	2APQ1006	2APQ1003	2APQ1004	2APQ6002	2APQ5001	2APQ1009

Fig. 2 Representation of the correct layout of chemical lockers

As can be seen, the correct order for the layout of the chemicals in the lockers has been determined to optimize the CR1 production line. In order to understand the impact of changing the lockers layout, the required production time in 2020 was calculated for article *XPTO*, if the chemicals were arranged in the lockers in the correct order, as illustrated in Fig. 2. With this new layout, the travel time required to fulfil a container of chemicals would be 20 seconds, as can be concluded by the following mathematical expression:

$$\text{Compounds pick-up time} = (2,5) + (2,5) + (2,5\times(4-2)) + (2,5\times4) = 20\ \text{s}$$

Thus, the total travel time in 2021 to produce the product *XPTO* would be:

$$6417\ blocks \times 3\ containers \times 20\ s \approx 106,95\ h \approx 4,46\ days$$

Considering the total time value of 13.37 days, consumed during the displacement of the trolley in the year 2020, in the production of the product *XPTO* using the initial arrangement of the lockers, then, through calculation: (13.37–4.46 days) ÷ 13.37 days, it can be concluded that, in the case of product *XPTO*, the time taken to move the trolley has decreased by 66.66%. In practical terms, this decrease means a great optimization for the line, due to the reduction of the time wasted in the operation of the dosage of chemicals. In addition, the time available for the operator to perform other tasks has increased significantly.

4 Conclusions

Regarding the results obtained with the study of the chemicals dosage, it was found that, in the case of *XPTO* production, class A product, the dosage time of the chemicals will be reduced approximately 67%, after the arrangement of the lockers is reorganized. Often, companies do not value small wastes of time, which, even though they seem small, affect the entire process. In this case, the time savings obtained are so significant that it may imply that other adjacent processes need to be revised in order to keep up with the speed now obtained in this sub-process. Thus, it can be seen that the mathematical formulation considered in the resolution of this problem was adequate, promoting a significant time saving in the process of collecting different chemical components to fill the necessary recipe for the production of composites in cork and rubber.

Acknowledgments F.J.G. Silva, L. P. Ferreira and M. T. Pereira acknowledge the INEGI research support.

References

1. Silva, F.J.G., Gouveia, R.M.: Cleaner Production – Toward a Better Future. Springer Nature, Cham (2020). ISBN: 978-3-030-23165-1
2. Gil, L.: A cortiça como material de construção – Manual Técnico (In Portuguese). Available online: https://www.academia.edu/9138885/A_cortiça_como_material_de_construção (2007). Accessed on 14 Apr 2021
3. Marinho, P., Pimentel, D., Casais, R., Silva, F.J.G., Sá, J.C., Ferreira, L.P.: Selecting the best tools and framework to evaluate equipment malfunctions and improve the OEE in the cork industry. Int. J. Ind. Eng. Manag. **12**(4), 1–13 (2021). https://doi.org/10.24867/IJIEM-2021-4-295
4. Sousa, E., Silva, F.J.G., Ferreira, L.P., Pereira, M.T., Gouveia, R., Silva, R.P.: Applying SMED methodology in cork stoppers production. Procedia Manuf. **17**, 611–622 (2018). https://doi.org/10.1016/j.promfg.2018.10.103
5. Gil, L.: Cork composites: a review. Materials. **2**(3), 776–789 (2009). https://doi.org/10.3390/ma2030776
6. Gürgena, S., Sofuoğlu, M.A.: Smart polymer integrated cork composites for enhanced vibration damping properties. Compos. Struct. **258**, 113200 (2021). https://doi.org/10.1016/j.compstruct.2020.113200

7. Costa, M.J.R., Gouveia, R.M., Silva, F.J.G., Campilho, R.D.S.G.: How to solve quality problems by advanced fully-automated manufacturing systems. Int. J. Adv. Manuf. Technol. **94**, 3041–3063 (2018). https://doi.org/10.1007/s00170-017-0158-8
8. Wikarek, J., Sitek, P., Nielsen, P.: Model of decision support for the configuration of manufacturing system. IFAC PapersOnLine. **52**(13), 826–831 (2019). https://doi.org/10.1016/j.ifacol.2019.11.232
9. Martins, M., Godina, R., Pimentel, C., Silva, F.J.G., Matias, J.C.O.: A practical study of the application of SMED to electron-beam machining in automotive industry. Procedia Manuf. **17**, 647–654 (2018). https://doi.org/10.1016/j.promfg.2018.10.113
10. Rafael, A.: Programação Linear e Algumas Extensões (In Portuguese). MSc dissertation in Mathematics for Teaching. Mathematics Department, Faculty of Sciences, University of Porto, Portugal, p. 76 (2014)
11. Burgard, S.: Análise e Melhoria dos Processos Produtivos numa Unidade Industrial de Cortiça e Borracha (In Portuguese). MSc dissertation in Mechanical Engineering, Department of Mechanical Engineering, Faculty of Engineering, University of Porto, p. 63. (2016)
12. Chen, D., Zhong, Y., Liao, Y., Li, L.: Review of multiple criteria and multiple constraint-level linear programming. Procedia Comput. Sci. **17**, 158–165 (2013). https://doi.org/10.1016/j.procs.2013.05.022
13. Serrano-Hernandez, A., Cadarso, L., Faulin, J.: A strategic multistage tactical two-stage stochastic optimization model for the airline fleet management problem. Transp. Res. Proc. **47**, 473–480 (2020). https://doi.org/10.1016/j.trpro.2020.03.152
14. Marchetti, P.A., Cerdá, J.: An approximate framework for large multistage batch scheduling problems focusing on bottleneck resources. Comput. Aided. **21**, 1991–1996 (2006). https://doi.org/10.1016/S1570-7946(06)80340-8
15. Lee, D.-H., Nie, Y., Chen, A.: A conjugate gradient projection algorithm for the traffic assignment problem. Math. Comput. Model. **37**(7–8), 863–878 (2003). https://doi.org/10.1016/S0895-7177(03)00090-6

Impact of Regulatory Changes on the Economic Viability of Photovoltaic Systems in Brazil

Paula Donaduzzi Rigo, Jordana Graciano Rech dos Santos, Cléber Vinícius de Freitas, and Julio Cezar Mairesse Siluk

Abstract In Brazil, the accelerated growth of solar photovoltaic energy through Distributed Generation (DG) conducts National Electric Energy Agency (ANEEL) to revise the Electric Energy Compensation System to balance the electricity grid costs DG participants and non-participants. For this, ANEEL proposed five Alternatives with the entry of grid electricity use costs (TUSD and TE) for DG participants. This article aims to analyze TUSD's Alternatives' impact on the economic viability of PV systems in Brazil in three scenarios: residential, commercial, and industrial. We verify that the new legislation significantly impacts the residential investor. From Alternative 0 to Alternative 5, the Payback Time can take approximately three times in years for the residential class, from 4.22 to 13.07 years. TUSD Alternatives' impact for the commercial and industrial class occurs with less intensity since the proportion that TUSD rates have compared to the level of electricity consumption is modest.

Keywords Photovoltaic energy · Distributed Generation · Net Metering

1 Introduction

In recent decades, fossil fuel's enormous exhaustion has led to large quantities of toxic particles and greenhouse gases, climate change, and destroying wildlife [1]. Based on this perspective, it was necessary to change or diversify energy use and find a new way to obtain energy. The 100-years-old model of centralized generation, which was causing damages to the world, and one-way power flow, is being disrupted by renewable energy technologies that enable customers to generate their electricity [2]. However, an energy market transition requires innovation, policy review and realistic implementation [3].

P. D. Rigo (✉) · J. G. R. dos Santos · C. V. de Freitas · J. C. M. Siluk
Federal University of Santa Maria, Department of Production and Systems Engineering, Santa Maria, RS, Brazil

V. M. López Sánchez et al. (eds.), *Industrial Engineering and Operations Management*, Springer Proceedings in Mathematics & Statistics 400,
https://doi.org/10.1007/978-3-031-14763-0_4

Brazil has a large part of its matrix for generating renewable electric energy based on hydroelectric plants, representing 61.99% in the energy matrix, followed by fossil fuels with 16.36% [4]. However, the hydrological cycle changes increase hydroelectric generation vulnerability in Brazil [5]. For this reason, it seeks to reduce dependencies on the water source without resorting to non-renewable sources of electricity generation. Renewable sources, such as solar photovoltaic, wind, and biomass, are essential for diversifying the countries' energy matrix [6]. One of the options for inserting single sources is to invest in Distributed Generation (DG). Ackermann et al. [7] defined Distributed Generation as "a source of electricity generation connected directly to the distribution network". The concept also includes measurement, control, and command equipment that articulate the generators' operation and the eventual control of loads so that they can adapt to the energy supply [8]. In other words, DG consists of the generation of electricity from renewable sources with connection to the distribution grid close to the consumer, capable of meeting the local load [9].

In April 2012, the National Electric Energy Agency (ANEEL) published Normative Resolution 482/2012 (REN ANEEL 482/2012), creating a new class of generators/consumers on a small scale: distributed micro-generation and distributed mini-generation. The REN ANEEL 482/2012 established the corresponding interconnection standards optimized for these projects. Consequently, DG's growth was exponential. In January 2021, there were 387,850 Renewable Energy Micro and Mini-generators, totaling 4.76 GW, with the photovoltaic (PV) technology representing 97.2% of this installed power [10]. The DG option proved to be an exciting investment due to the Electric Energy Compensation System. The Electricity Compensation System, which REN ANEEL 687/2015 defines as "a system in which the active energy injected by a consumer unit with microgeneration or distributed mini-generation is provided, through a free loan, to a local distributor and subsequently compensated with the consumption of active electricity" is responsible for DG's investment attractiveness [11].

The DG growth in Brazil after REN ANEEL 482/2012 emphasizes the importance of policies, investments, and government support to promote renewable energies, especially solar PV [12]. However, the increasing insertion of DG in the Brazilian electricity grid has been concerned with Electrical utility companies and their monopoly. They have faced the perspective that customers will reduce their electricity purchases as they adopt solar PV generation and enter a positive feedback loop (spiral of death) [13, 14]. Therefore, through public consultations in 2019 and 2020, ANEEL developed a new compensation rule for DG, which adds costs of the Distribution System Use Tariff (TUSD) by the energy utilities to DG users [15]. This system pricing can occur through five Alternatives and, if approved, it can implement from 2021 onwards.

Based on the Alternatives proposed by ANEEL, this article aims to analyze the impact of TUSD's Alternatives on the economic viability of PV systems in Brazil in three scenarios: residential, commercial, and industrial. The choice of application in these scenarios refers to the fact that these subgroups represent 85% of the DG [10]. To set the feasibility calculations in a local, we selected the city of Rio de

Janeiro because it is the city with the highest quantity of DG connections and has PV systems with an average size adhering to the national average, which is 8 kWp for residential systems, 30 kWp for commercial structures and 50 kWp for industrial systems. This article is divided into four sections, the introduction, the study method, the results obtained, and finally, conclusions.

2 Method

The method section has three subsections: The first subsection presents the current electricity tariff scenario in Brazil and the Alternatives proposed by ANEEL to modify the DG Electricity Compensation System; The second subsection presents the study scenarios, with the respective technical specifications of the PV systems; The third subsection presents the construction of cash flows to calculate the economic viability of PV systems. We chose three techniques for analyzing the scenarios' economic viability: Net Present Value (NPV), Internal Rate of Return (IRR), and Discounted Payback Time. These economic techniques are the most used by investors in photovoltaic energy.

2.1 The Electricity Tariff in Brazil and Regulatory Changes

Based on REN ANEEL 482/2012, the Distributed Generation (DG) user can compensate all the components of Fig. 1 with their electricity generation system. Federal (PIS/COFINS) and State (ICMS) taxes are exempt for DG users. This compensation system is the current GD's national tariff scenario, which we will call Alternative 0. The application PIS/COFINS and ICMS taxes are under TUSD plus TE's total amount for consumers who do not have DG. The second section presents the study scenarios, with the respective technical specifications of the PV systems.

As discussed in the Introduction, ANEEL pointed to future changes in the DG's compensation system. This change reduces consumer subsidies with DG because it inserts the costs of TUSD and part of TE. For this, ANEEL proposed 5 Alternatives. These Alternatives range from Alternative 0, which means the scope of the current standard (Fig. 1), to Alternative 5, which most negatively impacts DG projects' viability. In summary, the Alternatives presented by [15] are:

- **Alternative 0 (Current scenario):** the compensation for electricity injected into the grid occurs through all components of TUSD and TE;
- **Alternative 1:** The **B-Wire Transport** component will be charged for all energy consumed from the grid. The other tariff components continue to be charged only on the difference between the energy consumed and the energy injected into the grid;

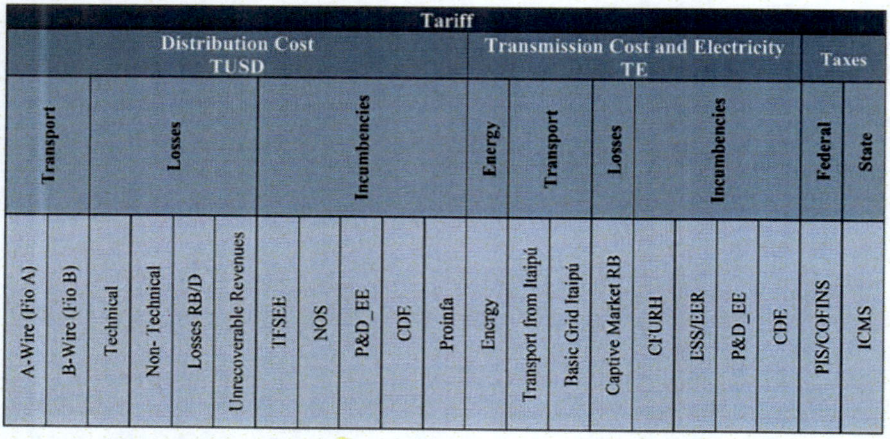

Fig. 1 Composition of the Electricity Tariff. (Source: [15])

- **Alternative 2:** The **B-Wire** and **A-Wire Transport** component will be charged for all energy consumed from the grid. The other tariff components continue to be charged only on the difference between the energy consumed and the energy injected into the grid;
- **Alternative 3:** The **B-Wire**, **A-Wire Transport,** and **TUSD Incumbencies** components will be charged for all energy consumed from the grid.
- **Alternative 4:** The **B-Wire**, **A-Wire Transport, TUSD Incumbencies,** and **Losses** components will be charged for all energy consumed from the grid. This Alternative corresponds to all TUSD.
- **Alternative 5**: The **B-Wire**, **A-Wire Transport, TUSD Incumbencies, Losses, and TE Incumbencies, Transport, and Losses** components will be charged for all energy consumed from the grid. This Alternative corresponds to all TUSD, only the TE Energy component would be set on the difference between the energy consumed and the energy injected into the grid. The other tariff components would affect all the energy consumed by the grid.

The ANEEL proposal document also discusses the method for establishing the Alternatives. Setting a fixed day to change the compensation system regime would be the best option due to the ease of controlling the DG expansion in Brazil. However, a fixed date could positively or negatively influence this diffusion, given the number of variables involved in energy generation and consumption. Indicating a fixed date can lead to an expansion and contraction of the sector, as the Normative Resolution provides 25 years of exemption for customers who adopt the DG until the standard changes.

Because of this, ANEEL recommends that the change in the compensation system occurs through the installed power levels, as it is possible to delimit the amount of power through the term "triggers" (levels for changes from one Alternative to another). It would also be possible to define the triggers using the amounts installed per Electrical utility companies, thus promoting DG penetration with greater dissemination and less concentration in regions with more favorable conditions from the point of view of solar incidence. According to [15], it would be possible to maintain Alternative 0 until the local DG market consolidates, with the installation of 3.4 GW in the whole country, and then to change to Alternative 1. With the triggers defined for the transitions between the Alternatives, the market will effectively monitor the DG dissemination.

2.2 *Study Scenarios*

We selected three photovoltaic power generation scenarios to apply the Alternatives. These three scenarios represent the main classes that install DG PV systems in Brazil: Residential, Commercial, and Industrial. Based on ANEEL data [10], we identified the average size in kW peak of each class's systems: Residential with 8 kWp, Commercial with 30 kWp, and Industrial with 50 kWp. Table 1 presents each scenario's characteristics and the reports/websites that we consult to determine the calculations' components.

The DG PV systems are connected to the low voltage grid, with the respective electricity tariffs for each scenario. We had to select a city to settle the viability estimations the city's electricity utility's rate. Therefore, we chose Rio de Janeiro's city because it has the highest quantity of Consumer Units (CU) with DG PV in Brazil, 4709 in January 2021. In Rio de Janeiro, the company *Light* performs electricity distribution [18]. We started with the regulatory classification of systems. Each class has a separate electricity tariff. The most common DG modality is Generation at CU, which consists of a system installed at the load site. After that, we selected the most used technology in Brazil for modules and inverters. Greener studies identified the most used brands in Brazil, *Canadian* for modules and *Fronius* for Solar Inverter [16]. The same study determined that Poli-PERC modules are the most significant trend in the year 2021. Given this, we collected the PV modules and inverter's sale prices in the Free Market of Brazil, calculated the number of modules, occupation area, and dimensioned the inverters. In addition to cabling, security devices, and installation service costs, we calculate the final price per Wp and compare it to the Greener study, totaling R$ 4.20 for residential systems and R$ 3.70 for commercial and industrial PV systems.

We project the size in m^2 of the system based on the number of modules. The module has area of 2.209 m^2. We used the system's total area to calculate the cost of predictive maintenance and electricity generation. The estimated cost of maintenance in Brazil is approximately R$ 20.00 m^2/year [17]. Finally, we identify the fees involved in electricity tariffs for each scenario. Considering that PV systems could only begin to influence the electricity tariff in Brazil from 2012, we have collected the value of electricity tariffs by classes of the Light company since 2012 – already included National Taxes (PIS / COFINS) and State Taxes (ICMS) – and we calculate the annual adjustment rate. In the city of Rio de Janeiro, a municipal tax

Table 1 Characteristics of each scenario

Characteristics		Scenario			References
Regulatory classification	Class	Residential	Commercial B3	Industrial B3	[10]
	Modality	Generation at the CU	Generation at the CU	Generation at the CU	
Modules	Model and quantify	20 Modules *Canadian* Poli-PERC 410 W	74 Modules *Canadian* Poli-PERC 410 W	122 Modules *Canadian* Poli-PERC 410 W	[16] e Mercado Livre Brazil
	Unit value	R$ 949.00	R$ 949.00	R$ 949.00	
	Amount	R$ 18,980.00	R$ 70,226.00	R$ 115,778.00	
Solar inverter	Model and quantify	*Fronius* 8,2 kW, Monophase 220 V, 2 Mppt (1 unit)	*Fronius* Eco 27.0 kW, Three-phase 380 V, (1 unit)	*Fronius* Eco 27.0 kW Three-phase, 380 V (2 units)	[16] e Mercado Livre Brazil
	Unit value	R$ 11,459.80	R$ 24,309.00	R$ 24,309.00	
	Amount	R$ 11,459.80	R$ 24,309.00	R$ 48,618.00	
Other equipment	Integration and cabling	R$ 3160.20	R$ 16,465.00	R$ 20,604.00	[16]
System final cost	Cost per Wp	R$ 4.20	R$ 3.70	R$ 3.70	
	Total cost	R$ 33,600.00	R$ 111,000.00	R$ 185,000.00	
System size	Size (kWp)	8 kWp	30 kWp	50 kWp	[10]
	Size (m^2)	44.2 m^2	163.5 m^2	269.5 m^2	
System maintenance	Maintenance (m^2/year)	R$ 20.00	R$ 20.00	R$ 20.00	[17]
	Maintenance (year)	R$ 883.67	R$ 3269.59	R$ 5390.41	
Values for calculations	Tariff (R$/kW)	R$ 0.83	R$ 1.01	R$ 1.01	[18]
	Tariff readjustment	6.30%	7.21%	7.21%	
	COSIP (R$/month)	R$ 27.95	R$ 52.43	R$ 131.83	
	COSIP readjustment	5.75%	5.78%	9.38%	
	National Inflation	3.01%	3.01%	3.01%	[19]
	WACC tax	7.32%	7.32%	7.32%	[20]

is also imposed, called COSIP. We did the same procedure to calculate the annual COSIP adjustment rate. COSIP is a value presented monthly by the city of Rio de Janeiro and is related to the unit's power consumption, which for the scenarios is between 500 and 1000 kWh for residential, between 1000 and 5000 kWh for commercial, and between 5000 and 10,000 kWh for industrial. To calculate an inverter's future purchase, we estimate Brazil's inflation to replace it every 10 years.

Table 2 Estimative of annual electricity generation

Characteristics		Value	References
Module efficiency	Module Poli-perc efficiency	17.88%	*Canadian*
	1° year degradation:	2.50%	*Canadian*
	Annual efficiency loss	0.70%	*Canadian*
	Other common losses	14.08%	[22]
	Module angle	20° to North	[22]
Inverter efficiency	Inverter DC to AC efficiency	96%	*Fronius*
Annual electricity generation	Solar radiation (kWh/m^2/day):	5.39	[22]
	Residential system 8 kWp (kWh/day)	12,791.20	Estimative
	Commercial system 30 kWp (kWh/day)	47,327.38	Estimative
	Industrial system 50 kWp (kWh/day)	78,026.23	Estimative

Finally, for economic viability techniques, it is necessary to use a Minimum Attractive Rate of Return (MARR). For this reason, we selected the MARR used by the electricity sector in Brazil, the Weighted Average Cost of Capital (WACC), the regulatory capital remuneration rate used to review the tariff or revenue of electricity distributors, transmitters, and generators [21]. The method for estimating the WACC is based on national parameters and uses public data sources. WACC treats the nature of the capital sources used in energy distributors differently, considering that equity is less flexible than third party capital. The Net Debt to regulatory EBITDA ratio defines distributors' capital structure, incorporating this ratio in the financial and credit markets.

We estimate electricity generation based on Table 2. The POLI-PERC modules' efficiency is 17.88%, and its degradation is greater in the first year of operation and approximately linear in the next years. To support our estimations, we consulted PVWatts, a photovoltaic power generation calculation platform. On PVWatts we collect the common loss rate of 14.08%, which corresponds to Soiling (2%), Shading (3%), Snow (0%), Mismatch (2%), Wiring (2%), Connections (0.5%), Light-Induced Degradation (1.5%), Nameplate Rating (1%), Age (0%), and Availability (3%). Considering that Rio de Janeiro is south of the equator line, PV modules ideal inclination is approximately 20° facing north. With these specifications, the solar radiation suffered by the Rio de Janeiro modules is on average 5.39 kWh/m^2/day. Solar radiation is the essential parameter needed for designing and developing solar PV systems [23]. We estimate the annual electricity generation by multiplying the average daily monthly radiation, the number of days in the month, and the system's size in m^2. Then, we multiplied the previous result by the module's efficiency and the inverter and discounted the common losses. We estimate the annual efficiency loss of the PV modules directly in the cash flow, 2.5% in first year and 0.7% in next years.

Year	System costs			Electricity grid costs			Power generation (kWh)	Recipe (R$)	Cash Flow (R$)	Discounted Cash Flow (R$)	Financial Balance (R$)
	Investment (R$)	Inverter replacement (R$)	Maintenance Cost (R$)	Tariff (R$/kW)	Minimum rate (R$)	COSIP (R$)					
0	33,600.00			0.83		335.40	12,791.18		-33,600.00	-33,600.00	-33,600.00
1			883.67	0.89	1,064.65	354.69	12,471.41	9,645.44	8,761.76	8,164.15	-25,435.85
2			910.27	0.94	1,131.73	375.08	12,384.11	10,172.72	9,262.45	8,042.00	-17,393.85
3			937.67	1.00	1,203.03	396.65	12,297.42	10,728.75	9,791.08	7,921.16	-9,472.69
4			965.90	1.07	1,278.82	419.45	12,211.33	11,315.11	10,349.22	7,801.62	-1,671.07
5			994.97	1.13	1,359.38	443.57	12,125.86	11,933.44	10,938.47	7,683.40	6,012.33
6			1,024.92	1.20	1,445.02	469.08	12,040.97	12,585.47	11,560.55	7,566.49	13,578.82
7			1,055.77	1.28	1,536.06	496.05	11,956.69	13,273.04	12,217.28	7,450.92	21,029.74
8			1,087.55	1.36	1,632.83	524.57	11,872.99	13,998.09	12,910.54	7,336.67	28,366.41
9			1,120.28	1.45	1,735.70	554.74	11,789.88	14,762.64	13,642.36	7,223.76	35,590.18
10		15,415.97	1,154.00	1.54	1,845.05	586.63	11,707.35	15,568.85	-1,001.13	-493.95	35,096.23
11			1,188.74	1.63	1,961.29	620.36	11,625.40	16,418.97	15,230.23	7,001.95	42,098.18
12			1,224.52	1.74	2,084.85	656.04	11,544.02	17,315.39	16,090.87	6,893.05	48,991.23
13			1,261.38	1.85	2,216.19	693.76	11,463.21	18,260.63	16,999.25	6,785.49	55,776.72
14			1,299.34	1.96	2,355.81	733.65	11,382.97	19,257.34	17,957.99	6,679.26	62,455.99
15			1,338.45	2.09	2,504.23	775.83	11,303.29	20,308.30	18,969.85	6,574.37	69,030.35
16			1,378.74	2.22	2,662.00	820.44	11,224.17	21,416.47	20,037.73	6,470.80	75,501.15
17			1,420.24	2.36	2,829.70	867.62	11,145.60	22,584.95	21,164.71	6,368.56	81,869.71
18			1,462.99	2.51	3,007.97	917.51	11,067.58	23,817.00	22,354.01	6,267.63	88,137.34
19			1,507.03	2.66	3,197.48	970.26	10,990.11	25,116.09	23,609.06	6,168.03	94,305.37
20		20,737.90	1,552.39	2.83	3,398.92	1,026.05	10,913.17	26,485.84	4,195.55	1,021.35	95,326.72
21			1,599.11	3.01	3,613.05	1,085.05	10,836.78	27,930.08	26,330.97	5,972.73	101,299.45
22			1,647.25	3.20	3,840.67	1,147.44	10,760.93	29,452.86	27,805.61	5,877.03	107,176.48
23			1,696.83	3.40	4,082.63	1,213.42	10,685.60	31,058.43	29,361.60	5,782.62	112,959.09
24			1,747.90	3.62	4,339.84	1,283.19	10,610.80	32,751.27	31,003.36	5,689.48	118,648.58
25			1,800.52	3.84	4,613.25	1,356.98	10,536.52	34,536.11	32,735.59	5,597.62	124,246.20

Fig. 2 Cash flow from residential system Alternative 0

2.3 *Cash Flow*

Figure 2 shows the cash flow from residential system Alternative 0, the actual scenario, where energy compensation happens in every component of TUSD and TE. The first cash flow section is related to the System costs, form by the: initial investment (system price), the inverters' replacement price, and the maintenance cost. We consider the inflation rate for the inverter's replacement in years ten (10) and twenty (20). Maintenance costs also suffer from the inflation rate.

REN 482 change will influence economic viability, particularly in the second cash flow section, Electricity grid cost. For PV systems connected to the distribution network, the cash flow entry is "not buying energy from the energy utility". The grid user with a PV system in Alternative 0 no needs to pay for energy, neither TUSD, TE, National, and State taxes. The user must pay only the Municipal tax (COSIP) and the Minimum rate. The Minimum rate refers to the single-phase, two-phase, or three-phase grid connection. For this study's scenarios, we consider a three-phase grid connection with a mandatory monthly contract of 100 kW. The Minimum annual rate is 1200 kW multiplied by the tariff. For example, in the first year, the electricity grid tariff is 0.88721 R$/kW multiplied by 1200 kW, resulting in R$1064.65 of minimum rate. The tariff cost is readjusted by 6.30% for residential systems and 7.21% for commercial and industrial. COSIP is subject to the adjustment rate for each scenario as presented in the previous paper's section.

In Alternatives 1–5, the change in the cash flow related to Alternative 0 is the Electricity grid costs parts that will not be compensated by the PV system, making the payment compulsory. The values for the three scenarios are the same, referring to [18], differing only in the tariff adjustment rate, following the residential example:

- **Alternative 1:** The B-Wire Transport cost of R$ 176.21 per month is added, totaling R$ 2247.73 per year, considering the readjustment rate.
- **Alternative 2:** The A-Wire Transport cost of R$ 29.73 per month is added, totaling R$ 379.24 per year, considering the readjustment rate
- **Alternative 3:** The TUSD Incumbencies cost of R$ 51.56 per month is added, totaling R$ 657.70 per year, considering the readjustment rate.
- **Alternative 4:** The TUSD Losses cost of R$ 112.68 per month is added, totaling R$ 1437.35 per year, considering the readjustment rate.
- **Alternative 5:** The TE Incumbencies, Transport, and Losses cost of R$ 71.72 per month is added, totaling R$ 914.86 per year.

The next column in Fig. 2 shows the value of power generation in kW. This value decreases annually at the efficiency loss rate. The Recipe column is the final cash flow entry, consisting of multiplying the generation in kW by the tariff, in R$/kW and deducting the mandatory payments from the Minimum rate and COSIP. For example, Eq. 1 demonstrates the Recipe for year 1.

$$Recipe = \left(12,471.41^{*}\ 0.88721\right) - 1064.65 - 354.69 = 9645.44 \quad (1)$$

Subtracting the Systems cost from the Recipe, we have the cash flow column. As all values are projected for 25 years, it is necessary to make the discounted cash flow, bringing each value to the present, using the WACC as MARR. We calculate the final balance by subtracting the discounted cash flow for the current year from the previous year's balance. We estimate Payback Time using the financial balance column to observe when the balance becomes favorable for the investor. The NPV and IRR calculation is based on the cash flow column because its formula already contains the WACC rate. We build cash flows for each Alternative and each scenario, totaling 18 cash flows.

3 Results and Discussion

We selected three photovoltaic power generation scenarios to apply the We based the economic feasibility analysis on the techniques of Net Present Value (NPV), Internal Rate of Return (IRR), and Discounted Payback Time. Table 3 presents the results of these techniques for the six Alternatives in the three scenarios. We also calculate the conversion from NPV to dollars and the growth rate of Payback Time from one Alternative to the other to present the impact of the change in the standard on the metric most used by investors, which is Payback Time.

The Payback Time for PV systems has reduced significantly in recent years, reaching 4.22 years for residential systems installed until January 2021, 2.67 years for commercial, and 2.70 for industrial. Many PV installers companies emerged in Brazil and expanded their market share due to this attractive Payback Time. These

Table 3 System and Alternative's economic viability analysis

System and Alternatives		NPV (R$)	NPV ($[a])	IRR	Payback Time Time	Payback Time increase
Residential System with 8kWp	A0	124,246.20	23,267.08	30.78%	4.22	–
	A1	77,444.15	14,502.65	23.28%	5.77	36.70%
	A2	69,547.75	13,023.92	21.95%	6.15	6.60%
	A3	55,853.21	10,459.40	19.57%	6.94	12.87%
	A4	25,924.97	4854.86	13.80%	9.70	39.86%
	A5	6875.86	1287.62	9.30%	13.07	34.68%
Commercial System with 30kWp	A0	830,821.06	155,584.47	46.76%	2.67	–
	A1	734,470.59	137,541.31	42.75%	2.94	10.13%
	A2	725,669.46	135,893.16	42.39%	2.97	0.98%
	A3	710,405.89	133,034.81	41.78%	3.01	1.09%
	A4	677,048.64	126,788.14	40.42%	3.10	3.10%
	A5	655,817.00	122,812.17	39.55%	3.15	1.52%
Industrial System with 50kWp	A0	1,348,246.36	252,480.59	46.30%	2.70	–
	A1	1,296,082.00	242,711.98	45.03%	2.76	2.25%
	A2	1287,280.87	241,063.83	44.82%	2.773	0.55%
	A3	1,272,017.30	238,205.49	44.45%	2.777	0.14%
	A4	1,238,660.05	231,958.81	43.63%	2.807	1.08%
	A5	1,217,428.41	227,982.85	43.11%	2.807	0.03%

[a]Dollar exchange rate of 5.34 reais on January 23, 2022

values have also contributed to the accelerated growth in the country's number of installations since 2015. The longer is the capital recovery time, the higher is the investor risk, and for this reason, the Payback Time is usually decisive for the investor.

For the current regulatory resolution (Alternative 0), the NPV of a residential system is R$ 124,246.20 for a total of 25 years of operation. This is a positive and attractive investment since the initial capital is R$ 33,600.00. However, already in Alternative 1 proposed for the new regulation, the NPV drops to R$ 77,444.15. For Alternative 5, NPV remains positive, but the amount of R$ 6875.86 is not attractive for any risks not measured in cash flow. If the NPV is negatively impacted, consequently, the return on investment increased with Alternatives' application. With the application of Alternative 1, Payback Time has grown by 36.70%. For Alternative 5, the accumulated growth exceeds 130%.

The change in regulation does not have the same impact on commercial and industrial systems when compared to residential. Because the commercial and industrial subgroups commonly have power generation systems with higher power than in the residential subgroup. As they are governed by rates and tariffs close to residential ones and with significantly higher consumption, the recipe expected for commercial and industrial systems is also higher, especially when PV systems meet 100% of their consumption.

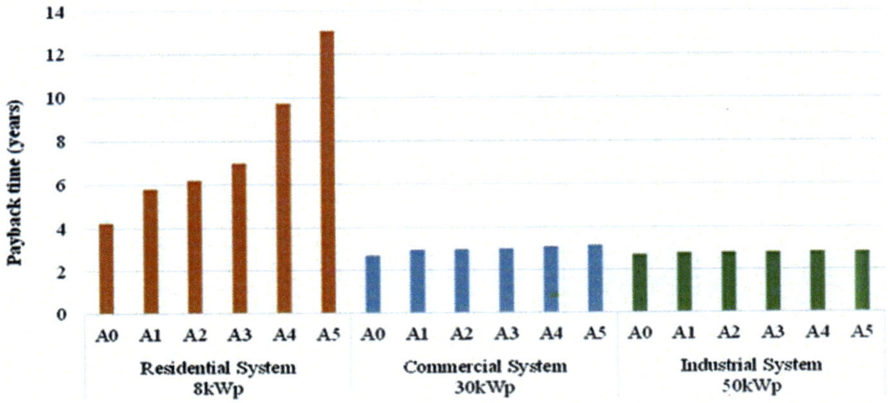

Fig. 3 Payback time by scenario and Alternative

Alternative 4 has a relevant feature when compared to the other Alternatives. In this alternative, occur the include values of TUSD Losses. Technical and non-technical losses are accounted for and passed on to consumers through TUSD. Utilities that have greater losses corroborate with more significant impacts on the PV systems' economic viability. Precisely the city of Rio de Janeiro has enormous losses in Brazil. These losses are due to the high number of energy fraud/deviations in the energy distribution system. Therefore, these scenarios have a significant impact when these losses enter as mandatorily charged in Alternative 4.

For the commercial and residential subgroups, the application of Alternatives implies minor impacts on the NPV values, adding up to a 16.82% growth in the Payback Time of commercial systems and only 4.06% for industrial systems. Figure 3 presents the Payback Time results to observe the differences between each Alternative and scenario.

According to the results in Table 3, NPV and IRR reinforce that the investment will continue to be economically viable, surpassing the WACC of 7.32% per year. The WACC rate as TMA can be considered a low rate, but that is consistent with the Brazilian economy's current period, since the Basic rate in Brazil, the SELIC rate, is 2% per year in 2021. However, the most significant impact for the future investor in photovoltaics is in the Payback Time of residential systems, or 8kWp systems. The observed growth of A0 (4.22 years) to A5 (13.07 years) reaches 130%. For commercial and industrial systems, or 30 kWp and 50 kWp, the Payback Time becomes only a few months longer from A0 to A5.

Evaluating the commercial and industrial subgroups scenarios, we can see a similarity and adherence in the evolution of the respective Payback Time, mainly due to the application rates that both have the same monetary value in the utility Light [18]. In this case, a more refined assessment by the Regulatory Agency of these two subgroups' tariff structure becomes essential, since most of the commercial points are inserted next to the residential neighborhoods. Thus, they share the same infrastructure as the energy utility, including facing the same impasses of technical and non-technical losses.

On the other hand, the industrial subgroup could receive separate incentives through a subdivision of activities because it demands energy consumption that is generally higher than the residential and commercial subgroups. Moreover, it is usually located in different geographic areas due to other municipal and state agencies' requirements. Currently, industrial subgroups receive incentives only when connected in higher voltage classes.

4 Conclusion

The electricity sector is undergoing an epochal transformation from the Bulk Grid (BG) to the Distributed Generation and Storage (DGS) model in which the role of energy utilities will change [24]. In Brazil, DG PV's accelerated growth conducts ANEEL to revise the Electric Energy Compensation System to balance the electricity grid costs for DG participants and non-participants. For this, ANEEL proposed five Alternatives with the entry of TUSD and TE costs for DG participants. Because today, the DG users do not need to pay any TUSD or TE costs. Based on the Alternatives, this article aims to analyze the impact of TUSD's Alternatives on the economic viability of PV systems in Brazil in three scenarios: residential, commercial, and industrial. The results help investors understand how much the new Electric Energy Compensation System impacts the future DG PV users.

Through the economic viability analysis, it was possible to verify that the new legislation significantly impacts the residential investor. From Alternative 0 to Alternative 1, the Payback Time increased by 36.70% for this class. This impact is because B-Wire – the first cost added in the regulatory change – has the highest cost. From Alternative 0 to Alternative 5, the Payback Time can take approximately three times in years for the residential class. For the residential class, the fees have a proportion in the total cost that is much higher than in the other two types, and therefore, the insertion of these costs significantly impacts the return on investment. TUSD Alternatives' impact for the commercial and industrial class occurs with less intensity since the proportion that TUSD rates have compared to the level of electricity consumption is modest.

It is a known fact that consumers who have distributed generation systems use the DG participants use the electricity grid more intensively than other consumers. However, this is a market heating the Brazilian economy and diversifying the Brazilian energy matrix through clean energy sources. This study shows that the new regulation proposal will intensely impact the residential class due to its lower electricity consumption. Therefore, ANEEL could develop other options for charging TUSD that consider electricity consumption, seeking a balance between the classes of residential, commercial, and industrial investors, and consumer units that did not adhere to DG. According to [25] and [26], there are numerous barriers to the distribution and diffusion of photovoltaic energy in Brazil, such as technical, economic, social, managerial, and political barriers. Among these impediments, financial issues stand out, such as the high cost of the initial investment, high Payback time of the investment, and the impact of TUSD, factors discussed in the present study. Although the regulatory change occurs gradually, the consequences will still impact the final consumer and PV market. As future studies, we recommended analyzing new tariff options to preserve the DG market's balance and potential actions by the government and ANEEL. The PV market niche must continue generating jobs and income growth in Brazil. Another point to deepen in future studies is comparing other countries and how this charge exists in contrast to Brazil's panorama.

References

1. Smith, K.M., Koski, C., Siddiki, S.: Regulating net metering in the United States: a landscape overview of states' net metering policies and outcomes. Electr. J. **34**(2), 106901 (2021)
2. Rosa, C.B., Rigo, P.D., Rediske, G., Moccellin, A.P., Mairesse Siluk, J.C., Michels, L.: How to measure organizational performance of distributed generation in electric utilities? The Brazilian case. Renew. Energy. **169**, 191–203 (2021)
3. Mabea, G.A.: Electricity market coupling and investment in renewable energy: East Africa Community power markets. Int. J. Sustain. Energy. **39**(4), 321–334 (2020)
4. de Aneel, A.N.E.E.: Capacidade de Geração do Brasil. *SIGA- Sist. informações Geração da ANEEL*. (2021)
5. Lima, M.A., Mendes, L.F.R., Mothé, G.A., Linhares, F.G., de Castro, M.P.P., da Silva, M.G., Sthel, M.S.: Renewable energy in reducing greenhouse gas emissions: reaching the goals of the Paris agreement in Brazil. Environ. Dev. **33**, 100504 (2020)
6. Barros, J.J.C., Coira, M.L., de la Cruz López, M.P., del Caño Gochi, A.: Comparative analysis of direct employment generated by renewable and non-renewable power plants. Energy. **139**, 542–554 (2017)
7. Ackermann, T., Andersson, G., Söder, L.: Distributed generation: a definition. Electr. Power Syst. Res. **57**(3), 195–204 (2001)
8. Pepermans, G., Driesen, J., Haeseldonckx, D., Belmans, R., D'haeseleer, W.: Distributed generation: definition, benefits and issues. Energy Policy. **33**(6), 787–798 (2005)
9. Camilo, H.F., Udaeta, M.E.M., Gimenes, A.L.V., Grimoni, J.A.B.: Assessment of photovoltaic distributed generation – issues of grid connected systems through the consumer side applied to a case study of Brazil. Renew. Sust. Energ. Rev. **71**, 712–719 (2017)
10. de Aneel, A.N.E.E.: Geração Distribuida. (2021)
11. de Aneel, A.N.E.E.: Resolução Normativa n°687, de 24 de novembro de 2015. 25. (2015)
12. Kabir, E., Kumar, P., Kumar, S., Adelodun, A.A., Kim, K.-H.: Solar energy: potential and future prospects. Renew. Sust. Energ. Rev. **82**, 894–900 (2018)
13. Muaafa, M., Adjali, I., Bean, P., Fuentes, R., Kimbrough, S.O., Murphy, F.H.: Can adoption of rooftop solar panels trigger a utility death spiral? A tale of two U.S. cities. Energy Res. Soc. Sci. **34**, 154–162 (2017)
14. do Nascimento, F.M., Mairesse Siluk, J.C., de Souza Savian, F., Bisognin Garlet, T., Renes Pinheiro, J., Ramos, C.: Factors for measuring photovoltaic adoption from the perspective of operators. Sustainability. **12**(8), 3184 (2020)
15. de Aneel, A.N.E.E.: Revisão das regras aplicáveis à micro e minigeração distribuída – Resolução Normativa n° 482/2012. 60. (2020)
16. Greener: Strategic Market Research Distributed Generation: Solar PhotoVoltaic (PV) Market. 105. (2020)

17. Rezende, V.G.S. de, and Zilles, R.: Análise Das Perdas De Produtividade Em Geradores Fotovoltaicos Por Efeito De Sujidade. *Cebens*, 10. (2018)
18. Light: Composição da Tarifa. (2021)
19. IBGE, I.B. de G. e E.: ´indice Nacional de Preços ao Consumidor Amplo. (2021)
20. Jornal do Comércio: Aneel propõe WACC de 7,71% para distribuidoras entre 2018 e 2020. (2020)
21. de Aneel, A.N.E.E.: Diretoria da ANEEL aprova nova metodologia para WACC. (2020)
22. PVWattes: Solar Resource Data. (2019)
23. Muhaisen, N., Khan, S., Habaebi, M.H., Ahmed, N.A., Ahmed, M.M.: Feasibility analysis of implementing PV street lighting system in an arid region. Int. J. Sustain. Energy. **41**, 1–22 (2021)
24. Houldin, R.W., Yang, B.: Going viral: the impacts of covid19 on retail electricity. Electr. J. **34**(2), 106903 (2021)
25. Garlet, T.B., Ribeiro, J.L.D., de Souza Savian, F., Siluk, J.C.M.: Paths and barriers to the diffusion of distributed generation of photovoltaic energy in southern Brazil. Renew. Sust. Energ. Rev. **111**, 157–169 (2019)
26. Rigo, P.D., Siluk, J.C.M., Lacerda, D.P., Rosa, C.B., Rediske, G.: Is the success of small-scale photovoltaic solar energy generation achievable in Brazil? J. Clean. Prod. **240**, 118243 (2019)

Energy Management Maturity Models: Literature Review and Classification

Nathália Jucá Monteiro, Renata Melo e Silva de Oliveira, Sérgio Eduardo Gouvêa da Costa, Fernando Deschamps, and Edson Pinheiro de Lima

Abstract Energy management is a concerning subject due to the rising energy prices and environmental concerns. In this scenario, the maturity model measures an organization's competences with benchmarking actions and performance improvement. Energy Management Maturity Models (EMMM) are already discussed in the literature. Still, few works focus on the classification of the existing models to identify gaps. This article aims to identify the main EMMM of the literature and propose a classification scheme aiming to identify gaps in the construction of future EMMM.

To achieve the goal, a literature review used an adaptation of the ProKnow-C method and snowball. Eleven maturity models were analyzed, eight directly related to energy management and three not directly related to energy management but with contributions to the research. The classification focused on the description of the type of the model, model structure, number of maturity levels, methodology of analysis, reference to international standards, mode of assessment, results of the evaluation, and guide to improvement. Results showed that most of the models are descriptive, which means they analyzed the current situation. Still, with no guide for improvement, here is also a preference for utilizing five maturity levels. ISO 50001 appears as an international reference standard, which is explicable by the area of study, which is energy. As for the presentation of the results, graphics and visual tools are common to present the results. Since the classification of the maturity

N. J. Monteiro (✉)
Universidade do Estado do Para (UEPA), Belém, PA, Brazil

Pontifical Catholic University of Parana (PUC-PR), Curitiba, PR, Brazil
e-mail: nathalia.monteiro@uepa.br

R. M. e. S. de Oliveira
Universidade do Estado do Pará, Belem, Para, Brazil

S. E. G. da Costa · E. P. de Lima
Universidade Tecnológica Federal do Paraná, Curitiba, Parana, Brazil

F. Deschamps
Pontifícia Universidade Católica do Paraná, Curitiba, Parana, Brazil

V. M. López Sánchez et al. (eds.), *Industrial Engineering and Operations Management*, Springer Proceedings in Mathematics & Statistics 400,
https://doi.org/10.1007/978-3-031-14763-0_5

models has resulted from a literature review, this research is limited by the period, keywords, and area of study. For future works, it is suggested that the expansion of the research area to identify relationships between EMMM and other maturity models is also offered the construction of a new EMMM, which fulfills the gaps identified in this research.

Keywords Energy management · Maturity models · ISO 50001

1 Introduction

Energy management is a rising concern to manufacturing since this activity is responsible for the consumption of 35% of the entire global electricity supply [1]. The high energy prices and the limited resources also reinforce the importance of energy efficiency in this scenario [2]. The study of energy efficiency changed from energy conservation from energy management [3]. Energy management can be understood as the sum of all measures and activities executed or planned to minimize the energy consumption of a company or institution [4].

Sola and Mota [5] defined energy management as a tool to improve the energy performance and reduce the gas emissions of an organization. Energy management can be divided into five activities: planning, operation, control, organization, and culture [6].

In this scenario, the industries optimize their energy efficiency, which can be controlled and systematized through an Energy Management System (EMS) [4]. An EMS allows an organization's control, evaluation, and decision-making [6, 7]. Despite its applicability, the development of an EMS is complex because industrial energy management is part of a particular context, so it can be challenging to replicate solutions in a different industrial sector or location [8]. To access an as-is situation of an industry and to prioritize improvement solutions, the maturity models (MM) are helpful tools [9].

Becker et al. [9] define a maturity model as a sequence of levels for a class of objects. These models are designed to assess the maturity of a domain based on a set of criteria [10]. Maturity is linked to energy management as the ability of an organization to manage its energy needs based on best practices and the utilization of efficient technologies [11].

The research of maturity models for energy management is not new, and some works have already developed MM for energy management [11, 12]. The applications are varied, like compressed air systems [13], sustainability in energy management [14], small and medium-sized enterprises [15], and multi-site organizations [16]. Despite these works, there is a lack of research in identifying and classifying the maturity models for energy management. So, the objective of this work is to find and organize the maturity models for energy management to identify current trends and research gaps for the construction of new models. To achieve this goal, a literature review was conducted based on the ProKnow-C method and snowball, followed by the classification of the models and identification of gaps for the construction of future models.

The achieve its goal, this article is divided as follows: Section 1 presents the contextualization of the work and its objective; Section 2 presents the methodology that led to the selection of the maturity models; Section 3 shows the results, classifying the MM founded and discussing its limitations, and in Sect. 4 the conclusions are presented.

2 Methodology

To achieve the objective of this research, a literature review was conducted utilizing an adaptation of the ProKnow-C method, which is already used by Monteiro et al. [1] and Vieira et al. [17]. The snowball method was also applied in the literature review to cover all the related works with the theme.

This research was conducted in five steps as described below:

- **Step 1:** Definition of the keywords and the databases for the study.
- **Step 2:** Application of the ProKnow-C method adapted.
- **Step 3:** Based on the works found in Step 2, the snowball method was utilized to select another research related to the theme but who are not founded in the previous step.
- **Step 4:** Classification of the maturity models for energy management.
- **Step 5:** Discussion of the results.

In step 1, the keywords and the databases were selected. For the keywords, two main areas were selected: energy and maturity. The keywords were combined in pairs and utilized in two databases: Scopus and Science Direct. These bases were considered the most relevant for the study area, so they were selected. The search protocol is presented in Table 1.

For Steps 2 and 3, the articles' selection process was conducted. Figure 1 presents the process with the number of articles found in each phase.

Eleven articles remained in the final analysis that was conducted in step 4. Step 4 (classification of MM) was considered some classifications proposed by Introna et al. [11] and De Bruin et al. [10]. Eight categories were selected to classify the maturity models of energy management. These categories are described below:

Table 1 Search protocol

Search Axes	Keywords
"Energy"	"Energy management"
	"ISO 50001"
	"Energy efficiency"
"Maturity"	"Maturity models"
	"Assessment models"

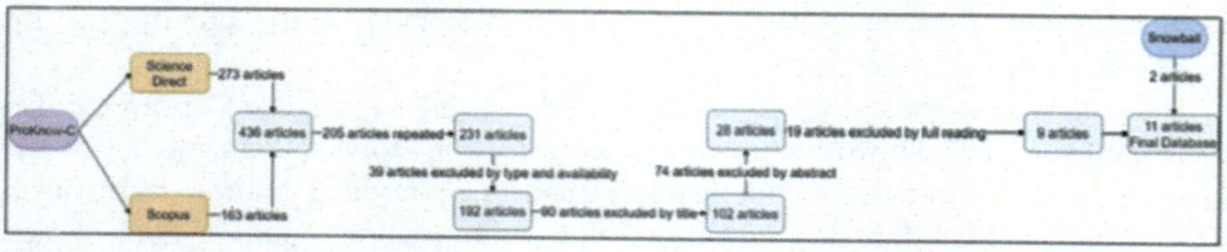

Fig. 1 Process of selection of articles

- **Model structure:** How the model is structured – Staged (well-defined stages for achievement), Continuous (Not defined stages), or Hybrid (Mix of both previous classifications).
- **A number of maturity levels:** The organization can achieve a number of maturity levels.
- **Type:** Purpose of the model – Descriptive (Analyses the current situation, without propositions of improvement), Prescriptive (Roadmap for improvement, presenting concrete actions for improvement), and Comparative (Benchmarking between various organizations or regions to comparison);
- **Methodology of analysis:** How the maturity is determined.
- **Reference to international standards:** If the MM is based on any international standard.
- **Mode of assessment:** How the process of assessment is technically conducted.
- **Results of assessment:** How the results of the MM are presented to the stakeholders.
- **Guide to improvement:** If the MM presents recommendations for improvement and how they are given.
- The classification of the MM analyzed is presented and discussed in Sect. 3.

3 Results

The classification of the eight articles related to energy management in Sect. 2 is presented in Table 2.

Table 3 presents the same classification as Table 2, but only for the MM not directly related to energy management. These articles were founded during the steps of the ProKnow-C method, and despite not having a MM for energy management, this benchmarking is interesting because the type of model, the levels, and other insights can be obtained.

Analyzing Table 2, it is possible to observe that most of the models are structured in staged phases; the total of the MM is also in a staged structure, which represents a common practice for the maturity area. The utilization of hybrid structures is low, but its application is encouraged because it presents a more efficient form of assessment. After all, you have stages to achieve, and the scores utilized show how much is still missing to pass to another level.

Table 2 Classification of the MM related to energy management

Model	Model structure	Number of maturity levels	Type	Methodology of analysis	Reference to international standards	Mode of assessment	Results of assessment	Guide to improvement
Energy Management Maturity Model [12]	Staged	1–5	Prescriptive	Nonspecified	ISO 50001	Ontology mapping with ISO 50001	Table presenting the comparison between the model and the ISO 50001	No guide for improvement
Compressed Air Systems Energy Efficiency Maturity Model (CASEEMM) [13]	Staged	1–5	Prescriptive	Benchmarking of best practices in the area	No specified, consistent with ISO 50001 sections	Survey	Bar graphs and radar diagram	Attached to assessment results
Energy Management Maturity Model for Multi-site Industrial Organisations with a Global Presence [16]	Hybrid	1–5	Prescriptive AND Comparative	Evaluation framework	–	Survey	Graphical comparisons between the site and external peers	Less specific recommendations for scores below 2.0 based on PDCA
Energy Management Maturity Model (EMMM) [11]	Staged	1–5	Descriptive	Five dimensions	ISO 50001	Survey	Graphical comparison between the current situation and the objective	Less minor structure recommendations based on the gaps
Energy Management Maturity Model for China (EMMM – China) [18]	Continuous	1–5	Prescriptive	Requirements of the ISO 50001	ISO 50001	Case study AND Survey	Radar diagram for the maturity levels	The article presents a table for gap analysis, which lead to improvements
ISO 50001-based Energy Management Maturity Model (EMMM50001) [19]	Continuous	1–5	Descriptive	Requirements of the ISO 50001	ISO 50001, PDCA cycle, and Capability Maturity Model Integration	Survey	Tables for the results of the survey	No guide for improvement
Energy and Utility Management Maturity Model (EUMMM) [14]	Staged	1–5	Descriptive	Matrix of characteristics based on CMMI	Capability Maturity Model Integration (CMMI)	Case of study AND Survey	Exerts from the interview survey	Less specific structure in a matrix of process areas from the model
Energy Efficiency Maturity (EEM) Assessment Framework [15]	Continuous	1–5	Prescriptive	Seven dimensions based on the literature	–	Case study AND Survey	Tables and radar graphs for the maturity results and diagrams for the analysis of the process	Proposals for energy efficiency improvement projects

Table 3 Classification of the MM is not related to energy management

Model	Model structure	Number of maturity levels	Type	Methodology of analysis	Reference to international standards	Mode of assessment	Results of assessment	Guide to improvement
Intelligent Manufacturing Capability Maturity Model [20]	Hybrid	1–5	Descriptive	Capability elements, maturity levels, and maturity requirements	–	Case study	Bar and round graphs for the maturity scores	No guide for improvement
Maturity Model for Assessing Industry 4.0 Readiness and Maturity of Manufacturing Enterprises [21]	Continuous	1–5	Descriptive	Nine dimensions and 62 items	–	Case study AND Survey	Radar diagram for the maturity scores	No guide for improvement
Sustainability Maturity Model for micro, small and medium enterprises (SMMM) [22]	Staged	1–4	Descriptive	Three factors	Nonspecified	Survey	Bar graphs for the maturity results	No guide for improvement

According to De Bruin et al. [10], the number of stages can vary between the models. Still, the Capability Maturity Model (CMM), the first model to introduce the concept of maturity, reinforces the utilization of 5 maturity levels, being one the lowest and five the highest. This practice is also identified in this work, where apart from the SMMM, all the models present five maturity levels, following the concept of CMM.

It is essential to mention the maturity levels that all the models analyzed do not consider a stage zero, where the companies do not know of energy management practices. The model of Finnerty et al. [16] assumes the existence of no knowledge but at level 1; the others consider level 1 with no implementation of processes, forgetting the information necessary before the beginning of the performance.

Maturity models are used for benchmarking to compare the state of a company [23, 24] in the MM analyzed in this article, only the model of Finnerty et al. [16] for multi-site organizations utilizes the benchmarking in the construction of the model; the CASEEMM also utilizes the benchmarking, but for the methodology of analysis. Antunes et al. [12] affirm that one of the benefits of using energy management maturity models for an organization is the benchmarking of current energy practices of other organizations. Still, as seen in the analysis, this practice is little considered, being a gap to be mentioned.

The model type is also something to consider, and the analysis shows a balance between the descriptive and the prescriptive models. The prescriptive's model presents a more interesting approach for the organizations since it presents concrete actions for improvements and, consequently, increases the maturity levels and scores. The MM analyzed did not offer concrete improvement actions, except for the CASEEMM of Benedetti et al. [13]. This situation suggests an opportunity for improvements since the prescriptive models should enhance their recommendations for concrete actions of improvements regarding energy management maturity models. Also, it should be pointed out that comparative models are scarce, only encountered in Finnerty et al. [16], which utilizes the benchmarking intra-company comparing the situations of the various industrial plants of the organization.

Regarding international standards, the ISO 50001 is the most mentioned. The ISO 50001 was created on June 15, 2011 by the International Organization for Standardization, and it establishes a process for the systematic management of energy and improvements in energy efficiency [25]. It is important to associate the standard with the maturity models because ISO 50001 presents the requisites for an energy management system but does not establish goals for improving energy efficiency [26]. On the other hand, the MM focuses on continuous improvement, presenting a set of processes that the organizations should implement for enhancement [12]. So, the combinations of these two tools can represent a better way to achieve the purpose.

The mode of assessment was also examined, and the MM of this work utilized case studies and surveys to evaluate the organizations' maturity levels and the proposed model. De Bruin et al. [10] mentioned that the mode of assessment is part of the methodology for building a maturity model. It is necessary for validation of the model and to measure its reliability in presenting accurate results. These authors also mention the survey as a tool of assessment, and the case study can help in the standardization of the model or its specification. The MM analyzed has specific domains, and this is a research gap that needs to be fulfilled by new maturity models for energy management.

As for the results, a graphical perspective is adopted, utilizing radar or bar graphs. This practice allows a comparative analysis between the ideal and the current situation, permitting the academic community and practitioners to have easy access to the most important results and values that need to be improved.

To summarize, the research gaps identified are presented below:

- Poor utilization of a hybrid model structure should be encouraged due to its better visualization of how much needs to be improved.
- There is no utilization of a level zero considering that the organizations do not know energy management practices.
- Few benchmarking practices fragilizing the core competence of the models.
- Few models focused on a prescriptive point of view without presenting recommendations for improvement.
- MM with specific domains, imposing some difficulties in standardizing energy management practices in a broader area.

With the discovery of the research gaps, new models can be built, improving the research around energy management and maturity models, helping the organizations to improve their energy efficiency practices.

4 Conclusions

Energy management is a critical area for organizations. In this scenario, the maturity models can provide a tool to guide the actions of improvements. The objective of this article was to identify current trends and research gaps for energy management maturity models. The aim of this work was achieved utilizing a literature review and a qualitative analysis of the maturity models for energy management presented in the literature.

Eleven MM were selected, being eight directly related to energy management. After the analysis of the models, it was possible to list some research gaps, like the low utilization of benchmarking practices, the absence of guides for improvements in the models, and the specification of the MM, resulting in models for specific domains.

It also examined the current trends, like the significant presence of ISO 50001 as an international reference standard, leading the companies in pursuit of certification, and how the graphical presentations of the results are very utilized, allowing numerical and accessible comparations about the current and the ideal situation.

This work is limited by its methodology since it was a literature review, limiting the amount of the work results ended on the combination of the keywords and by the period when the research was executed. This work is a qualitative analysis but from a theoretical point of view, so no practitioners in the area were consulted during the classification of the models.

To surpass these limitations, some future works are selected to expand the research to find other maturity levels to create a more standard analysis regarding the maturity area of study. The practitioners of the site should be consulted to validate and expand the results focusing on identifying gaps in the application and the relevance of the models. Also, new energy management maturity models should be built and validated to close all the gaps identified in this research and expand the area's knowledge.

References

1. Monteiro, N.J., da Costa, S.E.G., de Lima, E.P., Deschamps, F.: Energy and industry 4.0: an application of ProKnow-C method. International Joint Conference on Industrial Engineering and Operations Management. (2020)
2. Eder, M., Ketenci, A., Auberger, E., Gotthard, M., Ramsauer, C.: Integration of low-cost digital energy meters in learning factory assembly lines. Procedia Manuf. **45**, 202–207 (2020)
3. Perroni, M.G., Gouvea da Costa, S.E., Pinheiro de Lima, E., Vieira da Silva, W.: The relationship between enterprise efficiency in resource use and energy efficiency practices adoption. Int. J. Prod. Econ. **190**, 108–119 (2017)
4. Fiedler, T., Mircea, P.M.: Energy management systems according to the ISO 50001 standard – challenges and benefits. In 2012 International Conference on Applied and Theoretical Electricity, ICATE 2012 – Proceedings. (2012)
5. Sola, A.V.H., Mota, C.M.M.: Influencing factors on energy management in industries. J. Clean. Prod. **248**, 119263 (2020)
6. Schulze, M., Nehler, H., Ottosson, M., Thollander, P.: Energy management in industry – a systematic review of previous findings and an integrative conceptual framework. J. Clean. Prod. **112**, 3692–3708 (2016)
7. Sivill, L., Manninen, J., Hippinen, I., Ahtila, P.: Success factors of energy management in energy-intensive industries: development priority of energy performance measurement. Int. J. Energy Res. **37**, 936–951 (2013)
8. Vikhorev, K., Greenough, R., Brown, N.: An advanced energy management framework to promote energy awareness. J. Clean. Prod. **43**, 103–112 (2013)
9. Becker, J., Knackstedt, R., Pöppelbuß, J.: Developing maturity models for IT management. Bus. Inf. Syst. Eng. **1**, 213–222 (2009)
10. de Bruin, T., Health, Q., Kulkarni, U., Rosemann, M.: Understanding the Main Phases of Developing a Maturity Assessment Model. Australasian Conference on Information Systems. (2005)
11. Introna, V., Cesarotti, V., Benedetti, M., Biagiotti, S., Rotunno, R.: Energy management maturity model: an organizational tool to foster the continuous reduction of energy consumption in companies. J. Clean. Prod. **83**, 108–117 (2014)
12. Antunes, P., Carreira, P., Mira da Silva, M.: Towards an energy management maturity model. Energy Policy. **73**, 803–814 (2014)
13. Benedetti, M., Bonfà, F., Bertini, I., Introna, V., Salvatori, S., Ubertini, S., Paradiso, R.: Maturity-based approach for the improvement of energy efficiency in industrial compressed air production and use systems. Energy. **186** (2019)
14. Ngai, E.W.T., Chau, D.C.K., Poon, J.K.L., To, C.K.M.: Energy and utility management maturity model for sustainable manufacturing process. Int. J. Prod. Econ. **146**, 453–464 (2013)

15. Prashar, A.: Energy efficiency maturity (EEM) assessment framework for energy-intensive SMEs: proposal and evaluation. J. Clean. Prod. **166**, 1187–1201 (2017)
16. Finnerty, N., Sterling, R., Coakley, D., Keane, M.M.: An energy management maturity model for multi-site industrial organisations with a global presence. J. Clean. Prod. **167**, 1232–1250 (2017)
17. Vieira, E.L., da Costa, S.E.G., de Lima, E.P., Ferreira, C.C.: Application of the Proknow-C methodology in the search of literature on performance indicators for energy management in manufacturing and Industry 4.0. Procedia Manuf. **39**, 1259–1269 (2019)
18. Jin, Y., Long, Y., Jin, S., Yang, Q., Chen, B., Li, Y., Xu, L.: An energy management maturity model for China: linking ISO 50001:2018 and domestic practices. J. Clean. Prod. **290**, 125168 (2021)
19. Jovanović, B., Filipović, J.: ISO 50001 standard-based energy management maturity model – proposal and validation in industry. J. Clean. Prod. **112**, 2744–2755 (2016)
20. Hu, J., Gao, S.: Research and application of capability maturity model for Chinese intelligent manufacturing. Procedia CIRP, 794–799 (2019). Elsevier B.V
21. Schumacher, A., Erol, S., Sihn, W.: A maturity model for assessing Industry 4.0 readiness and maturity of manufacturing enterprises. Procedia CIRP, 161–166 (2016). Elsevier B.V
22. Vásquez, J., Aguirre, S., Puertas, E., Bruno, G., Priarone, P.C., Settineri, L.: A sustainability maturity model for micro, small and medium-sized enterprises (MSMEs) based on a data analytics evaluation approach. J. Clean. Prod. **311**, 127692 (2021)
23. Pessl, E., Sorko, S.R., Mayer, B.: Roadmap Industry 4.0 – implementation guideline for enterprises. In: 26th International Association for Management of Technology Conference, IAMOT 2017, pp. 1728–1743. International Association for Management of Technology Conference (IAMOT) and the Graduate School of Technology Management, University of Pretoria (2020)
24. Dikhanbayeva, D., Shaikholla, S., Suleiman, Z., Turkyilmaz, A.: Assessment of Industry 4.0 maturity models by design principles. Sustainability. **12**(9927), 9927 (2020)
25. Chiu, T.Y., Lo, S.L., Tsai, Y.Y.: Establishing an integration-energy-practice model for improving energy performance indicators in ISO 50001 energy management systems. Energies (Basel). **5**, 5324–5339 (2012)
26. Organisation for Economic Co-operation and Development: An Introduction to Energy Management Systems: Energy Savings and Increased Industrial Productivity for the Iron and Steel. (2015)

Profitability Analysis and Commercial Strategy in Liquified Petroleum Gas Distribution

Sirlei Valim Melo, Leila Figueiredo Dantas, and Antonio Márcio T. Thomé

Abstract Companies have realized that certain customers, despite consuming on a large scale, are not profitable and demand great service efforts. Adopting methods that make it possible to differentiate customers according to their share of the company's results contributes to improving the management process. The present study aims to present the use of the Cost of Serving (CTS) methodology in the analysis of customer profitability and its use as a commercial strategy tool, demonstrating how we can apply it in a consumer portfolio of a Liquefied Petroleum Gas distributor (LPG) in the state of Rio de Janeiro. The work discusses how to calculate the CTS of each client, adopting concepts derived from the activity costing method and its impact on the contribution margin. Criteria for customer segmentation are discussed, and all these concepts support managerial decisions considering the market forces present in the sector it operates. The literature review supports the empirical study and is based on data obtained from the LPG company. The CTS analysis seeks to improve management efficiency, reduce costs and increase the company's profitability. The conclusions offer an instrument for the management of the company's portfolio and address logistical, commercial, financial and competitive aspects that affect the LPG distribution market.

Keywords Cost to Serve · Profitability Analysis · Commercial Strategy

1 Introduction

Companies need to generate revenue, earn profits and distribute dividends to their shareholders. In order to achieve their financial goals, it is necessary to serve the consumer market with a provision of services that efficiently meets customer demands.

S. V. Melo (✉) · L. F. Dantas · A. M. T. Thomé
Pontifícal Catholic University, PUC-Rio, Rio de Janeiro, Brazil

V. M. López Sánchez et al. (eds.), *Industrial Engineering and Operations Management*, Springer Proceedings in Mathematics & Statistics 400,
https://doi.org/10.1007/978-3-031-14763-0_6

For Kaplan and Anderson [1], not all customers are profitable; there are customers that the more they demand deliveries, the more they reduce profitability due to the quantity and unpredictability of orders. Understanding the costs that each customer generates for the company, differentiating them into profitable or not, makes it possible to segment the portfolio, opening a new perspective for portfolio management.

The costs of companies have been calculated based on general domain criteria; however, this knowledge does not allow an analysis of profitability per customer. In this context, the cost of serving per customer (CTS) concept is a tool that explains the different costs that makeup customer service and thus favors the individualized calculation of the Contribution Margin (CM). Braithwaite and Samakh [2] emphasize that this type of analysis also helps improve processes.

The article aims to present the adoption of these concepts in a case study applied to a customer portfolio of an LPG distributor in the State of Rio de Janeiro. We will review the academic literature, demonstrate how the analysis of profitability can be applied in this sector of activity, and how factors related to the competition model of the LPG market influence the company's commercial strategy.

The work will observe the following sequence: (i) introduction, with the contextualization of the theme, justification, and objectives; (ii) theoretical foundation contemplating the theories found in academic publications; (iii) case study, with the methodology; (iv) analysis of results; (v) conclusions and bibliographic references.

2 Theoretical Foundation

2.1 Customer Profitability Based on CTS and CM

Kaplan and Cooper [3] developed a costing system that allows relating costs to resource-consuming activities, giving rise to the Activity-Based Costing (ABC). The main characteristic of this methodology is to associate the costs related to human, technological, and material resources, relating them to the stages of producing, delivering, and meeting demand, making it possible to track these costs.

According to Cokins [4], a relevant factor to be observed in adopting the ABC method is to identify how costs are allocated to activities and consumed throughout the processes through drivers that have the role of distributing the costs generated according to the use of resources. Thus, the resource drivers identify the resource source through an accounting item, and those of activity distribute the costs among the activities. According to the established apportionment criteria, the object drivers define how the activity costs are attributed to each customer.

Kaplan and Anderson [1], aiming to make the ABC implementation process faster and simpler, proposed an approach using the cost of the supply unit and the time spent in carrying out the task. In this way, they used time as the main

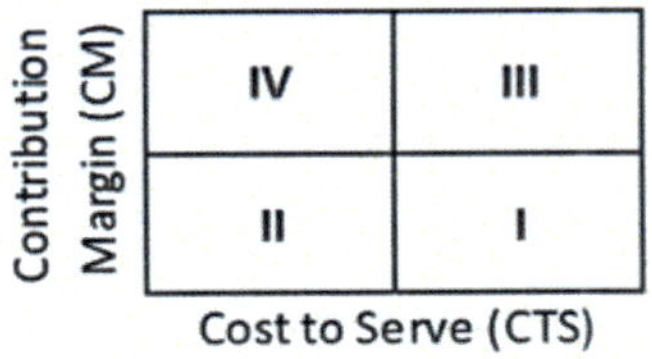

Fig. 1 Customer segmentation based on CM vs. CSC. (Source: Adapted from Mejía-Argueta et al. [6])

reference to measure the costs of activities and gave greater agility to the ABC method, becoming known as Time-Driven Activity-Based Costing (TDABC).

The CTS covers customer relationship activities and may include, in addition to logistical costs, those of a commercial nature, such as sales and marketing costs, as proposed by Guerreiro et al. [5]. In the LPG distributor studied were considered only the logistical costs for calculating the CTS.

The CTS is the sum of the costs involved in the different stages related to the service and delivery of the contracted product to the customer, based on cost drivers and apportionment criteria applied throughout the activities. In this way, these costs can be attributed to a final consumer, duly identified, thus determining the CM for each client as an expression of profitability.

The CTS methodology provides elements to carry out the portfolio segmentation based on the CM and to evaluate the possibility of optimizing the results with measurable data. Segmentation is a way used to understand customers' profitability, and a quadrant classification formed by the relationship of the CM with the CTS can be adopted, allowing a broader view of the distribution of the portfolio [3].

Mejía-Argueta et al. [6], adopted the distribution of customers by quadrants, classifying the segments by zones, namely (i) danger, with low CM and high CTS; (ii) construction, with low CM and low CTS; (iii) engineering, with high CM and high CTS; (iv) protection, with high CM and low CTS, attributing to each of the classifications a specific commercial strategy to increase profitability and add value to the company. According to Shin et al. [7], clients classified as generators of losses with low CM and high CTS can be subtly excluded from the portfolio through price increases. The Fig. 1 shows the division of customer segmentation quadrants.

Braithwaite and Samakh [2] emphasize the importance of CTS in measuring customer profitability and the use of Pareto analysis in this context. Kaplan and Narayanan [8] relate the accumulated percentage of CM with that of customers, generating an ABC curve, or whale curve, divided into three zones according to profitability, as follows: A, where is found the high-return customers; B, those with low profitability; and C, those with very low or negative CM.

Adopting the concepts presented allows the company to distinguish profitable from non-profitable customers, make decisions based on the trade-offs of the CTS and CM, define actions to protect its most profitable customers against competitors, and provide elements for the commercial strategy.

2.2 Factors That Interfere with Commercial Strategy

Based on CTS and CM, individual customer profitability is determined by the difference between revenues and costs. In this way, the prices practiced are of fundamental importance in determining the profit and must be established considering the CTS, the cost of the raw material, competitive characteristics of the sector in which the company is inserted and other factors.

According to Storbacka [9], segmentation considers the retrospective analysis of customer relationships and should be able to say something about their behavior. This approach allows the development of long-term market strategies, including pricing policy, new service concepts, among other factors that influence customer behavior. This information needs to be aligned with the competitive reality, the executives' intuition, and the perception of the business [10].

Customers sensitive to price adjustments tend not to accept increases, even if the CTS and profitability analysis indicate. There are situations in which profitability analysis may not be the only factor to be considered in customer relationship management decisions. Competitors looking to expand their market may charge lower prices to win customers from another company's portfolio. Understanding the dynamics of the sector in which the company operates, how it creates and shares value allows for a greater understanding of the market and how the competitive reality applies to the profitability of the customer portfolio.

To better understand the competitive environment in which the company operates, Porter [11] describes five forces that serve as an evaluation method in the sector. Each of the forces affects prices and costs to a different extent. The bargaining power of customers may demand lower prices or better service (costs). Suppliers can increase costs by demanding higher prices. Companies with substitute products can become direct competitors and affect revenues. When attracted to the sector, new entrants can pressure the price and modify the market share. The rivalry between players can be reflected in a price reduction or increased costs.

It should be considered that certain sectors of activity, such as LPG, have their prices influenced by the macroeconomic conditions of their markets. Uncertainty about currency and inflationary aspects leads companies to assume assumptions for prices and costs based on financial market expectations. Considering that prices are affected by factors beyond companies' control and interfere with their profitability, it is important to focus efforts on managing their costs.

3 Case Study

3.1 Case Study Methodology

The methodological definitions proposed by Cauchick-Miguel et al. [12] were adopted for the elaboration of the case study. For the theoretical foundation, we

Table 1 Description of the variables analyzed

Acronym	Variable name	Type of variable	Unit of measurement	Description of aggregate data from january to august 2021 (except price)
COD	Code	factor	-	Unique code generated by the customer's CNPJ
VOL	Measure	numerical	kilo (kg)	Measure sold to the customer in kilo
PRE	Price	numerical	Real/tonne (R$/ton)	Average unit price per ton paid for LPG per customer, in the period of 8 months
NV	Number of Visit	numerical	-	Number of visits made per customer
TV	Time in Visit	numerical	Hour (h)	Time spent on customer service
CP	Personal Cost	numerical	Real (R$)	Apportionments related to the Unit's personnel, administrative and distribution costs, per customer
CL	Rental Cost	numerical	Real (R$)	Apportionments related to the cost of leasing the delivery truck per customer
CC	Fuel Cost	numerical	Real (R$)	Apportionments related to delivery truck fuel cost per customer
CE	Cost of Stay	numerical	Real (R$)	Apportionments related to cost when there is a stay on the trip per customer
OC	Other Costs	numerical	Real (R$)	Apportionments related to electricity, internet and communication of the Unit per customer
CTS	Cost to Serve	numerical	Real (R$)	Sum CP, CL, CC, CE and OC per customer
CM	Contribution Margin	numerical	Real (R$)	Price multiplied by the Volume from which the CMV, taxes and CSC are deducted, obtaining CM

searched in the SCOPUS database, using the boolean OR and selecting the title, abstracts, and keyword fields with the pseudo-code in quotation marks: ["cost to serve" OR "customer profitability analysis" OR "segmentation customer"].

In planning the case study, we made the option for the LPG distributor because it is an exemplary case, has a consolidated brand, is representative of the market in which it operates, and is academically interested in the subject. The study was based on real data collected from the company operating throughout the national territory for more than 50 years. However, this study is restricted to the portfolio of customers in Rio de Janeiro State, as it is an important consumer center with dispersed deliveries, attending hospitals, condominiums, steel mills, gyms, among others.

The instruments and methods for data collection and the definition of unstructured interviews were defined. The questions addressed CM criteria for calculating the CTS, and feeding data into the System. Employees of the company from the commercial and logistics sectors were interviewed, visiting the operational sector to learn about the internal processes and validate the mechanisms for obtaining data.

The data collection stage examined the central management and accounting instruments related to costing and profitability, with the definition and authorization of access to their information sources. The data available referred to the usual purchases of 428 customers and brought aggregated information for 8 months (January to August 2021), except for the price that consisted of the average charged per customer.

The name of the company was anonymized. The actual numbers were modified using a proportionality factor to protect the confidentiality and security of strategic information. We chose the RStudio software to perform the treatment of the database and the analyses. Table 1 describes the 12 variables used in the analyses.

In analyzing the variables presented in Table 1, descriptive statistics were initially performed to detail the behavior of the observed numerical data. Next, Pearson's Correlation was performed, which measures the relationship between two numerical variables.

We constructed a graph considering the median, a metric of central position not very sensitive to extreme points, of the CM vs. CTS that provided customer segmentation [6]. Management indicators were developed to verify the percentage relationships by comparing the average per segment of the variables analyzed with the average practiced in the company.

An analysis of the price variable was carried out based on the created indicators, comparing the average price per segment with the company's average price. In sequence, it was verified how much the price interferes with the profitability of the customer portfolio and which policies are most appropriate for pricing. Concerning the analysis of CM, an ABC curve was prepared, based on Pareto analysis, to verify the percentage share of customers in profitability.

After that, it was analyzed how Porter's forces interfered with the company's commercial strategy. A comparison was made with other studies to contrast the analyses results with other academic studies mentioned in Sect. 2. These actions allowed us to establish inferences and verify the consistency of the analyses developed with the conceptual basis researched in the theoretical foundation. We finally prepared the research report.

3.2 *Method of Calculating CTS and Profitability*

The methodology describes the calculation of CTS in an LPG distributor. It considers in its composition the accounting items related to Personnel Cost (CP), Rental Cost (CL), Fuel Cost (CC), Stay Cost (EC), and Other Costs (OC) consumed by clients during service visits. Each installment is calculated according to apportionment criteria defined by the company applied to these five accounts, according to the proportionality in which they are used.

The CTS tool considers the particularities of each customer since the LPG distribution trips present a significant variation of parameters that influence their final cost. Thus, for each cost apportionment, there is a multiplier factor considering the NV characteristics of each customer service trip with different drivers [4]. Table 2 presents the cost drivers per customer based on the CTS calculation components.

Table 2 Cost drivers per customer

Cost lines (resources/monetary values)	Segmentation of cost (activity)	Customer drivers (objects)
Personal Cost (CP-46% of CTS)	• Distribution-Road • Distribuição-Visita • Administrative-Visit • Administrative – Volume	→ Time-Distance → Time → Time/Number of Visit → Time/Volume
Rental Cost (CL~27% of CTS)	• Fixed-Volume • Fixed-Road • Fixed-Visit Variable	→ Time/Volume → Time/Distance → Time Distance
Fuel Cost (CC~10 from CTS)	• Road	→ Distance/Consumption
Cost of Stay (EC~3% of CTS)	• Overnight stay	→ Time
Other Costs (OC~14% of CTS)	• Visit • Volume	→ Number of Visit → Volume

The cost lines present the average share (%) for each CTS component in the analyzed company.

To calculate profitability per customer, the data to be measured is the CM. Invoicing is obtained by multiplying the PRE by the VOL of a given customer's purchases. Taxes, cost of goods sold, and CTS are subtracted from the billing to calculate the CM. The value obtained from the sale, reflected in the billing, is a proxy of how much a customer is willing to pay for the company's product. This value is expected to be greater than costs for the customer to be considered profitable.

4 Analysis of Results

4.1 Initial Data Analysis

A descriptive analysis of the variables present in the customer sales cycle was carried out, as shown in Table 3.

It is observed that the mean and the median of the price present a small variation between them. It is because prices are impacted by market forces, such as intensity among competitors in the LPG segment. Pearson's correlation was developed to verify which variables are highly correlated, as shown in Table 4.

It is observed that the most significant correlation occurs between the variables VOL and CTS (0.85), followed by NV and CTS (0.82) and VOL and TV (0.78). The high correlation between NV and CTS exists because the CTS tends to increase as there is a need for more visits to the client. The influence of VOL on CTS occurs directly, as VOL determines the drop size, which is the maximum amount that we can deliver to the customer in one visit. In this way, the VOL impacts the NV, which is the main factor affecting the CTS.

Table 3 Description of the variables associated with customer service

	Measure (kg)	Contribution margin (R$)	Price (R$/tonne)	Number of visit	Time in visit (h)	Cost to serve (R$)
Minimum	2.231	- 1.630.473	388.275	97	9,06	16.479
1º Quartile	101.341	97.461	443.620	582	89,24	165.924
median	245.992	374.614	589.071	1.067	128,04	360.599
Mean	736.043	1.450.886	604.793	1.335	149,38	551.046
3º Quartile	532.670	1.136.016	730.813	1.649	167,81	625.953
Maximum	6.804.086	53.092.271	2.739.059	7.954	1.169	7.190.455
Standard Deviation	1.771.127	4.424.355	327.845	1.160	120	731.544

Table 4 Pearson's correlation of the analyzed variables

	Measure	Cost to serve	Price	Number of visit	Time in visit	Contribution margin
Measure	1,000	0,859	−0,001	0,623	0,781	0,570
Cost to serve	0,859	1,000	0,072	0,820	0,644	0,399
Price	−0,001	−0,072	1,000	0,201	0,168	0,259
Number of visit	0,623	0,820	0,201	1,000	0,475	0,274
Time in visit	0,781	0,644	0,168	0,475	1,000	0,563
Contribution margin	0,570	0,399	0,259	0,274	0,563	1,000

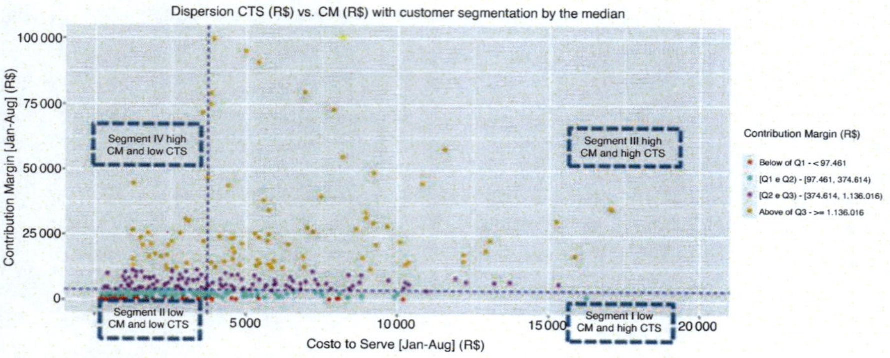

Fig. 2 CTS dispersion vs. CM with customer segmentation by the median

4.2 *Customer Segmentation: Analysis of CM Versus CTS*

The segmentation of customers using the analysis criterion CTS versus CM is adopted by the academic literature to carry out the analysis of profitability and guide the commercial strategy taking into account the position occupied by customers in each segment (see Fig. 1). This classification defines commercial policies for each grouping according to its characteristics. Figure 2 shows the dispersion of CTS by CM, with customer segmentation, considering the medians (see Table 3) of the CM dividing the "Y" axis into high and low CM and the CTS dividing the "X" axis into high and low CTS.

The CM classification is represented by the colors as follows, namely: (i) red, CM below the first quartile; (ii) blue, CM between the first quartile and the median; (iii) purple, between the median and the third quartile and (iv) yellow, above the third quartile. With the establishment of borders demarcated by the median of the CM and CTS, it was possible to perform the segmentation of customers, positioned according to these characteristics: (i) low CM and high CTS, contains 76 customers and corresponds to 18% of the customer portfolio; (ii) low CM and low CTS, covers 138 clients and 32% of the portfolio; (iii) high CM and high CTS, comprises 138 customers, also with 32% of the portfolio; (iv) high CM and low CTS, with 76 clients and 18% of the client portfolio.

Regarding to segment I, which concentrates low-return customers and/or loss generators, it is necessary to act on cost reduction and price increase. Regarding to segment II, this group, due to the low return it delivers and/or loss, points to adjustments in the price variable. As for segment III, which offer profitability, efforts should focus on CTS related aspects in order to make them even more profitable. As for segment IV, which has the best cost-benefit ratio, its main characteristics must be analyzed and reproduced in relation to other customers.

Table 5 Variables that impact CTS and CM by customer segment

	Mean per segment					
	Measure (kg)	Cost to serve (R$)	Number of visit	Time in visit (h)	Contribution margin (R$)	Price (R$/tonne)
Segment I (high CTS and low CM)	600.084	818.333	1.935	140	−22.039	591.695
Segment II (low CTS and low CM)	104.981	159.555	586	79	125.576	413.422
Segment III (high CTS and high CM)	1.720.935	986.626	2.120	245	3.705.526	724.878
Segment IV (low CTS and high CM)	229.524	203.710	673	114	1.236.345	747.441

Table 6 Managerial indicators of the segment average with the company average

	mean of segment / mean of company					
	Contribuition Margin/Revenue	Price	Measure	Cost to Serve	Number of Visit	Time in Visit
Segment I (high CTS and low CM)	−0,6%	98,0%	82,0%	149%	145%	94%
Segment II (low CTS and low CM)	28,1%	68,0%	14,0%	29%	44%	53%
Segment III (high CTS and high CM)	28,8%	120,0%	233,0%	179%	158%	164%
Segment IV (low CTS and high CM)	69,9%	124,0%	31,0%	37%	50%	82%

4.3 *Management Analysis*

We developed a comparative analysis between the variables that affect the behavior of the CTS and the CM, which are variables that support the company's profitability and commercial strategy. The evaluation consists of the construction of indicators that can contribute to guiding the actions to optimize the results. Table 5 shows the average of these variables by segment and aims to present the differences to understand the management analysis and improvement proposition better.

For a comparative evaluation between the segments, some indicators were established. We adopted the percentage relationship between the average of the variables for each segment and the general average of the company's customer portfolio to verify the differences and guide the managerial measures of improvement, according to the percentage relationship presented in Table 6. An additional indicator for each segment was created to verify the relationship between CM and revenue.

Regarding segment, I, it is observed a CTS 49% higher than the company's average and an NV 45% higher, which justifies a high CTS, as identified by the correlation analysis (see Table 4), signaling that the NV must be monitored to decrease the number of supplies, evaluating the possibility of increasing the VOL per delivery. In respect of the price, which is in a range of compliance with the company's average, we can observe that the commercial action should preferably seek a reduction in CTS to increase the CM.

In segment II, the NV is 44% of the company's average, while the TV is 53%, which confirms its classification as a low CTS and shows that there is little room for improvement in the scope of service costing. On the other hand, the price variable is 68% of the company's average price, so it should be the object of attention in future contract reviews; this suggests that the price tag is the factor that impacts the CM of this group. Customers in this segment may be price-sensitive and have low prices due to market pressures exerted by competition.

Regarding segment III, it has a CTS 79% higher than the company's average, which can be explained because the NV is 58% higher, signaling that the NV should

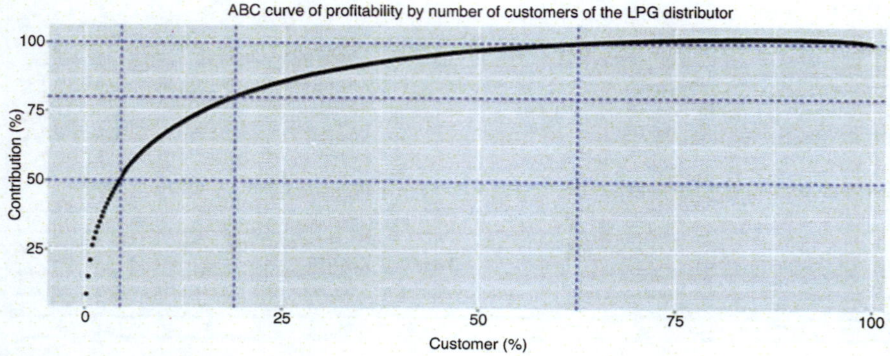

Fig. 3 ABC curve of profitability by number of company customers. (Source: Adapted de Kaplan and Narayanan [8])

be the focus of action for a reduction in the CTS. Considering that the VOL is expressive in this segment, and the NV, the reduction of the CTS must be given by the readjustment of the drop size (storage capacity) to deliver the same VOL with a lower NV. There is a good CM/revenue ratio and price 20% higher than the company's average, which explains its good profitability.

As for segment IV, with low CTS, the NV is 50% of the company's average, which shows little room for improvement in the scope of service costing. It can be seen in Table 6 that this group has the best CM/revenue ratio, approximately 70%, and a price 24% higher than the average practiced by the company, which is therefore favorable in terms of price. Presenting the ideal combination of CM vs. CTS serves as a reference to be followed. It is suggested to adopt loyalty actions for customers in this segment, with the objective of maintaining them in the portfolio, may be adopted as an internal benchmark for the company for the relationship with customers.

Another analysis performed, considering the CM, is the ABC curve of profitability by the number of customers of the LPG distribution company. In line with the Kaplan and Narayanan [8], the ABC curve presented in Fig. 3 shows the distribution of the number of customers that participate in the company's profitability.

We can observe that section A represents the most expressive portion of the CM and a much smaller amount of clientele. This region of the ABC curve includes part of the members of segments III and IV, with high CM, and corresponds to 82% of the CM and 18% of the customer basis, characterizing a classic illustration of the Pareto principle. Region B has 16% of the CM and 45% of the clients in the portfolio, including parts of segments II, III, and IV. Finally, in region C of the ABC curve, part of segment II and the entirety of segment I are located and contribute only 2% of profitability but represent 37% of customers.

4.4 Factors of Commercial Strategy

As evidenced in the managerial analysis, the price has a fundamental role in the CM. If it suffers significant reductions due to competition factors, it can reassess the commercial strategy and market positioning as a whole. The analyzed LPG distributor practices a managerial analysis of profitability based on the CTS with defensive pricing strategies regarding market share, according to the characteristics of the customers and the movement of the competition.

Suppose the attempt to reduce the CTS does not generate results in the medium term. In that case, it is up to the company to assess whether it wants to keep customers with these characteristics or implement price readjustments, even with losing some customers. Some clients would only be worth keeping in their wallets at much more competitive prices, as suggested by Shin et al. [7] for high CTS customers.

The analyzed elements took into account the prices charged by the company in the current contracts, and we suggested actions regarding pricing to improve profitability. In a competitive market such as LPG, in which the five main players compete for the consumer, it is also essential to analyze the price policy through the forces of Porter [11], according to the theoretical foundation session of this paper.

In this context, an expressive group of customers exerts pressure to reduce prices, using as a commercial argument the pricing of competitors. In addition, there are new entrants in the sector, and there is also the risk of natural gas expanding its distribution, presenting itself as a more advantageous substitute among other energy sources such as oil, wood chips, and firewood.

To deal with possible movements in the levels of prices practiced, it is advisable to protect supply contracts with longer terms and termination penalties that discourage or neutralize competition actions about changes in suppliers caused by price movements. Thus, the pricing strategy needs to be defined with an eye on the CM, but considering the strengths of Porter's [11] market forces in this sector, as well as the market experience of the company's executives and the macroeconomic aspects that affect it, such as exchange rate, inflation, global demand in the international oil market, among others.

4.5 Comparison with Other Studies

For Guerreiro et al. [5], logistics and commercial costs are included in the CTS calculation. The company considers the costs related to transport, supply, distribution, delivery, and service, not including marketing costs. The adoption of quadrants that relate CM and CTS for classification into segments is found in the works of Mejía-Argueta et al. [6] (see Fig. 1) and Shin et al. [7], aimed at basing managerial decisions in line with the proposed strategies in the same way as this work does.

Another widely referenced aspect, mainly in the works of Helgesen [10] and Guerreiro et al. [5], concerns the Pareto principle. This concept is also present in this work in terms of the number of clients responsible for the most expressive portion of the CM and was also endorsed as a valuable source of information to assess the degree of concentration of the portfolio's profitability and dependence on a smaller portion of customers.

5 Conclusion

The company case benefits from the CTS methodology and customer profitability analysis. This is perceptible at the operational level by bringing a process view that allows optimizing customer service and reducing costs. The tactical aspect guides customer segmentation, providing intelligence for pricing policy and commercial actions. In strategic terms, it provides elements for decision-making regarding market positioning, market share, and competition analysis. In this aspect, Porter's forces, mainly the bargaining power of customers associated with the intensity of competitors, exert pressure on prices and profit margin.

This article aims to contribute to theory and practice. It makes a practical contribution by presenting the CTS analysis method for increase the profitability and to achieve strategic and operational benefits, based on statistical and managerial analyzes. Some indicators are created to establish comparisons between customer segments, supporting the commercial strategy. The theoretical contribution, depending on the type of case study methodology chosen, for the elaboration of theories: (i) contributes to the calculation of the CTS, suggesting, based on the literature review and field observations, that these calculations include the marketing costs; (ii) it combines the profitability calculation with the strategic analysis of Porter's market forces, to have a balanced view of CTS optimization and market share, taking into account the bargaining power of suppliers and customers, substitute products and the intensity of competition in an oligopolistic sector, as is the LPG sector.

References

1. Kaplan, R., Anderson, S.R.: Time-driven activity-based-costing. **82**(11) (2004). Havard Business Publishing. Homepage, https://hbr.org/2004/11/time-driven-activity-based-costing, last accessed 27 Nov 2021
2. Braithwaite, A., Samakh, E.: The cost-to-serve method. Int. J. Logist. Manag. **9**(1), 69–84 (1998). Homepage, https://www.emerald.com/insight/content/doi/10.1108/09574099810805753/full/html, last accessed 18 Jan 2022
3. Kaplan, R., Cooper, R.: Cost and Performance: Manage your Costs to Be more Competitive, 1st edn. Futura, São Paulo (1998)
4. Cokins, G.: Activity-Based Cost Management: An Executive's Guide. John Wiley & Sons, New Jersey (2001). Homepage, http://www.untag-smd.ac.id/files/Pepustakaan_Digital_1/

ACCOUNTING%20ActivityBased%20Cost%20Management%20%20An%20Executive%27s%20Guide.%5B2001%5D.pdf, last accessed 16 Nov 2021
5. Guerreiro, R., Bio, S.R., Merschmann, E.V.V.: Cost to-serve measurement and customer profitability analysis. Int. J. Logist. Manag. **19**(3), 389–407 (2008). Homepage, https://www.emerald.com/insight/content/doi/10.1108/09574090810919215/full/html?fullSc=1&mbSc=1&fullSc=1&fullSc=1, last accessed 30 Jan 2022
6. Mejía-Argueta, C., Higuita-Salazar, C., Hidalgo-Cavajal, D.: Methodology for the offer of differentiated service through the analysis of the cost of service. Manag. Stud. **31**(137), 441–454 (2015). Homepage, http://www.scielo.org.co/scielo.php?pid=S0123-59232015000400010&script=sci_abstract&tlng=pt, last accessed 30 Nov 2021
7. Shin, J., Sudhir, K., Yoon, D.-H.: When to "fire" customers: customer cost-based pricing. Manag. Sci. **58**(5), 932–947 (2012). Homepage, https://pubsonline.informs.org/doi/abs/10.1287/mnsc.1110.1453, last accessed 24 Jan 2022
8. Kaplan, R., Narayanan, V.G.: Measuring and managing customer profitability. J. Cost Manag. **15**(5), 5–15 (2001). Homepage, https://www.researchgate.net/publication/291939528_Measuring_and_managing_customer_profitabilit, last accessed 27 Nov 2021
9. Storbacka, K.: Segmentation based on customer profitability: retrospective analysis of retail bank customer bases. J. Mark. Manag. **13**(5), 479–492 (1997). Homepage, https://www.tandfonline.com/doi/abs/10.1080/0267257X.1997.9964487, last accessed 29 Nov 2021
10. Helgesen, Ø.: Do customer accounting and profitability analytics provide managers with new decision support? Evidence from Norwegian fish exporters. Int. J. Food Syst. Dyn. **12**(2), 95–107 (2021). Homepage, http://centmapress.ilb.uni-bonn.de/ojs/index.php/fsd/article/view/78, last accessed 29 Oct 2021
11. Porter, M.: The five competitive forces that shape strategy. **86**(1), 58–77 (2008). Harvard Business Publishing. Homepage, https://hbr.org/2008/01/the-five-competitive-forces-that-shape-strategy, last accessed 2 Jan 2022
12. Cauchick-Miguel, P.A. (coord.), Fleury, A.: Research Methodology in Production Engineering and Operations Management, 3rd edn. Elsevier, Rio de Janeiro (2018). Homepage, https://eu-ireland-custom-media-prod.s3-eu-west-1.amazonaws.com/Brasil/Downloads/14-10/mztodologia.pdf, last accessed 28 Sept 2021

The Triple Bottom Line in Sustainable Supply Chain Management Frameworks and Their Gaps in the Period 2014–2022

Priscilla Cristina Cabral Ribeiro, Paulo Roberto Borges Amaral, Carlos Francisco Simões Gomes, and Marcos dos Santos

Abstract This paper aims to offer a classification of frameworks, as well as a critical reading of gaps in Sustainable Supply Chain Management (SSCM) published from 2014 to 2022. SSCM leads companies to produce with minimum environmental impact. Besides, this production becomes economically feasible and socially responsible. Many authors present frameworks joining the SSCM subjects, following three dimensions of Triple Bottom Line (TBL). The research had three search phases in Scopus and Web of Science databases. After the definition of the research question and the strings, some filters were applied to select papers for the sample. An analysis of titles and abstracts reduces the number of papers. Within the papers, some gaps emerged. They are related to sectors, countries, the partners in a supply chain, information and communication technologies implementation, suppliers' assessment, Circular Economy in SSCM, social subjects, and risk. The economic dimension appears in a higher number of articles than other dimensions. Most frameworks integrate the three dimensions of Triple Bottom Line (TBL). Most of the papers that join the TBL dimensions in their frameworks were published from 2018 to 2022, and less than 10% of selected articles presented new dimensions.

Keywords Sustainable supply chain management · Triple Bottom Line · Frameworks

1 Introduction

The importance of Circular Economy and sustainability needs to be spread out, as those concepts provide benefits in many areas [1]. Recently the concept of

P. C. C. Ribeiro (✉) · P. R. B. Amaral · C. F. S. Gomes
Universidade Federal Fluminense, Niterói, RJ, Brazil
e-mail: priscillaribeiro@id.uff.br

M. dos Santos
Instituto Militar de Engenharia (IME), Rio de Janeiro, RJ, Brazil

V. M. López Sánchez et al. (eds.), *Industrial Engineering and Operations Management*, Springer Proceedings in Mathematics & Statistics 400,
https://doi.org/10.1007/978-3-031-14763-0_7

Sustainable Supply Chain Management (SSCM) has emerged and become an essential part of supply chain management literature [2]. SSCM is the integration of the traditional supply chain with social, environmental, and economic activities [3]. Even though this is the understanding of a real SSCM, the integration of three dimensions of Triple Bottom Line (TBL) proposed by Elkington [4], some frameworks found in this study do not integrate them. Although the feasibility of TBL is undeniable, there is also limited research considering all three dimensions of sustainability [5, 6], though there are many frameworks based on sustainability subjects. Besides, some frameworks do not present this integration, because the one-dimensional framework that focuses on one dimension is the simplest and most commonly used [7]. This situation motivates this research and leads it to the research question: What kind of frameworks have been presented in Sustainable Supply Chain Management (SSCM) papers published from 2014 to 2022?

The paper is organized as follows. Section 2 presents the Method followed by an analysis of the publications about the SSCM and frameworks, presenting the gaps, future research, the dimensions added to TBL, and the papers which join the three dimensions of TBL. The paper concludes in Sect. 4, with the theoretical and managerial contributions.

2 Method

The aim of the research method is to plan the paper to answer the main research question. For this purpose, secondary questions will be answered by some actions in Table 1.

The first step of this research was to search in two databases, Scopus and Web of Science. We chose these databasis because WOS was the only database which covers all domains of science for many years, while Scopus is built on a similar breadth and scale, with a superior number of journals but with lower impact and limited to recent articles [8] using strings as 'sustainability and supply and chain' because the first search was more general. The authors limited the search to 'article title, abstract, keywords', the date range was from 2014 to 2018 (the research began in 2018 and authors intended to search the last 5 years), and the document type chosen was just 'articles'. The keywords selected to decrease the number of the papers were Sustainable, Supply Chain, Environmental, Sustainability Performance,

Table 1 Actions to answer the secondary questions

Secondary questions	Actions
1. What is the most studied dimension of sustainability?	Execution step
2. What are the 'new' dimensions proposed by studies about sustainability?	
3. Are there research that study and develop a framework which integrate the three dimensions of sustainability?	Result analysis

Literature Review, Sustainability Indicator, Case study, Green Supply Chain, Triple Bottom Line.

After that, the areas chosen were Environmental Science, Engineering, Decision Sciences, Economics, Econometrics and Finance, and Multidisciplinarity (Scopus). The authors decided to limit in two languages (Portuguese and English), resulting in 209 papers. After cutting some repetitions, the sample had 132 papers.

To update the research, the authors did the second step extending the period from 2018 to 2020, but the databases were the same, filtering in articles, with a string 'framework and sustainability and supply chain', finding 1.040 papers in Web of Science, and 557 papers in Scopus. The idea in the second search was to limit to the aim of the paper, related to frameworks in SSCM.

The second filter was the subject areas (Scopus) and categories (WOS), and the total decrease to 711 and 353, respectively. The third filter was the journal's impact factors (IF), where the median of them was 3.9, and the journals selected had IF above this value, resulting in a total of 193 (WOS) and 93 (Scopus).

The fourth filter was composed by cutting repeated articles and after those that did not present a framework on SSCM. Hence, the total number was 70 papers in Scopus and 38 papers in WOS. Within these 108, we exclude the articles with a quantitative approach because their frameworks were not based in a deep literature as the papers which have qualitative approach, so we have 31 (Scopus) and 38 (WOS) papers in the fifth filter. Next, to update the research, there was another search in databases (same strings, categories, keywords), where we found 80 papers (2021–2022), and we selected nine papers. In the end, the sample had 78 papers.

The second step, the sample analysis was compounded by the reading of the titles and abstracts, summarizing the main idea of each article, their data (authors, year), the objective, the concepts presented, and what dimensions of sustainability (environmental, economic, and social) they discuss, the gap, and the result was 58 papers.

The third step, the result analysis, began with a reading of the 58 papers, their frameworks, finding the dimensions and their integrations, gaps, and future research.

3 Results

3.1 Frameworks and Main Topics

In this section, we present the papers regarding frameworks found in both databasis. They are organized in the main themes: general models; indicators; evaluation models; collaboration; and risk.

General models Frostenson and Prenkert [9] and Zhang et al. [10] purpose frameworks to allow a better understanding of SSCM and Circular Economy generate an understanding of the prerequisites for sustainable supply chain management. Hugé et al.[11] and Long et al. [12] structured frameworks focusing on the social and economic dimensions, such as economic, institutional, behavioral, and organizational.

Thompson [13] focused more on the environmental issue, developing a framework to assess corporate strategies and actions. In this managerial theme, Norris et al. [14] presented a framework that supports the analysis and management of value-creating relationships between the focal firm, suppliers, and stakeholders of suppliers.

Junaid et al. [2] and Rialti et al. [15] propose a framework relating sustainable supply chain integration and innovation, though the first authors discussed green innovation in the economic dimension and in the second ones the framework was broader, discussing economic and environmental dimensions.

Akbar and Ahsan [16] and Freudenreich and Schaltegger [17] proposed a framework in the apparel industry based on the discussion of social compliance implementation challenges, aiming to reduce social and environmental problems in this sector. Concerning social problems, Huyard [18] and Silva et al. [19] discussed different subjects (food waste and natural disasters), which lead to serious consequences in a high number of people in the world.

Some frameworks discuss logistic subjects, such as evaluation of the materials and processes applied, but also the trade-offs between the economic, environmental, and social pillars (3BL) in an automotive industry context [20]; strategies for the development of a sustainable supply chain in purchase activities [21]; assessing logistical locations and their production levels, measuring their sustainable performance [22]; adoption of sustainable initiatives for companies that work with transport and logistics [23, 24] (assessed SCM and SSCM practices in different sectors, in an arbitrary number of suppliers, manufacturers, distributors, and retailers).

Paul et al. [25] studied the contribution of Blockchain technology (BCT) in the motivation of actors to change their attitudes in the tea supply chain, becoming more competent and improving their sustainable performance.

Indicators A permanent subject in the Supply Chain stream of research is key performance indicators (KPI), which lead to measurements, but this article has a subsection a part of evaluation models because of the unique characteristic of frameworks about the theme. Some of them developed a framework to identify sustainable practices [26]. Other authors intended to analyze the level of sustainability, identifying the main dimensions involved and the issues to be further explored [27], including in cities [28] and human resource management [29].

Some authors studied, in different ways, implementation of GSCM, presenting sustainable practices to influence and be influenced by the 3BL [30–32].

Another set of authors focuses on evaluating the suppliers' performance [33] about industrial sustainability [34] and the furniture industry [35]; sustainable tourism [36].

Evaluation models The models and their frameworks differ in aim, locus of research, and sectors, but they focus on different dimensions, or two or just one.

The set of authors who developed their frameworks on environmental dimension had different works such as in canteen production chain [37]; manufacturing industry [38]; digital manufacturing [39]; production of raw materials, transport, emissions from processing (direct and indirect), end-use applying Land Use Change Greenhouse Gas (GHG) or (LUC) and Life Cycle Optimization (LCO) [40, 41], who structure a framework for redesigning the food supply chain to improve environmental sustainability.

In the social dimension, Suganthi [42] develops her framework about the adoption of social responsibility initiatives, sustainable practices, and the effect on the performance of some workers. Islam et al. [43] aim to measure, through cultural attributes of companies, the level of corporate sustainability achieved. D'Eusanio et al. [44] present a methodological framework, using Sustainable Life Cycle Assessment (S-LCA), and showed several gaps in the implementation phase due to difficulty in evaluating the typical issues of the social dimension.

Other authors integrated two dimensions, such as Panda et al. [45], who develops research about social and environmental, measuring customer altruism, purchase intent, customer loyalty, and indoctrination, to observe consumer behavior and how changes in productive chain these characteristics produce. Angeles-Martinez et al. [46] developed a computational framework for simulation and optimization of the production and distribution, integrating the technological and environmental point of view. Aivazidou et al. [47] integrate environmental with economic foster supply chain sustainability, through water fresh (WF) management for the ex-ante evaluation of eco-efficient supply chain water management policies.

A set of authors integrated three different dimensions, which are not included in TBL, as Aboelmaged [48] who classified them as drivers (technological, organizational, and environmental) on sustainable production practices and their influence on these practices in the market. Xia [49] develop a paper where they discuss three different dimensions (environmental performance and firm performance), having technological (green) in common with Aboelmaged [48].

Some of them integrated the TBL 'classical' dimensions, such as Zailani et al. [50] and Azevedo et al. [51] (in the upstream part of the supply chain); Svensson et al. [52] shed light on the structural properties focusing on the direct effects and the indirect effect (i.e., mediation) between the TBL elements; and Coşkun [3] address the problem of sustainable supplier development through supplier evaluation, discussing suppliers' development. Kazançoglu et al. [6] identify 14 criteria for companies in diffusing sustainability standards throughout multi-tier supply chains to manage the food supply chain and Tapia-Ubeda et al. [53] introduce the Circular Economy in their novel SSCM framework, which assesses the sustainability of a productive process to improves the sustainable development.

Depending on the work, the authors focused on one dimension outside TBL, such as Guo et al. [54] present the framework that shows the energy-saving technological routes and identifies the most energy-consuming stages in the complete supply chain and key single technologies in each stage. Dos Santos et al. [35] developed a theoretical Transaction Costs (TCs) framework of the prefabricated house (PH) supply chain, developing the economic dimension in their work.

Collaboration Leigh and Li [55] develop a conceptual framework to embrace the integration and identify opportunities for companies to work collaboratively, in industrial symbiosis, as Herczeg et al. [56] proposed, identifying the main collaboration aspects and performance impacts. This concern was discussed in other frameworks related to the collaboration theme [57, 58]. The integration and their mutual benefits allowed by collaboration were discussed by Veleva and Bodkin [59]. The theme was explored in tactical and operational levels of the chain by Allaoui et al. [60]. Kaur et al. [61] concluded that focus on agility for target customers through collaboration and information sharing in SSCM will support business continuity.

Risk In this subtheme, Torres-Ruiz and Ravindran[62] proposed a framework, which quantifies the potential risks to the sustainability of the supply chain for different supplier segments, including TBL's dimensions. The framework structured by Ozturkoglu et al. [63] consider, in the ship recycling industry, some constructs, which relate to the work with TBL.

Valinejad and Rahmani [64] in their framework showed an intersection between themes such as sustainability, risk, and supply chain, presenting two more dimensions, such as institutional and technical.

While authors integrate all TBL's dimensions and found others, Shad et al. [65], proposed a framework that focused just on the economic dimension for oil and gas companies.

There were 58 papers, 33 join the three dimensions (four of them add one or more dimensions), 15 analyze two dimensions, and 10 just one. The theme 'Risk' concentrates the higher number of papers discussing all of TBL dimensions, compared to other constructs, 100%, followed by 'Indicators' (80%), General models (53%), Collaboration (43%), and Evaluation models (40%). Related to years, 2018 had, in this sample, the higher number of papers published that integrate the TBL's dimensions (10), followed by 2019 (7), 2015 (5), 2021 (4), 2016, 2020, and 2022 (2 each), and 2017 (1). Economic and environmental was the combination that appeared in a larger number of papers (9), than social and environmental (4), and social and economic (3). Even though the previous research Ribeiro et al. [66, 67] economic dimension was the most studied, in this study, social was the dimension that was most studied alone in the sample, in six papers followed by environmental (3) and economic (2). There are eight papers published from 2014 to 2017 within the 24 papers that built their analysis considering the 3BL, and 25 from 2018 to 2020. In three papers other four dimensions were presented: operational (2016), institutional and technical (2018), industrial sustainability (2019), and productivity (2021) in just four papers.

The main gaps/future research suggested by authors expanded the research to other countries and continents, sectors, and specific productive processes; all stakeholders in the different supply chain; economic feasibility in other business models; assess a wide set of sustainable performances related to information and communication technology (ICT); legal and ethical implications of ICT adoption in the SSCM; industrial symbiosis, product circularity metrics, and assessment, performance implications of Circular Supply Chain Management (CSCM); quantitative analysis and longitudinal data records; identification of opportunities for improvements regarding water consumption, use of lands, and impacts on local communities; and risks. Table 2 summarizes the main results.

4 Conclusion

There are many frameworks developed by different authors from various institutions from 2014 to 2022 about SSCM and different constructs. After a huge search for papers and analysis of 58 frameworks, this study concluded that the main gaps are about the locus of research, agents in the supply chain, ICT, assessment, CE in SSCM, social concerns, and risk.

There are other answers which came from the study. First, the economic dimension of TBL appeared more in the frameworks than environmental and social. The 'new' dimensions proposed by studies about sustainability are operational, institutional, technical, industrial sustainability, and productivity. Finally, most papers integrate the three dimensions of sustainability (TBL). The late period (2018–2022) of research has been developing more research that integrates the TBL's dimensions. These results show that there are many gaps to study in SSCM stream of research related to the two 'main' dimensions (social and environmental), the authors are encouraged to find (or create) new dimensions, and there is a tendency to integrate the TBL dimensions SSCM in frameworks.

This research supports academics who develop articles in SSCM and frameworks because it offers a summary of the main themes, the authors who are working on these themes, what dimension(s) could be studied, and what gaps exist to be explored. There are some suggestions for future research, such as develop a framework in one of these gaps, integrating all dimensions of TBL; develop studies about these gaps and/or new dimensions; develop similar research in a larger period; and study specific subjects of one dimension, as environmental, from some frameworks. Practical implications are related to the application of sustainable performance indicators related to the use of information and communication technology (ICT); industrial symbiosis; product circularity metrics; assessment of Circular Supply Chain Management (CSCM) operations; analysis of water consumption, use of lands, and impacts on local communities.

Table 2 Frameworks and the integration of TBL's dimensions

Themes	Dimensions	Authors
General models	Environmental and social	Frostenson and Prenkert [9]
	Social	Hugé et al. [11]
	Economic and environmental	Thompson [13]
	TBL	Long et al. [12]
	Social	Akbar and Ahsan [16]
	Social and environmental	Freudenreich and Schaltegger [17]
	TBL	Huyard [18]
	TBL	Silva et al. [19]
	Economic and environmental	Centobelli et al. [23]
	TBL	Tajbakhsh and Shamsi [22]
	TBL	Stoycheva et al. [20]
	TBL	Chowdhury et al. [21]
	TBL + productivity	Paul et al. [25]
	TBL	Norris et al. [14]
	TBL	Zhang et al. [10]
	Economic	Coşkun et al. [3]
	Economic and environmental	Rialti et al. [15]
Indicators	TBL	Kannegiesser et al. [30]
	TBL + operational	Luthra et al. [31]
	TBL	Jabbour and de Sousa Jabbour [32]
	TBL	Khan et al. [33]
	TBL	D'Souza et al. [26]
	TBL	Petit et al. [27]
	TBL	Brito et al. [28]
	Social	Mousa and Othman [29]
	TBL and industrial sustainability	Cagno et al. [34]
	Economic and environmental	Dos Santos et al. [35]
Evaluation models	TBL	Zailani et al. [50]
	TBL	Xia et al. [49]
	TBL	Azevedo et al. [51]
	TBL	Svensson et al. [52]
	Economic and environmental	Aivazidou et al. [47]
	Environmental	Krishnan et al. [41]
	Environmental and social	Schaubroeck et al. [37]
	Economic and environmental	Guo et al. [54]
	Economic and social	Wu et al. [36]
	Social	D'Eusanio et al. [44]
	Economic and environmental	Unterfrauner et al. [39]
	Environmental	Garcia et al. [40]
	Social	Suganthi [42]
	Social	Islam et al. [43]
	Environmental and social	Panda et al. [45]

(continued)

Table 2 (continued)

Themes	Dimensions	Authors
	Economic and environmental	Angeles et al. [46]
	TBL	Aboelmaged [48]
	TBL	Coşkun et al. [3]
	TBL	Kazançoglu et al. [6]
	TBL	Tapia et al. [53]
Collaboration	TBL	Herczeg et al. [56]
	TBL	Liu et al. [58]
	TBL	Allaoui et al. [60]
	Environmental	Leigh and Li [55]
	Economic and environmental	Wong et al. [57]
	Social and economic	Veleva and Bodkin [59]
	Economic	Kaur et al. [61]
Risk	TBL	Torres-Ruiz and Ravindran [62]
	TBL + institutional + technical	Valinejad and Rahmani [64]
	TBL	Ozturkoglu et al. [63]
	TBL	Shad et al. [65]

Em environmental, *Ec* economic; *S* social

References

1. Ribeiro, P.C.C., Carpilovsky, L.D., Gomes, C.F.S.: Circular economy and supply chain management: Publications and main themes. In: Tavares Thomé, A.M., Barbastefano, R.G., Scavarda, L.F., Gonçalves dos Reis, J.C., Amorim, M.P.C. (eds.) Industrial Engineering and Operations Management. IJCIEOM 2021 Springer Proceedings in Mathematics & Statistics, vol. 367. Springer, Cham (2021)
2. Junaid, M., Zhang, Q., Syed, M.W.: Effects of sustainable supply chain integration on green innovation and firm performance. Sustain. Prod. Consump. **30**, 145–157 (2022)
3. Coşkun, S.S., Kumru, M., Kan, N.M.: An integrated framework for sustainable supplier development through supplier evaluation based on sustainability indicators. J. Clean. Prod. **130287** (2022)
4. Elkington, J.: Accounting for the triple bottom line. Meas. Bus. Excell. **2**(3), 18–22 (1998)
5. Varsei, M., Soosay, C., Fahimnia, B., Sarkis, J.: Framing sustainability performance of supply chains with multidimensional indicators. Supply Chain Manag. **19**(3), 242–257 (2014)
6. Kazançoglu, Y., Ozturkoglu, Y., Mangla, S.K., Ozbiltekin-Pala, M., Ishizaka, A.: A proposed framework for multi-tier supplier performance in sustainable supply chains. Int. J. Prod. Res., 1–23 (2022)
7. Ibrahim, Y., Arafat, H.A., Mezher, T., AlMarzooqi, F.: An integrated framework for sustainability assessment of seawater desalination. Desalination. **447**, 1–17 (2018)
8. Chadegani, A.A., Salehi, H., Yunus, M.M., Farhadi, H., Fooladi, M., Farhadi, M., Aleebrahim, N.: A comparison between two main academic literature collections: Web of Science and Scopus databases. Asian Soc. Sci. **9**(5), 18–26 (2013)
9. Frostenson, M., Prenkert, F.: Sustainable supply chain management when focal firms are complex: A network perspective. J. Clean. Prod. **107**, 85–94 (2015)
10. Zhang, A., Wang, J.X., Farooque, M., Wang, Y., Choi, T.M.: Multi-dimensional circular supply chain management: A comparative review of the state-of-the-art practices and research. Transp. Res.Part E Logist. Transp. Rev. **155**, 102509 (2021)

11. Hugé, J., Mac-Lean, C., Vargas, L.: Maturation of sustainability in engineering faculties–From emerging issue to strategy? J. Clean. Prod. **172**, 4277–4285 (2018)
12. Long, T.B., Blok, V., Coninx, I.: The diffusion of climate-smart agricultural innovations: Systems level factors that inhibit sustainable entrepreneurial action. J. Clean. Prod. **232**, 993–1004 (2019)
13. Thompson, B.S.: Institutional challenges for corporate participation in payments for ecosystem services (PES): Insights from Southeast Asia. Sustain. Sci. **13**(4), 919–935 (2018)
14. Norris, S., Hagenbeck, J., Schaltegger, S.: Linking sustainable business models and supply chains—Toward an integrated value creation framework. Bus. Strateg. Environ. **30**(8), 3960–3974 (2021)
15. Rialti, R., Marrucci, A., Zollo, L., Ciappei, C.: Digital technologies, sustainable open innovation and shared value creation: evidence from an Italian agritech business. Br. Food J. (2022)
16. Akbar, S., Ahsan, K.: Workplace safety compliance implementation challenges in apparel supplier firms. J. Clean. Prod. **232**, 462–473 (2019)
17. Freudenreich, B., Schaltegger, S.: Developing sufficiency-oriented offerings for clothing users: Business approaches to support consumption reduction. J. Clean. Prod. **247**, 119589 (2020)
18. Huyard, C.: Fostering sustainable diets through a comprehensive approach of value-adding operations: What can vegetable box schemes contribute? J. Clean. Prod. **256**, 120615 (2020)
19. Silva, M.E., Pereira, S.C., Gold, S.: The response of the Brazilian cashew nut supply chain to natural disasters: A practice-based view. J. Clean. Prod. **204**, 660–671 (2018)
20. Stoycheva, S., Marchese, D., Paul, C., Padoan, S., Juhmani, A.S., Linkov, I.: Multi-criteria decision analysis framework for sustainable manufacturing in automotive industry. J. Clean. Prod. **187**, 257–272 (2018)
21. Chowdhury, M.M.H., Agarwal, R., Quaddus, M.: Dynamic capabilities for meeting stakeholders' sustainability requirements in supply chain. J. Clean. Prod. **215**, 34–45 (2019)
22. Tajbakhsh, A., Shamsi, A.A.: facility location problem for sustainability-conscious power generation decision makers. J. Environ. Manag. **230**, 319–334 (2019)
23. Centobelli, P., Cerchione, R., Esposito, E.: Evaluating environmental sustainability strategies in freight transport and logistics industry. Bus. Strateg. Environ. **29**(3), 1563–1574 (2020)
24. Tajbakhsh, A., Hassini, E.: Performance measurement of sustainable supply chains: a review and research questions. Int. J. Product. Perform. Manag. **64**(6), 744–783 (2015)
25. Paul, T., Mondal, S., Islam, N., Rakshit, S.: The impact of blockchain technology on the tea supply chain and its sustainable performance. Technol. Forecast. Soc. Chang. **173**, 121163 (2021)
26. D'Souza, C., McCormack, S., Taghian, M., Chu, M.T., Mort, G.S., Ahmed, T.: An empirical examination of sustainability for multinational firms in China: Implications for cleaner production. J. Clean. Prod. **242**, 118446 (2020)
27. Petit, G., Sablayrolles, C., Yannou-Le Bris, G.: Combining eco-social and environmental indicators to assess the sustainability performance of a food value chain: A case study. J. Clean. Prod. **191**, 135–143 (2018)
28. Brito, V.T., Ferreira, F.A., Pérez-Gladish, B., Govindan, K., Meidutė-Kavaliauskienė, I.: Developing a green city assessment system using cognitive maps and the Choquet integral. J. Clean. Prod. **218**, 486–497 (2019)
29. Mousa, S.K., Othman, M.: The impact of green human resource management practices on sustainable performance in healthcare organisations: A conceptual framework. J. Clean. Prod. **243**, 118595 (2020)
30. Kannegiesser, M., Günther, H.O., Autenrieb, N.: The time-to-sustainability optimization strategy for sustainable supply network design. J. Clean. Prod. **108**, 451–463 (2015)
31. Luthra, S., Garg, D., Haleem, A.: The impacts of critical success factors for implementing green supply chain management towards sustainability: An empirical investigation of Indian automobile industry. J. Clean. Prod. **121**, 142–158 (2016)

32. Jabbour, C.J.C., de Sousa Jabbour, A.B.L.: Green human resource management and green supply chain management: Linking two emerging agendas. J. Clean. Prod. **112**, 1824–1833 (2016)
33. Khan, S.A., Kusi-Sarpong, S., Arhin, F.K., Kusi-Sarpong, H.: Supplier sustainability performance evaluation and selection: A framework and methodology. J. Clean. Prod. **205**, 964–979 (2018)
34. Cagno, E., Neri, A., Howard, M., Brenna, G., Trianni, A.: Industrial sustainability performance measurement systems: A novel framework. J. Clean. Prod. **230**, 1354–1375 (2019)
35. Dos Santos, B.M., Godoy, L.P., Campos, L.M.: Performance evaluation of green suppliers using entropy-TOPSIS-F. J. Clean. Prod. **207**, 498–509 (2019)
36. Wu, K.J., Zhu, Y., Chen, Q., Tseng, M.L.: Building sustainable tourism hierarchical framework: Coordinated triple bottom line approach in linguistic preferences. J. Clean. Prod. **229**, 157–168 (2019)
37. Schaubroeck, T., Ceuppens, S., Luong, A.D., Benetto, E., De Meester, S., Lachat, C., Uyttendaele, M.: A pragmatic framework to score and inform about the environmental sustainability and nutritional profile of canteen meals, a case study on a university canteen. J. Clean. Prod. **187**, 672–686 (2018)
38. Cai, W., Lai, K.H., Liu, C., Wei, F., Ma, M., Jia, S., et al.: Promoting sustainability of manufacturing industry through the lean energy-saving and emission-reduction strategy. Sci. Total Environ. **665**, 23–32 (2019)
39. Unterfrauner, E., Shao, J., Hofer, M., Fabian, C.M.: The environmental value and impact of the Maker movement—Insights from a cross-case analysis of European maker initiatives. Bus. Strateg. Environ. **28**(8), 1518–1533 (2019)
40. Garcia, D.J., Lovett, B.M., You, F.: Considering agricultural wastes and ecosystem services in Food-Energy-Water-Waste Nexus system design. J. Clean. Prod. **228**, 941–955 (2019)
41. Krishnan, R., Agarwal, R., Bajada, C.: Arshinder, K.: Redesigning a food supply chain for environmental sustainability–An analysis of resource use and recovery. J. Clean. Prod. **242**, 118374 (2020)
42. Suganthi, L.: Examining the relationship between corporate social responsibility, performance, employees' pro-environmental behavior at work with green practices as mediator. J. Clean. Prod. **232**, 739–750 (2019)
43. Islam, M.S., Tseng, M.L., Karia, N.: Assessment of corporate culture in sustainability performance using a hierarchical framework and interdependence relations. J. Clean. Prod. **217**, 676–690 (2019)
44. D'Eusanio, M., Serreli, M., Zamagni, A., Petti, L.: Assessment of social dimension of a jar of honey: A methodological outline. J. Clean. Prod. **199**, 503–517 (2018)
45. Panda, T.K., Kumar, A., Jakhar, S., Luthra, S., Garza-Reyes, J.A., Kazancoglu, I., Nayak, S.S.: Social and environmental sustainability model on consumers' altruism, green purchase intention, green brand loyalty and evangelism. J. Clean. Prod. **243**, 118575 (2020)
46. Angeles-Martinez, L., Theodoropoulos, C., Lopez-Quiroga, E., Fryer, P.J., Bakalis, S.: The Honeycomb model: A platform for systematic analysis of different manufacturing scenarios for fast-moving consumer goods. J. Clean. Prod. **193**, 315–326 (2018)
47. Aivazidou, E., Tsolakis, N., Vlachos, D., Iakovou, E.: A water footprint management framework for supply chains under green market behaviour. J. Clean. Prod. **197**, 592–606 (2018)
48. Aboelmaged, M.: The drivers of sustainable manufacturing practices in Egyptian SMEs and their impact on competitive capabilities: A PLS-SEM model. J. Clean. Prod. **175**, 207–221 (2018)
49. Xia, D., Chen, B., Zheng, Z.: Relationships among circumstance pressure, green technology selection and firm performance. J. Clean. Prod. **106**, 487–496 (2015)
50. Zailani, S., Govindan, K., Iranmanesh, M., Shaharudin, M.R., Chong, Y.S.: Green innovation adoption in automotive supply chain: The Malaysian case. J. Clean. Prod. **108**, 1115–1122 (2015)
51. Azevedo, S.G., Carvalho, H., Ferreira, L.M., Matias, J.C.: A proposed framework to assess upstream supply chain sustainability. Environ. Dev. Sustain. **19**(6), 2253–2273 (2017)

52. Svensson, G., Ferro, C., Høgevold, N., Padin, C., Varela, J.C.S., Sarstedt, M.: Framing the triple bottom line approach: Direct and mediation effects between economic, social and environmental elements. J. Clean. Prod. **197**, 972–991 (2018)
53. Tapia-Ubeda, F.J., Isbej Muga, J.A., Polanco-Lahoz, D.A.: Greening factor framework integrating sustainability, green supply chain management, and circular economy: The Chilean case. Sustainability. **13**(24), 13575 (2021)
54. Guo, Y., Yu, Y., Ren, H., Xu, L.: Scenario-based DEA assessment of energy-saving technological combinations in aluminum industry. J. Clean. Prod. **260**, 121010 (2020)
55. Leigh, M., Li, X.: Industrial ecology, industrial symbiosis and supply chain environmental sustainability: A case study of a large UK distributor. J. Clean. Prod. **106**, 632–643 (2015)
56. Herczeg, G., Akkerman, R., Hauschild, M.Z.: Supply chain collaboration in industrial symbiosis networks. J. Clean. Prod. **171**, 1058–1067 (2018)
57. Wong, S.Y., Susilawati, C., Miller, W., Mardiasmo, D.: Improving information gathering and distribution on sustainability features in the Australian residential property market. J. Clean. Prod. **184**, 342–352 (2018)
58. Liu, L., Zhang, M., Hendry, L.C., Bu, M., Wang, S.: Supplier development practices for sustainability: A multi-stakeholder perspective. Bus. Strateg. Environ. **27**(1), 100–116 (2018)
59. Veleva, V., Bodkin, G.: Corporate-entrepreneur collaborations to advance a circular economy. J. Clean. Prod. **188**, 20–37 (2018)
60. Allaoui, H., Guo, Y., Sarkis, J.: Decision support for collaboration planning in sustainable supply chains. J. Clean. Prod. **229**, 761–774 (2019)
61. Kaur, A., Kumar, A., Luthra, S.: Business continuity through customer engagement in sustainable supply chain management: Outlining the enablers to manage disruption. Environ. Sci. Pollut. Res. **1-19** (2021)
62. Torres-Ruiz, A., Ravindran, A.R.: Multiple criteria framework for the sustainability risk assessment of a supplier portfolio. J. Clean. Prod. **172**, 4478–4493 (2018)
63. Ozturkoglu, Y., Kazancoglu, Y., Ozkan-Ozen, Y.D.: A sustainable and preventative risk management model for ship recycling industry. J. Clean. Prod. **238**, 117907 (2019)
64. Valinejad, F., Rahmani, D.: Sustainability risk management in the supply chain of telecommunication companies: A case study. J. Clean. Prod. **203**, 53–67 (2018)
65. Shad, M.K., Lai, F.W., Fatt, C.L., Klemeš, J.J., Bokhari, A.: Integrating sustainability reporting into enterprise risk management and its relationship with business performance: A conceptual framework. J. Clean. Prod. **208**, 415 (2019)
66. Ribeiro, P.C.C., Amaral, P.R.B., Rodrigues, L.B.: O Estado da arte dos temas Economia Circular e Avaliação do Ciclo de Vida na gestão da cadeia de suprimentos. In: 26th Simpósio de Engenharia de Produção, pp. 1–13, Bauru (2019)
67. Ribeiro, P.C.C., Amaral, P.R.B., Benites, M.V.A.: Sustentabilidade e gestão da cadeia de suprimentos: o estado da arte dos temas, 39th Encontro Nacional de Engenharia de Produção, pp. 1–14, Santos (2019)

Applying Lean Tools to Improve the Performance of a Small and Medium-Sized Cutlery Company

Helena Macedo, Inês Moniz, José Dinis-Carvalho, and Rui Sousa

Abstract Today's VUCA (Volatile, Uncertain, Complex and Ambiguous) world and the continuous search for new products and services requires companies to pursuing excellence and continuous improvement. This paper aims to describe a lean implementation project developed in a Portuguese cutlery SME. Action-research was adopted as research method. The production processes of the most representative products (spoons, forks and knives) were analyzed with adequate tools (e.g. VSM – value stream mapping – and indicators monitoring). The main problems identified were excess of inventory (of both finished products and WIP – work-in-process), and consequent long throughput times and low productivity. The root causes for those problems were investigated. Then, improvement proposals, involving the creation of a system to monitor and control production orders, the implementation of daily kaizen meetings, the redefinition of safety stocks and reorder point of finished products and the promotion of training and creation of work instructions regarding equipment setups, were developed and implemented to mitigate the identified problems. The implementations had quite positive impacts such as WIP reduction of 255 K€, 38.5% decrease in finished products stock, throughput time reduction from 17.3 to 4.6 days for spoons and forks, and from 12.4 to 6.8 days in the case of knives, and, finally, an overall productivity enhancement of 36.5%.

Keywords Lean manufacturing · Lean tools · Continuous improvement · Cutlery industry

H. Macedo (✉) · I. Moniz · J. Dinis-Carvalho · R. Sousa
University of Minho, Guimaraes, Portugal

V. M. López Sánchez et al. (eds.), *Industrial Engineering and Operations Management*, Springer Proceedings in Mathematics & Statistics 400,
https://doi.org/10.1007/978-3-031-14763-0_8

1 Introduction

Production competitiveness is one of the most important factors for companies, and is the fuel to survive in an extremely competitive economic market [24]. Companies once exposed to the competitive pressure of market changes, face the challenge of reducing the costs of production processes [4]. In addition to this, for a company to be competitive must possess enormous adaptability to its environment, and must be able to adopt a philosophy of continuous improvement, analyzing and improving processes in order to meet the goal of reducing costs and waste. Aware of these circumstances/needs, the referred cutlery company intends, in general terms, to adopt the lean manufacturing philosophy to increase its competitiveness, especially since it has identified some difficulties in managing the flow of materials. Hence, the global objective of the project described in this paper is to improve the overall performance of the company by applying principles and tools/techniques inherent to the lean manufacturing paradigm. The specific objectives are: (i) reduction of WIP by 50%; (ii) reduction of finished product stocks by 30%; (iii) reduction of lead-time by 50%, and; (iv) increase of productivity by 20%.

Action Research was the research strategy employed in this project. This methodology comprises the involvement of the researcher with the company's employees, enabling collaboration, and creation of critical knowledge, as well as empowering social change [20]. The methodology adopted followed the following phases:

- Diagnosis – In this phase, a Value Stream Map was built depicting the current state of the production process with emphasis on work in process, cycle times and takt time.
- Future state definition – After identifying the existing problems and collecting the necessary data, the VSD (Value Stream Design) of the desired state for the end of the project was defined.
- Action planning – the necessary actions to reach the future state were planned.
- Implementation of the actions – Then the improvement actions were implemented according to the plan.
- Evaluation – Results evaluation, impact evaluation and stabilization of the new standard.

The paper structure was organized accordingly. Section 2 briefly presents the theoretical background of the project, mainly focusing on the principles and tools/techniques of lean manufacturing. Section 3 presents the analysis of the current production process, identifying the main problems and corresponding root causes. Section 4 describes the development and implementation of the improvement actions required to mitigate the identified problems. The results achieved are presented and discussed in Sect. 5, and finally, conclusions are outlined in Sect. 6.

2 Theoretical Background

The Lean Manufacturing paradigm presented by a researcher from the MIT [22] is based on the practices of the Toyota Production System (TPS), which seeks to reduce waste (Muda) as much as possible [14]. The same author considers waste to be all the activities that do not add value to the product, from the final customer's point of view, and identifies seven types of waste: overproduction, waiting, transportation, inventory, over-processing, motion, and defects. However, Jeffrey Liker [9] identified an eighth waste related to not taking advantage of human potential, considering that it will possibly lead to lost time, ideas, improvements, and growth opportunities because workers may not be engaged in their daily work.

Lean Manufacturing evolved and gave rise to Lean Thinking, which is based on five principles that underlie the tools and techniques that eliminate waste [21]: (i) defining value from the customer point of view; (ii) identifying the value stream; (iii) creating a continuous flow; (iv) implementing a pull production, and; (v) pursuing perfection (continuous improvement). The effective management of materials and information flow is clearly an important component of the lean philosophy as well as in other popular excellence models such as Shingo Model [12] and Toyota Way [10]. A very popular tool to represent and assist the improvement of flow is the Value Stream Map [19] but other alternative tools are also available such as Waste Identification Diagram [1, 3]. Flow improvements are crucial since result in less general production waste, less work in process, shorter throughput times and shorter value added ratios.

Continuous improvement is another powerful concept present in most excellence models, some of them referred earlier. This concept is so important that in some cases is confused with an excellence model, the Kaizen model [8]. An important component of continuous improvement is the human factors related to respect for people, the development of people and letting the people grow to their full potential [2]. Very much connected with continuous improvement are the teamwork and the continuous feedback given by performance indicators monitoring and visual management. Introduced by Toyota, visual management is an approach that transforms the physical work environment and makes it safer, simpler, standardized and more fluid, while ensuring that operational results are sustainable [5]. It creates a visual work environment, increasing the transparency of processes and using simple and accessible means (e.g., *andon* and dashboards) so that everyone can understand the status of the work being performed [19]. To effectively measure the success of Lean implementations and continuous improvement, it is necessary to establish a list of critical metrics or indicators that can be measured and quantified [15], the performance indicators, also referred to as Key Performance Indicators (KPI). Following the principles of continuous flow and continuous improvement have a great impact on the path to excellence in the most diverse business areas [7, 17, 18, 23, 25].

3 Production Process Analysis and Diagnosis

Cristema, is a Portuguese company founded in 1993, in Guimarães. Currently with 56 employees, the company has developed its activity based on the production of cutlery products. With over 9.7 million items produced per year, Cristema recorded revenues of 3.5 million euros in 2021.

The initial state of the production process of dinner and dessert spoons and forks as well as tea and coffee spoons will be described and analized as one since they use the same raw material and the production processes are identical (see Fig. 1).

Following, the description of the production process of dinner and dessert knives is presented in Fig. 2.

The diagnosis of the current situation was carried out using the Value Stream Mapping tool, monitoring of key performance indicators, direct observation of the

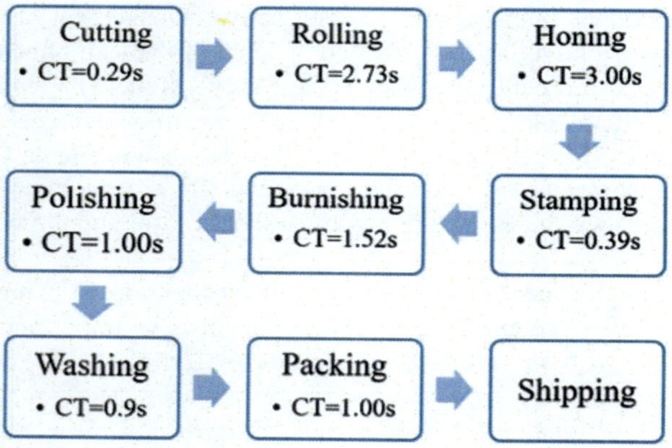

Fig. 1 Spoons and forks production process. *CT Cycle Time

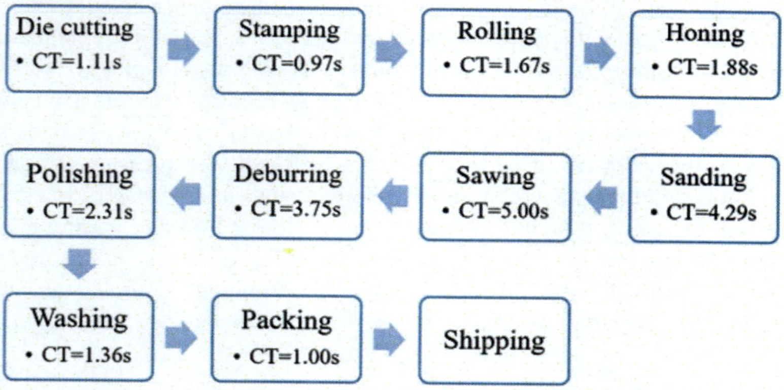

Fig. 2 Knives production process. *CT Cycle Time

Table 1 Stock of finished products (retrieved in February 2021)

Product	Stock of finished product (units)
Spoons	862,905
Forks	584,136
Knives	576,600

Table 2 Value Added Ratio for spoons, forks and knives production processes

Products	Processing time (seconds)	Throughput time (days)	Value-added ratio (%)
Spoons and forks	23.72	17.3	0.0025
Knives	37.12	12.40	0.0054

shop floor, and through conversations with the staff. Raw material was typically ordered for 3-month periods and the stock quantities of finished product is shown in Table 1.

Moreover, in order to identify the WIP quantity existing between processes, data collection was carried out during 2 days in which the company did not work, through direct observation and with the support of team leaders. The throughput time of the parts between processes is determined by dividing the existing WIP (130,370 knives and 537,045 spoons and forks) by the daily demand for each product type (10,500 knives and 31,000 spoons and forks), as presented in Eq. (1) adapted from the famous Little Law [11].

$$\text{Throughtput Time} = \frac{\text{WIP}}{\text{Daily Demand}} \tag{1}$$

The throughput time of 17.3 days was estimated for spoons and forks, and a throughput time of 12.40 days was estimated for knives. In addition to this indicator, the Value Added Ratio was measured for each type of cutlery, and was calculated using Eq. (2), where the Throughput Time represents the sum of all times without value added.

$$\text{Value Added Ratio (VAR)} = \frac{\text{Processing Time}}{\text{Throughput Time}} \times 100 \tag{2}$$

As can be seen in Table 2, from the moment the products enter the system as raw material to the moment they are shipped as finished products, 17.3 days will have passed, in the case of spoons and forks, and 12.40 days in the case of knives. Only 0.0025% and 0.0054% of the time was spent on activities that added value to the respective products. This low ratio of value-added activities, is related to the amount of WIP existing in the process, and some other problems described below.

The high amount of WIP and stock of finished products, inefficiency and lack of information in the production planning process, long setup times and waste in handling and transportation were some of the problems identified in the value stream (see Table 3).

Table 3 Problems identified through the value stream

Problem	Consequence	Waste
WIP	Occupation of the shop floor; Long throughput time: 17.25 and 12.40 days (spoons/forks and knives).	1. Overproduction 2. Inventory 3. Waiting
High amount of finished product stock	Occupation of the warehouse of finished products. Large amount of money invested in dead capital: 2023641 pieces in warehouse.	4. Inventory 5. Overproduction 6. Motion
High amount of raw material stock	Occupation of the warehouse of raw materials; Lack of standardization in the construction of raw materials cutters, generating a consequent need for more raw material references.	7. Inventory 8. Transportation 9. Defects
Inefficiency and lack of information in production planning	Lack of communication between team leaders and the production manager, leading to the production of non-complied with the specifications. Time loss, searching for information related to the production routing.	10. Waiting 11. Overproduction 12. Defects
Long setup times	Time loss, searching for information related to the production routing. Low productivity.	13. Overprocessing 14. Waiting
Excessive motion and transports	Time loss in unnecessary displacements when moving the parts to a subcontractor.	15. Motion 16. Transportation

3.1 Work-In-Process

The first type of waste identified was the high amount of WIP existing in different sections of the company. Through the VSM analysis, the highest amount of WIP was found between the polishing and washing processes of spoons and forks, due to the accumulation of some coffee spoons models that are washed in a specific machine that only works some days per week. Although the WIP existing between polishing and washing stations led to 8.35 of waiting, the WIP existing between the cutting and rolling station was also a matter of concern, since it contributed to increasing the parts' waiting time in the process by 2 days. This WIP can be explained when comparing the cycle time of cutting (0.29 s) and rolling (2.73 s). Since the cycle time of the first operation is much shorter than that of the second, there will naturally be an accumulation of stock between these processes. When analyzing the knives production process, the highest WIP can be found before the sanding operation. This process has the highest cycle time of the value stream (4.29 s) and therefore, it is considered the bottleneck in the production process of knives. Inefficient planning and lack of follow-up of production orders (PO) were identified as root causes of this problem, since there is no planning of POs in all stations, from the beginning to the end of the production process, they are often forgotten and overtaken by urgent

Table 4 WIP cost estimation

Product type	Average cost of an unfinished part (€)	WIP (units)	WIP (€)
Spoons and forks	0.55 €	537,045	295,374.75 €
Knives	0.80 €	130,370	104,296 €
Total			399,670.75 €

parts. Another problem identified was the existence of POs of the same article in WIP and in different sections. This problem shows that, by not having a follow-up of the PO, overproduction of the article can occur.

Based on the average cost of each cutlery product type supplied by the company, an estimation of the inventory cost was made (see Table 4).

This analysis highlights the financial impact of the high values of WIP in the process. This money is simply tied up, so it is important to ensure a minimum necessary stock throughout the production process, in order to liberate this money, so that it can be invested in something that adds value to the company.

3.2 *Throughput Time*

Throughput time is very much connected to WIP, as they show very much the same problem: materials waiting in the shop floor. The throughput time is the effect of the WIP in the company's response to demand. Long throughput time is frequently responsible for long delivery time and in many cases puts companies in a poor competitive position. As shown earlier, the spoons and forks take 17.3 days to go through all processes from raw materials to finished products. It is almost shocking realizing that during those more than 17 days the products only spend 23.72 seconds in processing operations. The current throughput time value is noticeably high being a clear problem that must be addressed.

3.3 *Finished Products Stock*

Another obstacle identified with VSM was the existence of a high amount of stock of finished products (FP). This problem is mainly due to the fact that production is mostly made for stock, and not to order. In addition, the delay of stock's information in the production software program, leads to overproduction, which increases the already existing problem of high stock of FP. This problem, in addition to generating very high ownership costs, in this case more than one million euros, also generates the need for a larger storage area (see Table 5).

Table 5 Cost estimation of finished products stock

Product	Stock of finished product (units)	Stock of finished product (€)
Spoons	862,905	474,597.75 €
Forks	584,136	321,274.80 €
Knives	576,600	461,280.00 €
Total	2,023,641	1,257,152.55 €

Table 6 Average (avg) time spent in Setup activities per day

	Avg number of setups per day		Avg time per setup (min)		Avg time spent in setup activities per day (min)	
Operation	Spoons/forks	Knives	Spoons/forks	Knives	Spoons/forks	Knives
Cut	7	1	60	10	420	10
Rolling	1	2	30	25	30	50
Deburring	1	2	25	35	25	70
Stamping	7	2	20	30	140	60
Sanding	–	3	–	5	–	15
Knurling	–	2	–	15	–	30
Burnishing	3	2	80	45	240	90
Polishing	9	7	60	60	540	420

3.4 *Productivity*

Field observation and monitoring of performance indicators, allowed the researcher to identify that the company was struggling to deliver on time – symptom of high levels of WIP as well as low productivity. This key performance indicator (KPI), besides being affected by the high quantity of WIP existing in the process, is also affected by factors such as: high setup times, excessive motion, excessive transports, excessive waiting, and defects. Another problem is the communication failures between the production manager and the section managers, which in turn lead to delays in the execution of POs and even in the production of products with wrong specifications or forgotten POs responsible for the existence of "forgotten" WIP on the factory floor and in the finished products warehouse. The lack of communication and inefficient planning were also identified as reasons that were affecting the performance of setup activities. In this regard, often the operators did not know what product/model was going to produce next, so there was no previous preparation of the tools needed for the setup. This means that the operators have to go to the section manager to know what product the team should produce next. The fact that the tools were not specified in the POs, as well as the lack of their location, also complicated the process, since it forced the operator to look for someone who knew where the tools were stored or simply knew which tool to use.

Table 6 shows the time spent daily in setups, according to the company's average reference values.

Through the analysis of the table, it is possible to verify that there are cases in which more than half of the available daily production time is spent on setup activities. In the case of polishing, 9 hours are spent daily in setup activities for the production of spoons and forks, and 7 hours in the case of knives.

4 Improvement Actions Development and Implementation

According to the problems identified and described in the last section, some solutions and improvements were proposed through the application of lean concepts and techniques such as Visual Management, Standard Work, Daily Kaizen and KPI monitoring. Some improvements were also implemented in the production planning and control practices.

4.1 Production Orders Monitoring and Control System

Since the reduction of the WIP existing in the factory was one of the main objectives of this project, it was necessary to recognize and account all the product parts that were forgotten on the shop floor, define their production status and determine the missing operations. Thereupon, together with the production and section managers an action plan was outlined and performed to finish all the ongoing production orders.

In addition, a follow-up system for all the POs was also implemented. A weekly production plan was outlined for polishing and cutting stations, and the remaining operations were ruled by FIFO. Along with this, a daily production plan was created, and shared with team leaders, in order to improve the management of their daily work. As can be seen in Fig. 3, when the PO is colored in green means that the PO was completed, when colored in blue means that it still in progress, and when colored in gray means that it hasn't started yet.

4.2 Kaizen Meetings

In order to improve communication, increase focus and prioritize work, a daily kaizen meeting of 10 minutes was implemented. During this meeting, the daily production plan (see Fig. 3) was delivered to the team leader so he could be best prepared to manage the daily work.

Moreover, with a view to overcome communication problems between departments, a 1-hour meeting per week was implemented. This meeting aimed to analyze the ongoing orders at the moment and those that will be produced in the next 2 weeks, in order to align needs, especially with the purchasing department, in case it

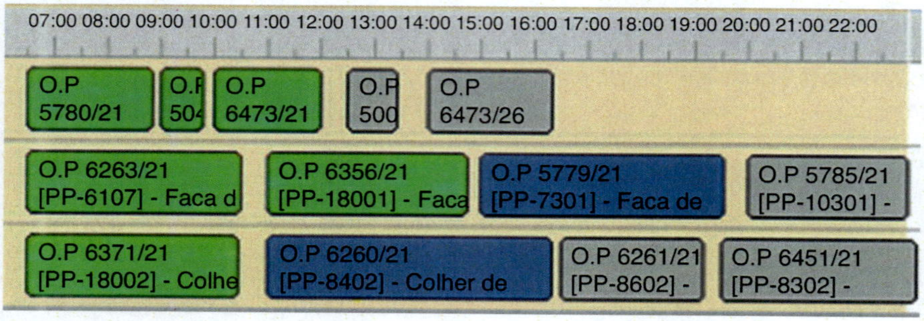

Fig. 3 Daily production plan

would be necessary to acquire some type of raw material. Both meetings followed a standard agenda, so the team could be focused and only the most important topics be discussed.

4.3 Redefinition of Safety Stocks and Re-order Point

Initially it was necessary to check and correct the inventory of each existing product's reference of the factory. Then the researcher created an excel document with all the references and the respective information such as demand and average and maximum lead-time, in order to define the Safety Stock (SS) and Re-order Point. The SS was determined based on Eq. 3:

$$SS = (\text{max daily demand} \times \text{max .lead time}) - (\text{avg daily demand} \times \text{avg lead time}) \quad (3)$$

The Re-order Point is determined using Eq. 4:

$$\text{Re-order Point} = (\text{avg daily demand} \times \text{avg delivery time}) + SS \quad (4)$$

Furthermore, an Excel document was created in order to support the production manager to oversee and manage the production alerts. This document analyzes the production needs and prioritize them based on the Re-order point and generates a corresponding color alert (green – low priority; red – high priority).

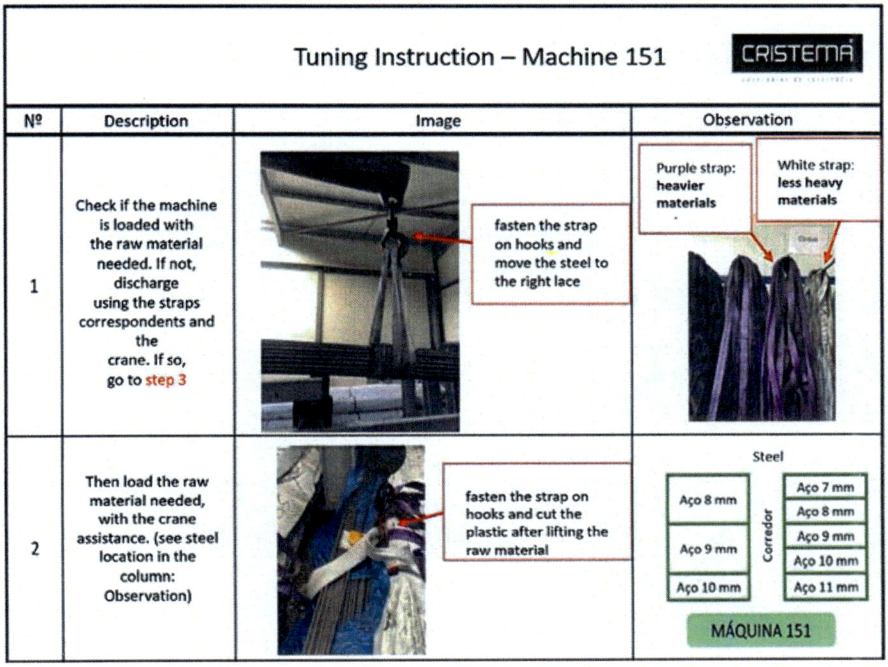

Fig. 4 Example of a tuning instruction

4.4 *Setup Training and Work Instructions*

Considering the long setup times and the difficulty of tuning most of the machines, tuning manuals were created, to support operators to perform the setup activities autonomously and decrease the total tuning time (see Fig. 4).

With the daily production plan defined in advance, the team leader was responsible for prepare the tools needed for setup and training the operators adopting the tuning manuals. With these improvements a decrease of 440 min per day in setup activities was verified.

4.5 *Team Boards*

The team boards implemented were designed by the researcher in a simple way for making them clear to workers, and their main objective is to enhance transparency, focus, monitor key performance indicators, promote communication, and reflection for all team members.

Fig. 5 Team board

As can be seen in Fig. 5, the team boards are divided into three sections: Team, KPI monitoring and improvement suggestions. In the team's section, the team's competencies matrix is presented and below that, there is a photo of each team member with a blank space underneath, so the production/section manager can write down tasks to be done by each member of the team. Regarding the KPI monitoring, the team in collaboration with the researcher defined the indicators that could relate best with their performance and started monitoring the setup times, number and time spent in stops and breakdowns and information related to the current production. Finally, a last section in the boards was reserved to the improvement suggestions that could be given by the employees. This initiative was created in an attempt to give space and make official the act of giving any suggestion that could improve the company's performance or the employees work conditions/environment and quality of life.

5 Results Discussion

This section presents the results obtained in this project, in order to evaluate the effectiveness of the implemented improvements. The completion and follow-up of all PO allowed a reduction of 255180.80€ in WIP costs, representing a decrease of approximately 66% in inventory, exceeding the 50% set as initial goal. The definition of safety stock and re-ordering point led to a decrease of 38.49% in finished products stock, also surpassing the initial goal of 30%. The creation of team boards and daily kaizen meetings promoted transparency and communication between workers, team leaders and production managers, the weekly kaizen meeting broke barriers existing between departments. This enhancement of transparency and communication, and the elimination of WIP led to a reduction in the Throughput Time from 17.3 to 4.6 days for spoons and forks (representing a reduction of 73.4%), and from 12.4 to 6.8 days in the case of knives (representing a reduction of 45.16%). The implementation of setup and tuning instructions and training of workers allowed a gaining of 440 productive minutes per day of work. With all the improvements implemented during this research project, the company's productivity increased by 36.46%, accomplishing and transcending the initial set goals of 20% for this work program.

The results of this project converge with the findings of several studies that state that most SMEs lean implementation initiatives focus on waste reduction on the shop floor, including reductions in inventory, space, time (e.g. changeover time and throughput time) and cost of products [7]. Zhou et al. [25], studied lean practices and related implementation issues in SMEs in the U.S. and clearly observed that lean tools and programs had positive impacts on the enterprises' performance whereas an increased productivity and efficiency, along with increased customer satisfaction and decline in manufacturing/inventory cost are the primary benefits brought by lean.

The main contributions of this work to lean research are: (i) Improving flow, a key lean principle, is crucial to overall production performance improvement, and; (ii) teamwork and visual management play a key role in continuous improvement effectiveness. Similar gains as the ones achieved in this work can be achieved in many traditional companies in Portugal as well as in any other country in the world as long as managers follow the type of methodology applied in this work.

This work resulted in a direct contribution to the operational excellence [13] of the company and its future sustainability [16]. The positive effect of lean effective implementations on sustainability is expected by the lean own nature [6]. In the work presented in this article, the company improved its economic sustainability because it reduced its costs and increased its competitiveness. The social component was improved because it created teamwork, improved communication between people and brought greater identification of each person's effort in everyone's performance. In the environmental component, the company began to produce more with less, reducing waste.

6 Conclusion

The objective of this research project was to analyze and improve the material flow management and productivity in Cutlery Company. With this objective in mind, lean tools and principles were applied. Through the analysis of the production process of products with the greatest impact on the company's income, it was possible to highlight some problems that were affecting the performance of the organization, such as the high amount of WIP, long throughput times, and high stock of finished product and raw material. It was also identified an inefficient and lack of information in production planning, long setup times and waste in motion and transport.

The implementation of lean concepts, routines and tools such as, Visual management, Standard work, Daily Kaizen meetings, and definition of Safety stock and Re-ordering point brought several benefits to the company, mostly in terms of productivity, inventory costs and improvement in information flow.

Confronting the objectives outlined initially, with the results obtained, productivity was increased by 36.46%, way above the defined goal of 20%. WIP was also reduced by 66.20%, which implies a reduction of 255180.80€ in inventory cost. Throughput time was also reduced by 73.4% for spoons and forks, and 45.16% for knives, allowing a faster response to end customers. Finally, the stock of finished products was reduced by 38.49%, reducing the needed space for storage, holding costs and leading to a possible restructuring of the warehouse layout, proposed as future work.

Acknowledgments This work has been supported by FCT – Fundação para a Ciência e Tecnologia within the R&D Units Project Scope: UIDB/00319/2020.

References

1. Dinis-Carvalho, J., Guimaraes, L., Sousa, R.M.R.M., Leao, C.P.C.P.: Waste identification diagram and value stream mapping: a comparative analysis. Int. J. Lean Six Sigma. **10**(3), 767–783 (2019). https://doi.org/10.1108/IJLSS-04-2017-0030
2. Dinis-Carvalho, J., Monteiro, M., Macedo, H.: Continuous improvement system: team members' perceptions. In: Lecture Notes in Networks and Systems, vol. 122, (2020). https://doi.org/10.1007/978-3-030-41429-0_20
3. Dinis-Carvalho, J., Moreira, F., Bragança, S., Costa, E., Alves, A., Sousa, R.: Waste identification diagrams. Prod. Plan. Control. **26**(3), 1–13 (2015). https://doi.org/10.1080/09537287.2014.891059
4. Dombrowski, U., Ebentreich, D., Krenkel, P.: Impact analyses of lean production systems. Proced. CIRP. **57** (2016). https://doi.org/10.1016/j.procir.2016.11.105
5. Galsworth, G.D.: Visual Workplace: Visual Thinking – Creating Enterprise Excellence Through the Technologies of the Visual Workplace. CRC PRess (2017)
6. García-Alcaraz, J.L., Morales García, A.S., Díaz-Reza, J.R., Jiménez Macías, E., Javierre Lardies, C., Blanco Fernández, J.: Effect of lean manufacturing tools on sustainability: the case of Mexican maquiladoras. Environ. Sci. Pollut. Res. (2022). https://doi.org/10.1007/s11356-022-18978-6

7. Hu, Q., Mason, R., Williams, S.J., Found, P.: Lean implementation within SMEs: a literature review. J. Manuf. Technol. Manag. **26**(7) (2015). https://doi.org/10.1108/JMTM-02-2014-0013
8. Imai, M.: Kaizen: The Key to Japan's Competitive Success. McGraw-Hill Education (1986). https://books.google.pt/books?id=q0rCTQlvNMoC
9. Liker, J.: Toyota Way: 14 Management Principles from the World's Greatest Manufacturer. McGraw-Hill Education (2004)
10. Liker, J., Franz, J.: The Toyota Way to Continuous Improvement : Linking Strategy and Operational Excellence to Achieve Superior Performance. McGraw-Hill Publishing (2011)
11. Little, J.D.C.: A proof for the queuing formula: $L = \lambda W$. Oper. Res. **9**(3) (1961). https://doi.org/10.1287/opre.9.3.383
12. Miller, R.: The Shingo Model for Operational Excellence, p. 45. Jon M. Huntsman School of Business (2016)
13. Naftanaila, I., Radu, C., Cioana, G.: Operational excellence - a key to world-class business performance. Studies in Business & Economics. **8**(3), 133–141 (2013)
14. Ohno, T.: Toyota Production System: Beyond Large-Scale Production (C. Press, Ed.; 3ª Edição). Productivity, Inc (1988).
15. Ortiz, C.A.: The TPM Playbook: A Step-by-Step Guideline for the Lean Practitioner. Productivity Press (2016)
16. Purvis, B., Mao, Y., Robinson, D.: Three pillars of sustainability: in search of conceptual origins. Sustain. Sci. **14**(3) (2019). https://doi.org/10.1007/s11625-018-0627-5
17. Ribeiro, P., Sá, J.C., Ferreira, L.P., Silva, F.J.G., Pereira, M.T., Santos, G.: The impact of the application of lean tools for improvement of process in a plastic company: a case study. Proced. Manuf. **38** (2019). https://doi.org/10.1016/j.promfg.2020.01.104
18. Rodrigues, J., Sá, J.C., Silva, F.J.G., Ferreira, L.P., Jimenez, G., Santos, G.: A rapid improvement process through "quick-win" lean tools: a case study. Systems. **8**(4) (2020). https://doi.org/10.3390/systems8040055
19. Rother, M., Shook, J.: Learning to see: value stream mapping to add value and eliminate muda. In: The Lean Enterprise Institute (1999). https://doi.org/10.1109/6.490058
20. Saunders, M., Lewis, P., Thornhill, A.: Research Methods for Business Students, 5th edn. Pearson Education (2009)
21. Womack, J., Jones, D.: Lean Thinking: Banish Waste and Create Wealth in Your Corporation. Fee Press (1996)
22. Womack, J., Jones, D., Roos, D.: The Machine that Changed the World. Free Press (1990)
23. Wong, Y.C., Wong, K.Y., Ali, A.: A study on lean manufacturing implementation in the Malaysian electrical and electronics industry. Eur. J. Sci. Res. **38**(4) (2009)
24. Yue, D., Brychko, A.: Competitiveness of production as a determining factor of effective management of the enterprise in market conditions. Ekon. Upr. APK. **2**(151), 10.33245/2310-9262-2019-151-2-67-75 (2019)
25. Zhou, B.: Lean principles, practices, and impacts: a study on small and medium-sized enterprises (SMEs). Ann. Oper. Res. **241**(1–2) (2016). https://doi.org/10.1007/s10479-012-1177-3

Digital Transformation in the Public Sector: Enabling Technologies and Their Impacts

Shayane Betiatto dos Santos, Fabio Capellin, Marcelo Trentin, Sandro Cézar Bortoluzzi, and Edson Pinheiro de Lima

Abstract Digital transformation has allowed advances in several areas, as it promotes the adoption of sophisticated digital technologies aimed at improving processes and reducing costs. Digital transformation technologies have the potential to encourage improvements and innovations in all types of institutions, both public and private. Unlike the evolution of private organizations, the public sector lacks adaptation to these innovations. Thus, to reap the benefits of digital transformation, the public sector needs to be more agile and more flexible and adaptive. Therefore, this study aims to raise contributions from enabling technologies for digital transformation in the public sector. We carried out a systematic review of the literature using the ProKnow-C method. Through content analysis, the focus of this article, the technologies were pointed out where and how they are being used in this sector, in addition to some benefits and difficulties of implementation. The survey results from the 18 articles in the bibliographic portfolio indicated that Europe and Asia have the highest utilization rate, especially the United Kingdom. In addition, Blockchain is the technology with the most significant applications. On the other hand, the Internet of Things, widely used in the private sector, had few appearances. However, public agencies in several countries have improved using these technologies despite the nascent maturity.

Keywords Digital Transformation · Smart Government · Public Sector · Enabling Technologies

1 Introduction

Information and Communication Technologies (ICTs) are considered one of the essential digital age topics, directly implying the development of human activities,

S. B. dos Santos (✉) · F. Capellin · M. Trentin · S. C. Bortoluzzi · E. P. de Lima
Universidade Tecnológica Federal do Paraná, Pato Branco, PR, Brazil
e-mail: shayane@alunos.utfpr.edu.br

V. M. López Sánchez et al. (eds.), *Industrial Engineering and Operations Management*, Springer Proceedings in Mathematics & Statistics 400,
https://doi.org/10.1007/978-3-031-14763-0_9

both personal, social, and work. The disruptive potential of ICTs requires the adaptation of markets, companies, and even governments to the new standards established by digitalization. Consequently, the adoption of ICTs is a fundamental aspect of the modernization of the public sector to improve administrative efficiency, centrality, and availability of services [1].

Considered a global event [2, 3], Digital Transformation (DT) has as its principle the adoption of sophisticated digital technologies to promote changes, disruptive or incremental [4], to improve public [5] and private organizations [3], allowing business models, operational processes and customer experiences create value for processes, products, and services [6].

Although there is still no unanimous definition for DT [7–10], there is a consensus among authors about their involvement in the use of ICTs [2, 3, 7], which impact the strategy [11], in the structure [12] and distribution of power [13]; to create new capabilities in businesses [7, 14], in governments [2, 3, 7] and manufacturing companies [15].

To enjoy the benefits of DT, such as cost reduction and service quality improvement [3], public institutions need to be able and flexible to cultural changes, thus allowing them to follow the evolutions and development of new sophisticated digital technologies [16–18]. These transformations focus on developing and applying digital solutions to streamline internal and external processes [18, 19] and improve the availability and delivery of services to society [16, 20, 21].

Emerging and disruptive technologies capable of enabling DT in the public sector are identified in the literature. Below, some of these technologies are presented and a brief definition with the characteristics associated with each one of them.

(i) Blockchain – technology used for secure financial transactions, such as Bitcoin and Huperledger-Fabric, and inventories or evidence of government charges [22]. Blocks are created to store part of the information for these transactions. These structures contain data to be mined, that is, processed until finding the result of the security key. They are chained linearly and chronologically to each other, ensuring the security sequence of the transaction through protocol consensus. This technology differs in that all blocks are secured; that is, to tamper with a transaction; all blocks must be manipulated, which requires a computationally complex operation [23].

(ii) Artificial Intelligence – refers to the use of autonomous and interdisciplinary systems, which can operate without human intervention and learn by themselves to identify patterns for decision-making [24]. In this way, AI can be described as a system that identifies the relationships that are taking place in the environment, interprets the observations, obtains conclusions, and determines actions to solve problems or perform an activity for which it was designed [25]. This technology has one of the most significant potential applications for improving the productivity of operations [26].

(iii) Internet of Things (IoT) – Considered a developing technology, the IoT aims to create value for users from intelligent resources that are shared and integrated [27]. Velsberg et al. [28] and Garcia et al. [27] describe IoT as electronic devices that process signals from sensors that transmit data through a wireless network. The IoT is divided into layers; authors classify between three and six layers, but all models present the level of perception, web, and application [29–32]. For this technology to meet its objective, it is necessary to guarantee data storage for temporal analysis, avoid system inoperability through well-defined architectures and standards, and ensure data privacy, security, and quality [33].

(iv) Business Intelligence and Analytics – By collecting, integrating, accessing, and analyzing data, in addition to transforming them into information with relevant content quickly and easily, Business Intelligence is a concept that uses technologies, applications, and tools to improve businesses to support decision making [34, 35]. The idea of Business Analytics (BA) seeks to present information quickly and transparently with information technologies so that attitudes are taken to strengthen economic agents, operational activities, and strategic businesses in supply chains [36]. The BA delivers "right decisions, to the right people, at the right time" [37], with the assignment of presenting information so that a company understands its market and knows its business [36].

(v) Big Data – To help decision-making and obtain competitive and economic advantages, Big Data can be understood as a set of architectures and technologies that collect large volumes of data, exploring its variety through analysis [22, 38]. There is no consensus on the definition of Big Data in the literature. Russom [39] defines this technology by the concept of 3V's – speed, volume, and variety. Lomotey and Deters [40] adopted the 5V's, adding veracity and value [40]. The authors complement each other, and the definitions of the V's agree.

(vi) Cloud Computing – Refers to quickly and with minimal effort or contact from service providers to access servers, services, applications, and other computing resources [41]. Its main benefit is to reduce costs in information technologies [42]. The consequence of this is the search for innovation in an intelligent, flexible, practical, and creative way, increasing productivity, reliability, and availability, reducing response times [43].

Although the literature clearly and directly addresses some technologies linked to the digital transformation of industries, it presents sophisticated digital technologies related to a few studies to DT in the public sector [44, 45].

Therefore, understanding and predicting DT characteristics becomes a vital aspect of being considered by policymakers, government executives, researchers, and all individuals who prepare, implement or evaluate government decisions [44, 46]. Given this, public sector organizations need to be more adept and flexible in adopting emerging technologies, deal with discontinuous changes and work collaboratively and efficiently with the new digital and intelligent options [1, 19].

This article aims to map and structure a theoretical framework on digital technologies' main contributions to the public sector's digital transformation to understand the problem better. Therefore, we performed a systematic literature review (SLR) using the ProKnow-C method [47]. In addition, through content analysis, we identify the insights and contributions of enabling technologies for digital transformation.

Despite the high expectations regarding DT, the failures in the transformation of public organizations (or at least in part of them) in recent years indicate a lack of understanding of the complexity of DT and the relationships between the adoption of technologies, the use of information, social contexts and institutional arrangements [44, 45]. Consequently, it is essential to understand these relationships and the changes that occur [44], as digital technologies can promote or satisfy the government's needs about DT, in addition to providing benefits in reducing operational costs and procedural time, promoting citizen participation, and improving the government's relationship with its stakeholders, enable the use of intelligent technologies and the development of cities, subsidize wise governments and governance, and promote transparency, accountability, and equity in the treatment of all [48].

2 Research Methodology

This article adopts an exploratory and descriptive approach to address the research problem. It seeks to investigate and identify the technologies that drive digital transformation in the public sector and describe their contributions in this area. In addition, a qualitative analysis is carried out regarding the research process, as it aims to understand DT technologies and how they relate to the development of the public sector [49].

This article uses the ProKnow-C method [47] for SLR to build a relevant and sufficient bibliographic portfolio (BP) about the research topic.

We delimited the research axes and the search terms for the first stage of BP selection. The research axes were three: (i) "**Digitalization and digital transformation**"; (ii) "**Analytics and assessment**"; and (iii) "**Public Sector**". The search terms were constructed to synthesize the objectives related to each research axis. In this way, the first axis (Digitalization and digital transformation had the following terms: "Industry 4.0"; "Digitization"; "Cyber-physical systems"; "Smart contracts"; "Blockchain"; "IoT"; "Corporate sustainability"; and "Big data analytics". Already your terms of the second axis (Analytics and assessment) were: "Analytics"; "Measurement"; "Assessment"; "Evaluation"; "Maturity model"; and "Measures". Moreover, the last axis (Public Sector) had its name as a search term. This resulted in 108 combinations.

We used Web of Science and Scopus databases without time limits, and between February 8 and 15, 2022, the search process presented 3057 articles as an initial gross portfolio. With the help of the Mendeley Desktop software, used to manage the search results, we eliminated 33 articles that were duplicates or that were published in books and conferences, thus maintaining 3024 articles.

The next step was to read the titles of the articles. To select those that would follow in the BP, the articles should have one of the following characteristics:

(i) public sector, public or state-owned company; (ii) digital technology; (iii) technological maturity; (iv) evaluation and measurement framework; and/or (v) public governance. Because of the filtering process, 440 articles were kept in the BP.

The sequence was based on the appreciation of the abstracts, maintaining those articles that met the characteristics of having some digital technology or evolution models and/or evaluation of digital transformation in the public sector. From this filtering, 100 articles remained.

For the final definition of the BP, those articles available in full for reading and that had DT contributions in the public sector remained. The following contributions are listed: (i) implementation of digital technology in the public sector; (ii) discussion on the modernization of public services in the face of digitalization; (iii) identification of challenges and limiting factors for the adoption of sophisticated digital technologies; and/or (iv) studies to assess the digital maturity of public companies. Thus, the final BP was composed of 18 articles cited in the references and highlighted by the identifier (*).

For content analysis, we used a grouping of similar characteristics of the studies, understanding how the investigated literature has evolved and is developing. In addition, the transformation of this information into a graphic representation makes it possible to synthesize the data extracted from the literature and demonstrate lines of construction of knowledge about the subject studied. In this sense, we first selected the central axes, verified by the recurrence of a topic or issue in the investigated studies. Subsequently, similar characteristics are grouped into areas of concern and ranked in order of priority or preference [50]. Consequently, we centralize strategic issues and allocate operational objectives at the ends, organizing the identified problems and topics hierarchically [7].

3 Results and Discussions

This section presents the results and discussions considering the impacts of the adoption of sophisticated digital technologies as enabling axes of digital transformation in the public sector. This study focuses on content analysis regarding the contributions of sophisticated digital technologies to public sector DT and not on bibliometric investigations.

The adoption of enabling technologies for digital transformation impacts the development of human activities, both personal, social, and work. Due to its disruptive potential, it requires governments to adapt to the new standards established by digitalization. Thus, we address below some cases found in the literature, highlighting contributions and difficulties in adoption.

These contributions are described in Table 1, which presents a synthesis of the technologies found in the systematic literature review. The authors pointed out the use of technologies, which countries already have some application reports, and which contributions or implementation difficulties these authors mention.

Table 1 Synthesis of enabling technologies for digital transformation in the public sector

Technology	Article	Parents	Contribution/Difficulty
Blockchain	[51–56]	Denmark, Honduras, Georgia, Dubai, Spain, Cyprus, Switzerland, USA, Estonia, Singapore, UK, Republic of Moldova	It reduces corruption, improves transparency and reliability, and increases investments of internal capital flows. Online management of traditional citizen services, such as authentication of individual identities for digital voting, to the management of identity of foreigners. It facilitates notary services, such as marriage certificates and management contracts.
Artificial Intelligence	[14, 57]	United Kingdom, United Arab Emirates, European Union	It helps in decision-making.
Internet of things	[28, 58]	Estonia	Efficiency in public services.
Business Intelligence and Analytics	[34, 59]	South Africa, Malaysia	Improvement in organizational performance.
Big Data	[60–65]	South Korea	Insights for policy decisions. Internal transparency and disclosure of information. High cost, lack of understanding, ability to process data, and a lack of theoretical foundation. The complexity of data integration, sharing, and security, plus the need to manage privacy and security risks. Ethical problems, difficulties in organizational change, lack of investment in information technologies, and the lack of qualified professionals.
Cloud Computing	[66]	UK, USA, Australia, Thailand, Saudi Arabia	Security issues.

As explained previously and observed in Table 1, Blockchain is the most used technology in public governance. It presents a more significant recurrence of authors and application cases, mainly in promoting the development of intelligent governments. Some of these applications can be cited, such as (i) the registration of data in the public health service in Estonia; (ii) the registration of public lands in Honduras, Georgia, and Dubai or, still; (iii) as a possible substitute for the national currency electronic payment system in Denmark for making small and micropayments [52, 53, 56].

Some applications propose using Blockchain as a facilitator of using virtual tokens in voting systems by electronic devices in a secure way [51].

Hyvärinen et al. [52] present a model for the use of Blockchain applied in the government of Denmark, in which the objective is to reduce fraud in dividend tax payments due to double taxation. In addition, Ølnes [54] exemplifies the use of Bitcoin in wise governments but points out that more functions than payment solutions are needed for this.

We also found that the UK has emerged as the main highlight in the modernization of public governance, adopting digital initiatives in Blockchain, Artificial Intelligence, and Cloud Computing, to improve the provision of public services such as authentication of individual identities for digital voting [55], automation of government services [14], as well as improvements in their information technology processes [66]. The UK has also encouraged public-private partnerships between government, industry, and academia to develop AI projects for the prosperity of society [57].

Most efforts found in the literature on DT in the public sector come from Europe and Asia. No relevant application or contribution was verified in South America. As the continent comprises emerging and underdeveloped countries, we understand that few initiatives are being developed to the detriment of cultural and social aspects. Still, regional or local DT initiatives may not have been incorporated in this analysis due to the delimitations of the sample field in the SLR.

We also note that one of the main goals of adopting sophisticated digital technologies is to improve the decision-making process [14, 34, 58–60]. As in private companies, these improvements aim to make the collection and analysis of data accurate and agile, transforming information into insights and forecasts that support assertive decision-making, focused on the citizen and legitimized by society. In this sense, Hartley and Seymour [59] gather information on how the implementation of Business Intelligence (BI) tools can benefit governments in obtaining information, analyzing data for recurring issues, in supporting the formulation of strategies and policies aimed at citizens. Service management and transparency. Abai et al. [34] present essential factors to be considered in adopting Business Intelligence and Analytics tools in the Malaysian public sector, especially related to professional skills for operationalization and effective adoption of the devices.

Consequently, Alassafi et al. [66] current government initiatives in Cloud Computing, such as (i) the United Kingdom with the adoption of cloud computing to improve IT services; (ii) the United States of America with the adoption of electronic public services; (iii) Australia from the transfer of public data to the cloud; (iv) Thailand through the development of a cloud-based e-mail platform; and (v) Saudi Arabia with investments in IT services.

Figure 1, presented below, depicts a conceptual framework for the uses of each of the technologies found in the fragment of the literature analyzed by this research.

We highlight in the current scenario as key enabling technologies of DT in the public sector: AI, Blockchain, Cloud Computing, Big Data, BI, and IoT. Its applications are as varied as possible. However, most of them focus on automating public

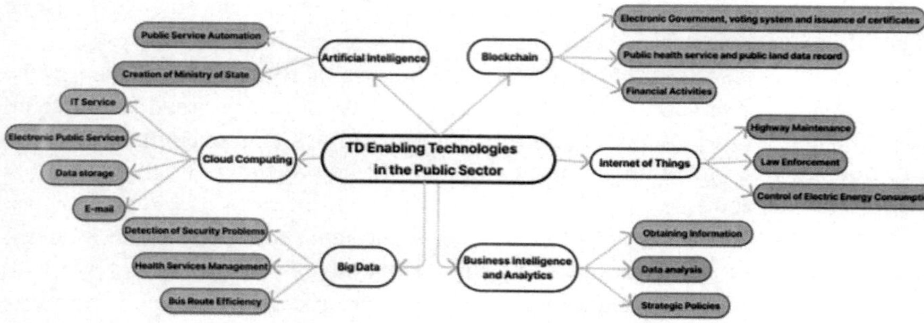

Fig. 1 Conceptual framework for the uses of enabling technologies for digital transformation in the public sector

services by adopting Artificial Intelligence, Cloud Computing, and Blockchain technologies.

Regarding Blockchain technology, we have evidenced benefits associated with improving data visibility, reducing the need for verification, allowing automation and data tracking. In addition, we emphasize its ability to ensure security in sharing information and producing operational improvements with increased volume of processed data and easy error detection in an end-to-end execution.

Big Data Business Intelligence and Analytics technologies are focused on data management in health services and strategic policies. Complementarily, Big Data technology can provide insights for policy decisions and better understand citizens' interests and opinions [61]. It can also positively impact internal transparency and the dissemination of information in the public sector.

In addition, Wahdain et al. [65] present the challenges of implementing Big Data in the public sector, highlighting that the solutions generally have a high cost, lack of understanding, and data processing capacity to a lack of theoretical foundation. In addition, Fraefel et al. [61] address the complexity of data integration, sharing, and security and the need to manage risks related to privacy and security. Malomo and Sena [62], Merhi and Bregu [63] point out in their studies the impossibility of applying Big Data in the public sector since they generate structured and static data, not meeting the volume requirements, variability, and variety. In addition to stating that the main barriers to implementation are related to ethical problems, difficulties in organizational change, lack of investment in information technologies, and the lack of qualified professionals.

The comments in the following paragraphs address the difficulties of implementing these technologies, complementing what was described in the studies.

The Internet of Things currently presents a low level of development in public governance applications, being restricted to collaborations in maintaining and controlling essential public services and applying laws.

As we highlighted earlier, the management process needs to understand better the issues addressed. Thus, the possibility of analyzing and managing large volumes of data extracted from the most diverse sources presents itself as a prominent opportunity to adopt these technologies to develop database management systems with predictive and prescriptive capabilities.

In general, the main difficulties in adopting enabling technologies for DT in the public sector are the change in culture, public investment, and lack of skills in collaborative teams.

The culture change directly influences the acceptance and effective use of proposed digital solutions. We can highlight the need to identify the impacts on human interaction with technology, modifying traditional relationships, and implying a strategic reengineering of functions and work teams. In addition, we note the lack of studies that investigate the digital experimentation of citizens in the use of digital services, identifying possible improvements in equity and access for all parts of the population.

In addition, another critical point is risk management as a strategy for evaluating and controlling cyberattacks. Therefore, efficient management demands changes in everyone's mentality and functionality in devices and personal activities.

We also highlight the need to identify gaps in analyzing and managing data collected through digital services. In this way, it is necessary to evolve in the collection and treatment of data and establish base structures that comprise dimensions such as accuracy, timeliness, consistency, and completeness.

Some limiting aspects, such as the obsolescence of processing devices and servers, are barriers to incorporating digital technologies that better understand citizens' needs.

The selected articles generally fail to investigate the relationship of industrialization or digital maturity of industries with the development of DT initiatives in the public sector.

In addition, the authors understand that the adoption of sophisticated digital technologies in the public sector must have a commitment to the person and society in the search for administrative, social, legal, and financial alternatives that improve the quality of life of the citizen. In this sense, addressing a sustainable agenda strategically aligned with the adoption of sophisticated digital technologies becomes necessary in the construction of governments and smart cities based on the democratization of public services, citizen participation in the formulation of public policies, and the perspective of practical environmental at a regional and global level.

4 Conclusion

DT exposes a new moment for public administration, services provision, and the relationship between interested parties. With the constant emergence of new technologies with disruptive potential and intelligent applications, tools capable of providing different solutions are presented, making the functional structure and processes more efficient.

We then observe that the enabling technologies of DT can be used in both the private and public sectors. In this sense, this article aimed to understand DT. It investigated how digital technologies are applied and found in the most diverse areas and at different technological levels.

As a result, we found technologies that enable digital transformation applied to the public sector, with some highlights, such as the use of Blockchain, which had the highest number of appearances. In addition to the applications, this study focused on understanding these technologies' contributions to the sector in question, with aid in decision-making being the most relevant.

Few factors were identified as limiting aspects of applying such technologies in the public sector. The main points addressed are the high cost of deployment, lack of understanding and processing of data, privacy and ethics, cultural disruption, legal frameworks, and data security.

Therefore, for the continuity of this study, we suggest investigating the DT process from databases in a regional way, aiming to understand the context of each region, identify, and classify how different desires and needs between regional contexts can be solved.

References

1. Pereira, G.V., et al.: Smart governance in the context of smart cities: a literature review. Inf. Polity. **23**(2), 143–162 (2018)
2. Schwer, K., et al.: Digital maturity variables and their impact on the enterprise architecture layers. Probl. Perspect. Manag. **16**(4), 141–154 (2018)
3. Lappi, T.M., Aaltonen, K., Kujala, J.: Project governance and portfolio management in government digitalization. Transform. Gov.: People Process Policy. **13**, 159–196 (2019)
4. Henriette, Emily, Feki, Mondher, Boughzala, Imed: Digital Transformation Challenges. MCIS 2016 Proceedings. **33**. Paphos – Chipre (2016). http://aisel.aisnet.org/mcis2016/33
5. Vial, G.: Understanding digital transformation: a review and a research agenda. J. Strateg. Inf. Syst. **28**(2), 118–144 (2019)
6. Morakanyane, R., Grace, A., O'Reilly, P.: Conceptualizing digital transformation in business organizations: a systematic review of literature. 30th bled econference: digital transformation – from connecting things to transforming our lives. Bled, Slovenia. (2017). https://doi.org/10.18690/978-961-286-043-1.30. ISBN 978-961-286-043-1
7. Dobrolyubova, E., Klochkova, E., Alexandrov, O.: Digitalization and effective government: what is the cause and what is the effect? In: International Conference on Digital Transformation and Global Society, pp. 55–67. Springer, Cham (2019)
8. Schallmo, D., Williams, C.A., Boardman, L.: Digital transformation of business models—best practice, enablers, and roadmap. Int. J. Innov. Manag. **21**(08), 1740014 (2017)
9. Fitzgerald, M., et al.: Embracing digital technology: a new strategic imperative. MIT Sloan Manag. Rev. **55**(2), 1 (2014)
10. Singh, A., Hess, T.: How Chief Digital Officers promote the digital transformation of their companies. MIS Q. Executive. **16**(1), 1–17 (2017)
11. Teichert, R., et al.: Digital transformation maturity: a systematic review of literature. Acta Univ. Agric. et Silvic. Mendelianae Brun. **67**(6), 1673–1687 (2019)
12. Matt, C., Hess, T., Benlian, A.: Digital transformation strategies. Bus. Inf. Syst. Eng. **57**(5), 339–343 (2015)

13. Wischnevsky, J.D., Damanpour, F.: Organizational transformation and performance: an examination of three perspectives. J. Manag. Issues. **18**, 104–128 (2006)
14. Reis, J., Santo, P.E., Melão, N.: Artificial intelligence in government services: a systematic literature review. In: World Conference on Information Systems and Technologies, pp. 241–252. Springer, Cham (2019)
15. Santos, R.C., Martinho, J.L.: An Industry 4.0 maturity model proposal. J. Manuf. Technol. Manag. **31**(5), 1023–1043 (2019)
16. Fishenden, J., Thompson, M.: Digital government, open architecture, and innovation: why public sector IT will never be the same again. J. Public Adm. Res. Theory. **23**(4), 977–1004 (2013)
17. Cordella, A., Tempini, N.: E-government and organizational change: reappraising the role of ICT and bureaucracy in public service delivery. Gov. Inf. Q. **32**(3), 279–286 (2015)
18. Magnusson, J., et al.: Closeness and distance: configurational practices for digital ambidexterity in the public sector. Transform. Gov.: People Process Policy. **15**(4), 420–441 (2021)
19. Mergel, I., Gong, Y., Bertot, J.: Agile government: systematic literature review and future research. Gov. Inf. Q. **35**(2), 291–298 (2018)
20. Bannister, F., Connolly, R.: ICT, public values and transformative government: a framework and programme for research. Gov. Inf. Q. **31**(1), 119–128 (2014)
21. Asgarkhani, M.: Digital government and its effectiveness in public management reform: a local government perspective. Public Manag. Rev. **7**(3), 465–487 (2005)
22. Wang, Y., et al.: Making sense of blockchain technology: how will it transform supply chains? Int. J. Prod. Econ. **211**, 221–236 (2019)
23. Makhdoom, I., et al.: Blockchain's adoption in IoT: the challenges, and a way forward. J. Netw. Comput. Appl. **125**, 251–279 (2019)
24. Tecuci, G.: Artificial intelligence. Wiley Interdiscip. Rev. Comput. Stat. **4**(2), 168–180 (2012). https://doi.org/10.1002/wics.200
25. Russell, S., Norvig, P., Intelligence, S.: Artificial. A Modern Approach. Prentice Hall, Upper Saddle River (2002)
26. Mittal, N., Kuder, D.S.H.: AI-fueled organizations: reaching AI's full potential in the enterprise. Deloitte Insights. (2019)
27. Gil-Garcia, J.R., Pardo, T.A., Gasco-Hernandez, M.: Internet of Things and the public sector. In: Beyond Smart and Connected Governments, pp. 3–24. Springer, Cham (2020)
28. Velsberg, O., Westergren, U.H., Jonsson, K.: Exploring smartness in public sector innovation-creating smart public services with the Internet of Things. Eur. J. Inf. Syst. **29**(4), 350–368 (2020)
29. ITU Internet Reports, International Telecommunication Union: The Internet of Things, 7th edn (2005). www.itu.int/internetofthings/on
30. Xiaocong, Q., Jidong, Z.: Study on the structure of "Internet of Things (IOT)" business operation support platform. In: 2010 IEEE 12th International Conference on Communication Technology, pp. 1068–1071. IEEE, Nanjing, China (2010). https://doi.org/10.1109/ICCT.2010.5688537
31. Han, K., et al.: Initially researches for the development of SSME under the background of IOT. Phys. Procedia. **24**, 1507–1513 (2012)
32. Khalid, A.: Internet of Thing architecture and research agenda. Int. J. Comput. Sci. Mob. Comput. **5**(3), 351–356 (2016)
33. Middha, K., Verma, A.: Internet of Things (IOT) architecture, challenges, applications: a review. Int. J. Adv. Res. Comput. Sci. **9**(1), 389–393, p. 5, (2018)
34. *Abai, N.H.Z., et al.: Integrating business intelligence and analytics in managing public sector performance: an empirical study. Int. J. Adv. Sci. Eng. Inf. Technol. **9**(1), 172–180 (2019)
35. Obeidat, M., North, M., Richardson, R., Rattanak, V., North, S.: Business intelligence technology, applications, and trends. International Management Review Journal. **11**(2), 47–56 (2015)
36. Lim, E., Chen, H., Chen, G.: Business intelligence and analytics: research directions. ACM Trans. Manag. Inf. Syst. (TMIS). **3**(4), 1–10 (2013)

37. Bronzo, M., et al.: Improving performance aligning business analytics with process orientation. Int. J. Inf. Manag. **33**(2), 300–307 (2013)
38. Sheng, J., Amankwah-Amoah, J., Wang, X.: A multidisciplinary perspective of big data in management research. Int. J. Prod. Econ. **191**, 97–112 (2017)
39. Russom, P., et al.: Big data analytics. TDWI best practices report, fourth quarter. **19**(4), 1–34 (2011)
40. Lomotey, R.K., Deters, R.: Towards knowledge discovery in big data. In: 2014 IEEE 8th International Symposium on Service Oriented System Engineering, pp. 181–191. IEEE, Oxford, UK (2014)
41. Mell, P., et al.: The NIST Definition of Cloud Computing. U.S. Dept. of Commerce, National Institute of Standards and Technology, Gaithersburg (2011)
42. Weng, F., Hung, M.: Competition and challenge on adopting cloud ERP. Int. J. Innov. Manag. Technol. **5**(4), 309 (2014)
43. Hayes, B.: Cloud computing. Commun. ACM. **51**, 9–11 (2008)
44. Gong, Y., Yang, J., Shi, X.: Towards a comprehensive understanding of digital transformation in government: analysis of flexibility and enterprise architecture. Gov. Inf. Q. **37**(3), 101487 (2020)
45. Tassabehji, R., Hackney, R., Popovič, A.: Emergent digital era governance: enacting the role of the 'institutional entrepreneur'in transformational change. Gov. Inf. Q. **33**(2), 223–236 (2016)
46. Janowski, T.: Digital government evolution: from transformation to contextualization. Gov. Inf. Q. **32**(3), 221–236 (2015)
47. Ensslin, L., et al.: ProKnow-C, knowledge development process-constructivist. Processo técnico com patente de registro pendente junto ao INPI. Brasil. **10**(4) (2010)
48. Cepik, M.A.C., Canabarro, D.R., Possamai, A.J.: Do novo gerencialismo público à era da governança digital. In: Governança de TI: transformando a administração pública no Brasil, pp. 11–35. WS, Porto Alegre (2010)
49. Miguel, P.A.C., et al.: Estudo de caso na engenharia de produção: estruturação e recomendações para sua condução. Revista Produção, SCIELO Brasil. **17**(1), 216–229 (2007)
50. Janovec, T.: Procedural Justice in Organizations: A Literature Map. Unpublished manuscript. University of Nebraska-Lincoln (2001)
51. *Benítez-Martínez, F.L., Hurtado-Torres, M.V., Romero-Frías, E.: A neural blockchain for a tokenizable e-Participation model. Neurocomputing **423**, 703–712 (2020)
52. *Hyvärinen, H., Risius, M., Friis, G.: A blockchain-based approach towards overcoming financial fraud in public sector services. Bus. Inf. Syst. Eng. **59**(6), 441–456 (2017)
53. *Khan, S.N., Shael, M., Majdalawieh, M.: Blockchain technology as a support infrastructure in E-Government Evolution at Dubai Economic Department. In: Proceedings of the 2019 International Electronics Communication Conference, pp. 124–130. Association for Computing Machinery, New York (2019)
54. *Ølnes, S. Beyond bitcoin enabling smart government using blockchain technology. In: International Conference on Electronic Government, pp. 253–264. Springer, Cham (2016)
55. *Warkentin, M., Orgeron, C.: Using the security triad to assess blockchain technology in public sector applications. Int. J. Inf. Manag. **52**, 102090 (2020)
56. *Sokolov, B., Kolosov, A.: Blockchain technology as a platform for integrating corporate systems. Autom. Control Comput. Sci. **55**(3), 234–242 (2021). Creswell, J.W.: A concise introduction to mixed methods research. SAGE Publications (2014)
57. *De Sousa, W.G., et al.: How and where is artificial intelligence in the public sector going? A literature review and research agenda. Gov. Inf. Q. **36**(4), 101392 (2019)
58. *Pool, R., et al.: The Internet of Things in a smart society: how government policy can help seize opportunities and mitigate threats. In: Beyond Smart and Connected Governments, pp. 25–48. Springer, Cham (2020)
59. *Hartley, M., Seymour, L.: Key success factors for implementing Business Intelligence in South African public sector organisations. In: 9th IDIA Conference, IDIA2015, pp. 249–265 (2015)

60. *Choi, Y., Lee, H., Irani, Z.: Big data-driven fuzzy cognitive map for prioritising IT service procurement in the public sector. Ann. Oper. Res. **270**(1–2), 75–104 (2018)
61. *Fraefel, M., Haller, S., Gschwend, A.: Big data in the public sector. Linking cities to sensors. In: International Conference on Electronic Government, pp. 276–286. Springer, Cham (2017)
62. *Malomo, F., Sena, V.: Data intelligence for local government? Assessing the benefits and barriers to use of big data in the public sector. Policy Internet **9**(1), 7–27 (2017)
63. *Merhi, M. I.; Bregu, K. Effective and efficient usage of big data analytics in public sector. Transform. Gov.: People Process Policy **14**(4), 605–622 (2020)
64. *Shin, D.: Demystifying big data: anatomy of big data developmental process. Telecomm. Policy **40**(9), 837–854 (2016)
65. *Wahdain, E.A., Baharudin, A.S., Ahmad, M.N.: Big data analytics in the Malaysian public sector: the determinants of value creation. In: International Conference of Reliable Information and Communication Technology, pp. 139–150. Springer, Cham, (2018)
66. *Alassafi, M.O., et al.: A framework for critical security factors that influence the decision of cloud adoption by Saudi government agencies. Telemat. Inform. **34**(7), 996–1010 (2017)

COVID-19 Impact on Ethanol Sales in Fuel Stations: An ITS Econometric Analysis

Henrique Duarte Carvalho and Thulio Eugenio Portes de Almeida Ricomini

Abstract This paper provides estimates of the impact of the COVID-19 outbreak on Brazilian Ethanol sales. To this end, weekly data on Ethanol sales volumes are analyzed through an ITS SARIMA model and a counterfactual analysis covering the 2019–2020. We find that the real effect of COVID-19 was a reduction above 77.97% in Brazil after the first COVID-19 death, in March 2020, and still a decrease of about 50.15% at the end of 2020. The empirical evidence confirms that the impact of the pandemic crisis, the counterfactual analysis allows estimating the real effect of COVID-19 is on average 3.76% greater than the observed against an index date reference. These results suggest that ethanol sales in Brazil were more affected than only when comparing previous results to the effects of the pandemic.

Keywords Ethanol sales · ITS Sarima model · COVID-19 pandemic crisis

1 Introduction

Brazil has been the second-largest ethanol producer globally, accounting for 28% of the world's production from 2016 to 2021 [1], of which 97% was consumed domestically [2]. As fuel ethanol can be used in a mix with gasoline (blend of anhydrous Ethanol) or directly as hydrous Ethanol; although the USA is the largest producer and consumer, there is no expressive market for hydrous Ethanol, a greenhouse product, in contrast to Brazil which light fuel market (gasoline and Ethanol) had an impressive hydrous ethanol market share of 37% in 2019 [2].

Since 2020 the world has suffered an unprecedented impact on its economy and supply chain of most products; the COVID-19 pandemic imposed a challenging crisis on a global scale as many markets, especially transport, faced new challenges in managing the available resources [1, 3]. The European air transport had a clear

H. D. Carvalho (✉) · T. E. P. de Almeida Ricomini
UNIFEI, Itabira, MG, Brazil
e-mail: henrique.carvalho@unifei.edu.br; thulioricomini@unifei.edu.br

V. M. López Sánchez et al. (eds.), *Industrial Engineering and Operations Management*, Springer Proceedings in Mathematics & Statistics 400,
https://doi.org/10.1007/978-3-031-14763-0_10

impact as most international traffic was suddenly interrupted, facing dilemmas in predicting the recovery and future scenarios [4], as well Tokyo's fuel market suffered the closing of many gas stations [5], the German transportation had variations from public transport to private cars due to the fear of the disease transmission [6], and USA suffered a decrease on human mobility from March 2020 onward [7].

Brazilian fuel market was not an exception, observing a decrease of 33% in the sales of light fuels from February 2020 to April 2020 [2], even though Brazil's first registered death occurred 2 months apart from the world's first death [8, 9] there was no time for the Brazilian economy to learn from international cases and prepare itself for the pandemic effect. The expected increase of 16% in light fuel demand from 2018 to 2020 faced a decrease of 9% in actual sales in Brazil, the hydrous Ethanol that had an average annual increase of 29.13% from 2017 to 2019 had a drop of 14.58% from 2019 to 2020 [2, 10].

With the present data from different gas stations across Brazil, there is enough data to evaluate and analyze how the pandemic affected the Brazilian hydrous ethanol market. No reseller or producer has the previous experience to prepare strategies to react to a disruptive macroeconomic event; therefore, it is necessary to supply information to the entire market on how a pandemic-like event can impact the demand.

In partnership with a Brazilian fuel station SaaS provider, this study collected historical data of 1123 across the entire country from 1st of January 2019 to 31st of December 2020 of hydrous ethanol sales, contemplating the previous behavior and the direct impact of the quarantine on demand. Using the interrupted time series (ITS) ARIMA model, this study aims to explore the key events of the pandemic that affected the ethanol market and the quantitative impact of the pandemic on demand, evaluating the postponed behavior of sales considering the key events.

As applied by [4] on air transport scenario, this study considers the COVID-19 pandemic as an exogenous shock using a single ITS model; this application may be used hence the shock affects the population and not only a group avoiding the need for a control group [11]. In the analysis applying an ITS, the contra factual trend is estimated using the time series [11, 12]. The effect of COVID-19 is assessed by comparing the observed data to the trend observed in the absence of the shock.

2 Empirical Methods

To estimate the effect of COVID-19 pandemic on Ethanol sales in Brazil we used data of sales across 1123 stores from 1st of January 2019 to 31st of December 2020. The data was treated to a time series with the average sales per week per store in a time series format. We applied an ITS structure as presented by Andreana et al. [4] with a few adjustments to our own time series structure where the SARIMA model is then defined by Eqs. 1 and 2 as follows:

$$y_t = b_0 + b_1 t + b_2 D + b_3 P + \beta x_t + e_t \quad (1)$$

$$\begin{aligned} e_t = {} & \varphi_1 e_{t-1} + \cdots + \varphi_p e_{t-p} + \varepsilon_t + \cdots + \theta_1 \varepsilon_{t-1} + \cdots + \theta_q \varepsilon_{t-q} \\ & + \Phi_1 \varepsilon_{t-s} + \Phi_1 e_{t-s} + \Phi_P e_{t-s-P} + \Theta_1 \varepsilon_{t-s-P} + \cdots + \Theta_Q \varepsilon_{t-s-Q} \end{aligned} \quad (2)$$

In Eq. 1 the variable t represents the time raging from the first to the last week in the time series, D is a dummy variable marking the first death as the intervention point, therefore marked as 0 pre-first death and as 1 post-first death. The last variable P marks the time elapsed after the intervention. The regressor x_t is given by the number of COVID-19 cases recorded in Brazil according to the John Hopkins University [13] database.

We estimate a log-linear model of Eq. (1), i.e., the dependent variable is the logarithmic transformation of the response variable. To identify the percentage variation in the volume of Ethanol.

To address some identification problems and have unbiased estimates of the coeficients some actions were made as follows, first the auto.arima() function in R was applied, which applies the Hyndman-Khandakar algorithm to set the best fitting coeficients to the ARIMA model [14]; secondly the SARIMA model's residuals were validated as white-noise or not. The post-estimation diagnostic analysis checked the residuals by plotting the autocorrelation function and by implementing a Ljung-Box test. Finally, the counterfactual time series used to compare the intervention effect was predicted with a SARIMA model using the Hyndman-Khandakar algorithm with the data limited to a month before the first death of COVID-19 in Brazil, predicting a trend as with the pandemic there happened. The counterfactual series predicted the data up to 31st of December 2020.

3 Data

The data used in this work come from the average volume of Ethanol sold per gas station between 2019 and 2020. Information was collected from 1123 gas stations throughout the Brazilian territory. In addition to the average volume of Ethanol, the number of new cases of Covid 19 observed in Brazil was used in this research, according to data from the John Hopkins University [13]. The data were grouped by week, and the descriptive statistics are presented in Table 1.

For a country with the dimensions of the territory as Brazil, it is reasonable to think that the country's regions were affected in different ways and intensities. The same can be said of decisions about starting and ending lockdowns across the country.

Thus, establishing a single date, a reference for the beginning or end of a lockdown period, is not a sensible strategy for analyzing data or assessing the impact of an event on the volume of Ethanol sold.

Table 1 Descriptive statistics

Statistic	Weekly average ethanol transactions (liters by gas station)	Covid – new cases
Min	1174	0
First quarter	2084	0
Median	2299	0
Mean	2228	106,548
Third quarter	2457	248,541
Max	3049	476,227

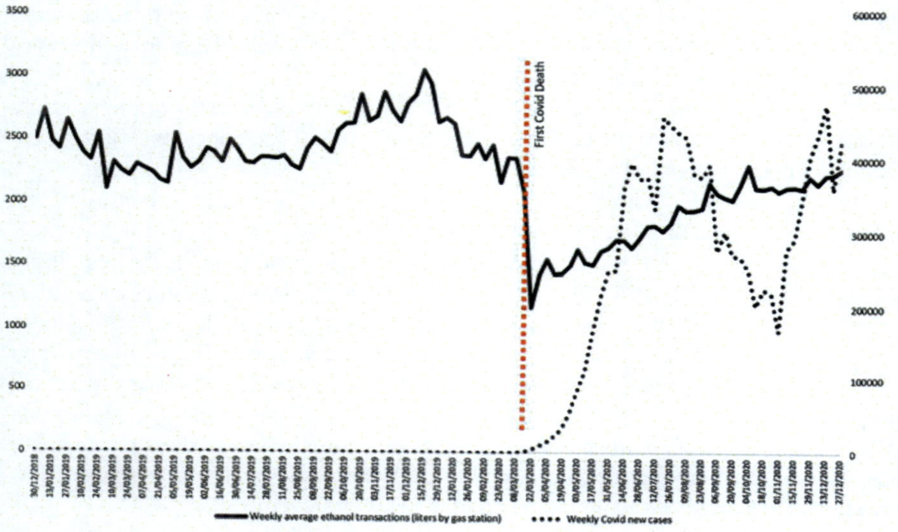

Fig. 1 Ethanol sales and COVID-19 time series

Thus, to promote an analysis of the impact of COVID-19 on the Brazilian ethanol retail trade, it was decided to use a reference date of the initial impact as the date of the first death by covid in Brazil, which occurred on the day 17 March 2020.

Figure 1 shows that the choice of this reference may well represent a scenario of structural change in the series; the vertical dashed line highlights the week in which the first death by covid occurred in Brazil. Until that date, throughout practically the entire year of 2019, there has been a slightly volatile behavior for the average volume of Ethanol sold, a slight increase at the end of 2019, and a trend towards a gentler decline at the beginning of 2020, the result of a seasonal consumption pattern, where sales usually are lower.

To understand the magnitude of this impact, albeit occasionally, when comparing the average volume of Ethanol sold in the same week of the previous year (an average of 2245 liters of Ethanol per gas station), in 2019, with the week after the 17th March 2020 (average of 1173 liters per gas station), there was a drop of 47.75%.

However, after the reference period, a positive trend characteristic in the average ethanol volume can be highlighted, clearly at a lower level than in 2019 and early 2020, but with a clear recovery, even in a scenario of clear expansion of people infected with the Covid-19 virus.

This characteristic structural change in the series, as well as the observation of the market recovery trend, requires a more careful analysis of the impact of covid on the commercialization of an asset as important to the Brazilian economy as ethanol, in addition to fuel resale services.

4 Results

To estimate the impact of the pandemic generated by COVID-19 on the average volume of Ethanol sold in Brazil, the ITS model, described in Eqs. 1 and 2, was used. A contrafactual series was generated, and the difference between the observed trend and the contrafactual scenario was estimated.

The variable of the logarithm of the average volume of Ethanol sold was considered here as a dependent variable of the model. As seen in Fig. 1, the data series does not have an evident seasonal component, with a more apparent trend only after the shock provided by the reference date (17 March 2020) used in this study as a structural change in ethanol sales.

In this work, we present two sets of results that aim to assess the scenario of the impact of covid 19 on ethanol sales made by gas stations and the subsequent partial recovery. Thus, the results obtained by analyzing a subsample that ends in the last week of March 2019 were described and thus capturing the immediate effect of the impact on the resale of Ethanol.

The second data set considers the complete series, in which the results of the ITS model were applied using the series until the end of 2020. This analysis can provide relevant information on the reaction of the ethanol supply through the change in the consumer's disposition according to the Covid 19. The results of the ITS model applied here are presented in Table 2.

The Hyndman-Khandakar algorithm [14] identified the coefficients of the ARIMA model in the two analyzes performed. The short-term results associated with the impact of covid 19 are presented in the second column of Table 2.

This algorithm iteratively searches over a series of potential SARIMA models for the one with the lowest AIC or BIC, with several constraints applied to avoid convergence problems. These include setting the maximum value of p and q to 5 and P and Q to 2, although the researcher can modify these settings if necessary. The model with the lowest information criteria selected by the algorithm was (2,0,0) × (0,0,1)52. In other words, the autocorrelation order of the model (p) was 2, the moving average order of the model (q) was 0, the autocorrelation order of the seasonal part of the model (P) was 0, and the moving average order of the seasonal component of the model (Q) was 1.

Table 2 SARIMA ITS model of the covid 19 effect in ethanol sales

Statistic	Observations until 31st March 2020	Observations until 31st December 2020
Dependent variable: *lethanol*		
Constant	7.803*** (0.000)	7.7809*** (0.000)
Time	0.001*** (0.000)	0.003*** (0.000)
FD Dummy	−0.7797*** (0.000)	−0.5015*** (0.000)
Time Since FD	0.0717** (0.003)	0.0115*** (0.000)
lcovid	−0.013 (0.134)	−0.021** (0.005)
ARIMA error model		
L.AR	0.4159*** (0.000)	1.7271*** (0.000)
L2.AR	0.327** (0.007)	−0.7943*** (0.000)
L3.AR		
L.MA		−1.3144*** (0.000)
L2.MA		0.2350 (0.2370)
L3.MA		0.3521** (0.0024)
Seasonal Effects		
L.AR		
L.MA	0.260* (0.096)	
N	66	105
AIC	−858,17	−320.9

p value in parentheses. $p < 0.1$, $^{*}p < 0.05$, $^{**}p < 0.01$, $^{***}p < 0.001$

Specifically, about the coefficients of the ITS model, the effect of the estimated "Time" variable shows that the estimated trend before the first death by covid in Brazil is positive and significant, equivalent to a seminal growth of 0.1%.

The coefficient associated with the variable "FD Dummy" indicates the immediate impact after the first death on the average volume of Ethanol sold in Brazil. A direct negative and significant impact of −77.97% represents a drastic change in the observed series.

The variable "Time Since FD" indicates that the trend changed immediately after the impact of the first kill. Promoting a relevant positive and significant effect, meaning that, for each of the 3 weeks that passed after the shock, an average growth of 7.17% per week was observed.

The estimated coefficient for the variable representing the cases of COVID-19 (lcovid) is not significant.

The results of the ITS model applied to the entire sample are presented in the third column of Table 2. The "Time" coefficient indicates the performance after the reference of the first death date; although significant, the contribution is not very expressive, only 0.01% per week.

The model with the lowest information criteria selected by the algorithm was an ARIMA(2,0,3). In other words, the autocorrelation order of the model (p) is equal to 2, and the moving average order of the model (q) is equal to 3.

The magnitude of the positive trend after the period measured by the variable "Time since FD" is not so expressive, with an impact of 1.15% per week; however, more relevant than the previous variable. These combined effects show that the recovery that started after the reference date was not strong enough to change the trend or, otherwise, overcome the effects promoted by the structural change of COVID-19 on the commercialization of Ethanol.

The "FD dummy" variable shows the model's sensitivity after the impact of COVID-19 for the entire series. Here, the immediate effect is negative and significant, indicating that covid promoted an impact of −50.15%. Although we do not observe an inversion in the sign of this variable here, when analyzing the entire series, the magnitude of the impact is significantly smaller when compared to the first model.

The previous variable, analyzed together with the result for the "Time since FD" coefficient, indicates that the recovery that started at the end of March 2020 was not strong enough to recover the previous levels of average volume of Ethanol traded and, thus, that the effects promoted by COVID-19 have not yet been fully absorbed.

The estimated coefficient for the variable representing the cases of COVID-19 (lcovid) presents itself as significantly and negatively correlated with the sales of Ethanol; here, we can observe an associated negative relation of 2.1% of new cases of COVID-19 with Ethanol sales.

To validate the results presented in Table 2 and verify if there could be any problems in the analysis, diagnostic tests were performed. The diagnosis consists of plotting the residual autocorrelation functions and implementing the Ljung-Box test.

Table 3 shows that the two analyzes performed do not reject the null hypothesis that the errors are independently distributed.

The ITS model makes it possible to derive a counterfactual trend and compare it with the observed data to obtain a better estimate of the impact of COVID-19 on the resale of Ethanol in Brazil. Figure 2 Presents this comparison. The black dots represent the observed time series, while the white triangles present the counterfactual time series.

In Fig. 2, a seasonal influence can be observed on 2019's demand and the data predicted for the counterfactual scenario. This format suggests a peak in demand for

Table 3 SARIMA and ARIMA ITS model diagnostic tests

	Observations until 31st March 2020	Observations until 31st December 2020
Q*	3.3452	3.8454
p-value	0.2345	0.2786
Ho	N.R.	N.R.

Q_* Ljung-Box statistics, N.R. H0 not rejected, R. H0 rejected

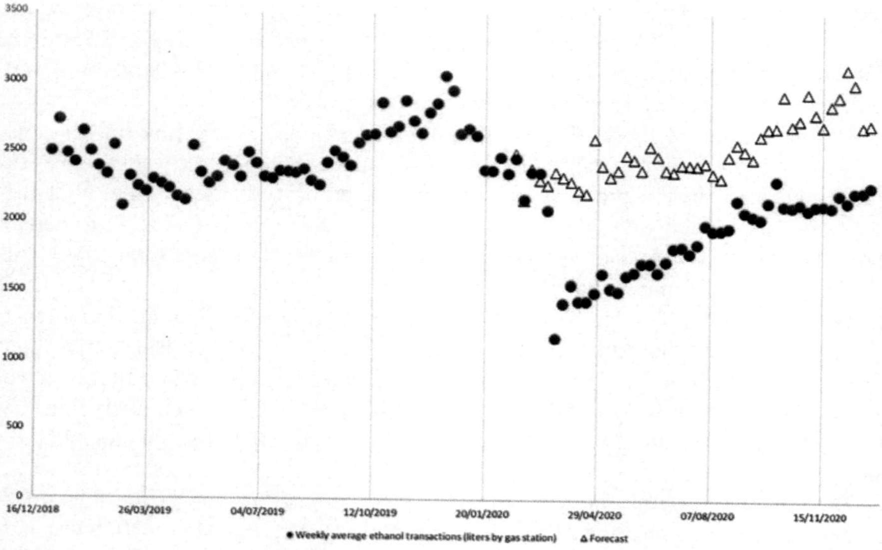

Fig. 2 Counterfactual analysis: no COVID-19 cases and no subsequent effect

Ethanol at the end of the year, followed by a drop in the initial months and a more evident recovery in the second semester of the year.

The actual effect of COVID-19 is represented by the difference between the observed and the counterfactual series, which can be broken down into two parts; the first being the difference between observed data and past data, and the second the difference between observed past data and predicted data in a scenario of the absence of COVID-19 effects on sales of Ethanol.

To highlight the effect of these components, the same strategy observed in Andreana et al. [4] was used; that is, the week of 16 February 2020, a period 1 month before the occurrence, was identified as a base equal to 100 of the first confirmed case of COVID-19 in Brazil.

After that, the difference between the observed levels and the database was initially estimated, and then the difference between the database for forecasted sales for the dates of interest. With this information in hand, the actual effect was then obtained as the sum of the two differences on each specific date of interest. All indices and results are presented in Table 4.

Table 4 The observed and real impact of lockdown on Ethanol sales

	Results			
Index counterfactual time series Base = 100 16th February 2020	1st March	95.59	5th April	91.91
	3rd May	96.59	7th June	95.28
	5th July	94.90	2nd August	97.43
	6th September	101.05	4th October	107.88
	1st November	117.67	6th December	124.87
Difference between counterfactual and 100	1st March	−4.41	5th April	−8.09
	3rd May	−3.41	7th June	−4.73
	5th July	−5.10	2nd August	−2.57
	6th September	1.05	4th October	7.88
	1st November	17.67	6th December	24.87
Index observed time series Base = 100 16th February 2020	1st March	95.92	5th April	63.21
	3rd May	66.44	7th June	69.49
	5th July	73.90	2nd August	80.66
	6th September	84.46	4th October	93.71
	1st November	85.24	6th December	87.48
Difference between observed and 100	1st March	−4.08	5th April	−36.79
	3rd May	−33.56	7th June	−30.51
	5th July	−26.10	2nd August	−19.34
	6th September	−15.54	4th October	−6.29
	1st November	−14.76	6th December	−12.52
Real COVID-19 effect	1st March	0.33	5th April	−28.7
	3th May	−30.15	7th June	−25.78
	5th Jul	−21.00	2nd August	−16.77
	6th September	−16.59	4th October	−14.17
	1st November	−32.43	6th December	−37.39

The real impact of the COVID-19, the difference between counterfactual analysis and the observed data, in Apr 2020 is −28.7 basis points, around further 8 basis points greater than those observed (i.e., −36.79 when compared with the 100 index); this improvement is undoubtedly due to the downward trend in forecasted data over the first half of the year. In December, the real effect is −37.39, while the observed series has only a −12.52 reduction. The loss has been about 25 additional points compared to that observed historical data.

It is also possible to highlight the real impacts of −30.15 and −32.43 for May and November.

The counterfactual analysis shows that Ethanol Sales performed a higher loss in the volumes of activity than that computed using observed data, on average equal to 3.76% from April to December. This evidence confirms that the impact of COVID-19 has been stronger than compared with the last observed value before the COVID-19 impact and amplifies the warnings regarding the Ethanol sales.

5 Conclusion

The aim of this work is to quantify the disruptive impact that the pandemic crisis caused by COVID-19 has had on Ethanol sales in Brazil. To do this, a time series ITS SARIMA econometric model is estimated, and a counterfactual analysis is performed. This approach allows measuring the impact of the pandemic by comparing the observed trend of business volumes to a counterfactual time series that represents the development trend in the absence of the crisis.

The empirical evidence confirms that the impact on Ethanol sales during the pandemic crisis has been relevant. In Brazil, the estimated real effect is a reduction in Ethanol sales greater than 77% after the first death date.

The counterfactual analysis allows estimating the real effect of COVID-19 is on average 3.76% greater than the observed against an index date reference.

These results suggest that ethanol sales in Brazil were more affected than only when comparing previous results to the effects of the pandemic. Such a strategic product, with a significant present and future market potential, experienced great market pressure in 2020, throughout the pandemic, however, the recovery throughout the year already shows signs of the product's relevance to the market, however, still in deficit conditions compared to the counterfactual scenario.

This scenario of gradual sales recovery suggests that any government intervention on the product might not be necessary, but this study did not assess this scenario, leaving this assessment, as well as the estimate of the change in demand for Ethanol, or the analysis for other fuels, as suggestions for future research.

References

1. Renewable Fuels Association: Annual Ethanol Production. Retrieved April 19, 2022, from https://ethanolrfa.org/markets-and-statistics/annual-ethanol-production (2022)
2. Agência Nacional do Petróleo; Gás Natural e Biocombustíveis: Painéis Dinâmicos da ANP. Retrieved April 19, 2022, from https://app.powerbi.com/view?r=eyJrIjoiMDVmMjkwMmMtZjE5OC00M2QwLThiZDEtYTA2Y2E1ZmM1ZmQzIiwidCI6IjQ0OTlmNGZmLTI0YTYtNGI0Mi1iN2VmLTEyNGFmY2FkYzkxMyJ9 (2022)
3. Chang, C.L., McAleer, M., Wang, Y.A.: Herding behaviour in energy stock markets during the Global Financial Crisis, SARS, and ongoing COVID-19. Renew. Sust. Energ. Rev. **134**, 110349 (2020)
4. Andreana, G., Gualini, A., Martini, G., Porta, F., Scotti, D.: The disruptive impact of COVID-19 on air transportation: an ITS econometric analysis. Res. Transp. Econ. **90** (2021)
5. Aruga, K.: Changes in human mobility under the COVID-19 pandemic and the Tokyo fuel market. J. Risk Financ. Manag. **14**(4), 163 (2021)
6. Eisenmann, C., Nobis, C., Kolarova, V., Lenz, B., Winkler, C.: Transport mode use during the COVID-19 lockdown period in Germany: the car became more important, public transport lost ground. Transp. Policy. **103**, 60–67 (2021)
7. Archer, C.L., Cervone, G., Golbazi, M., Al Fahel, N., Hultquist, C.: Changes in air quality and human mobility in the USA during the COVID-19 pandemic. Bull Atmos. Sci. Technol. **1**(3), 491–514 (2020)

8. Qin, A., Hernández, J. C.: China reports First Death from new virus. The New York Times. Retrieved April 19, 2022, from https://www.nytimes.com/2020/01/10/world/asia/china-virus-wuhan-death.html (2020)
9. Ministério da Saúde: Coronavírus Brasil. Retrieved April 19, 2022, from https://covid.saude.gov.br/ (2022)
10. de Andrade Junior, M.A.U., Valin, H., Soterroni, A.C., Ramos, F.M., Halog, A.: Exploring future scenarios of ethanol demand in Brazil and their land-use implications. Energy Policy. **134**, 110958 (2019)
11. Baicker, K., Svoronos, T.: Testing the Validity of the Single Interrupted Time Series Design. National Bureau of Economic Research. Technical report (2019)
12. Bernal, J.L., Cummins, S., Gasparrini, A.: Interrupted time series regression for the evaluation of public health interventions: a tutorial. Int. J. Epidemiol. **46**(1), 348–355 (2017)
13. John Hopkins University. John Hopkins University coronavirus research centre (2020)
14. Hyndman, R.J., Khandakar, Y.: Automatic time series forecasting: the forecast package for R. J. Stat. Softw. **27**, 1–22 (2008)

Lean Maturity Models: A Scoping Review

Philipe da Silva Simões, Luiz Felipe Scavarda, Taciana Mareth, and Rodrigo Goyannes Gusmão Caiado

Abstract The Lean Six Sigma approach has been driving several organizations worldwide through quality and performance improvement. Its implementation process is complex and success depends, besides several factors, on the organization's current maturity level. Lean Maturity Models (LMM) can reveal guidelines to conduct organizations during the implementation process, ensuring its continuous improvement. In this context, this paper presents a Scoping Review to synthesize which are the Maturity Models (MM) already developed and understand their contributions to the successful implementation of lean approaches. The results reveal a diversity of MM being developed to assist these implementations. Most of them are hybrid models and are highly generalizable. In general, these artifacts are used to assess and indicate to organizations their lean maturity level and the points to be improved. The synthesis of existing MM in the literature presented in this paper can be a starting point for creating or adapting new models. Moreover, organizational managers can verify the existence of any replicable model to assist them in lean maturity.

Keywords Maturity model · Lean approach · Lean implementation · Six sigma

1 Introduction

The Lean Manufacturing approach, which was originated in the Toyota Production System [1], has been adapted to new production systems and merged with other approaches, giving rise to new aspects. Some of them are already well consolidated in the literature, such as Lean Healthcare, Lean Office, Lean Construction [2],

P. da Silva Simões (✉) · L. F. Scavarda · R. G. G. Caiado
Pontifical Catholic University of Rio de Janeiro (PUC-Rio), Rio de Janeiro, RJ, Brazil

T. Mareth
University of Vale Do Rio Dos Sinos, Sao Leopoldo, RS, Brazil

V. M. López Sánchez et al. (eds.), *Industrial Engineering and Operations Management*, Springer Proceedings in Mathematics & Statistics 400,
https://doi.org/10.1007/978-3-031-14763-0_11

Agilean, and Lean Six Sigma. Although they are divergent, there is a consensus that the main goal of these approaches is to eliminate waste and increase the value perceived by customers through continuous improvement, human resource management, pull production, and customer-oriented production management [3].

Although the use of these approaches has become a popular continuous improvement strategy, there are still many challenges in implementing them in new industries [4]. This introductory step depends on the current scenario of the organization and requires considerable financial investments, because, an unsuccessful implementation can result in a large loss. According to Yadav and Desai [5], it is for this reason that the interest in improving the implementation process has been gaining space among the subjects addressed by academics. The authors reinforce that a well-structured implementation increases the chances of bringing about cultural level changes in organizations, since lean approaches are not a set of methods or tools, but a systemic approach, and organizational thinking or philosophy.

Anthony and Antony [6] indicated that the lack of maturity (or readiness) of organizations when implementing continuous improvement approaches has been a driver of research on Maturity Models used in quality programs. The authors further argued that some organizations are implementing lean practices only in parts of their processes and using very simple techniques. Jørgensen et al. [7] highlighted the Lean Maturity Models (LMM) as frameworks capable of assessing the maturity level of an organization to help its managers during the implementation of the lean approach.

In this context, this paper seeks to answer the following questions: What is the existing LMM? How are they developed to assess the maturity level? From the answers to these questions, it is possible to discuss how these models assist in the implementation process of lean approaches. Thus, the main objective of this paper is to investigate how the existing maturity models contribute to the implementation process of lean approaches.

To this end, a scoping review [8, 9] is conducted to pursue the following goals: define the current state of the art of studies using LMM in the implementation process; synthesize the identified LMM; understand the relevance of the LMM in the implementation processes of lean practices. It is worth mentioning that, unlike a systematic review, the scope review does not assess the quality of the selected articles, it provides a narrative or descriptive report of available literature. According to Arksey and O'Malley [9], this limitation can lead to difficult decisions about the extent to which breadth is more relevant than depth. However, the scoping review provides a robust and transparent method to illustrate the field of interest of a subject, in line with the objectives of this study. This analysis makes it possible to summarize and disseminate the research results in a relatively short period of time compared to a systematic review.

This paper is structured in five sections, the first of which is the introduction. The second presents the concepts of the Maturity Model and the possible impacts on the implementation of lean approaches. The third section describes the research methodology adopted, while the fourth section presents the analysis of the results obtained. Finally, the last section offers the main conclusions and contributions of the article, with practical implications and directions for future research.

2 Theoretical Foundation

The origin of the first Maturity Models (MM) was in Quality Management, and their use was considered a major advance toward performance improvement approaches [10]. MM can be defined as artifacts that have elements arranged on an evolutionary scale with measurable transitions from one level to another and can be used for benchmarking, self-assessment, and continuous improvement [11]. The maturity level of a process or organization is the evaluative measure of the ability to meet pre-established objectives.

The Capability Maturity Model (CMM) is the best-known framework for achieving the maturity of a process [12]. Developed by the Software Engineering Institute, its initial goal was to evaluate software capability. However, from its success, several works ended up adapting the model to other areas, such as construction [13], industrial automation [14] and new product development [15]. Later, in 2002, the same institution refined the model, which they called Capability Maturity Model Integrated (CMMI), a model with greater ability to guide process improvement in areas other than software engineering [16].

Fraser et al. [17] conceptualize in their study the key aspects of maturity models. According to the authors, maturity models are composed by the arrangement of the components: series of quality levels; names of the levels; brief descriptions about each level, dimensions of the theme; series of elements that compose these dimensions; brief descriptions of these elements. According to the authors, every MM must have maturity levels (continuous or discrete) and a set of dimensions (or processes). In some cases, these dimensions can be broken down into elements (or activities). The authors have classified MM into three types: Maturity Grid, Hybrids and CMM – a brief description of each type is presented below.

The Maturity Grid has moderate complexity and consists of textual descriptions of activities for each maturity level, such as Crosby's Quality Management Grid [10]. According to Paula et al. [18], these frameworks provide a qualitative description of each process area, i.e., each grid cell contains descriptions of expected performance at different maturity levels. Thus, best practices in each process knowledge area are described within each maturity level, from the most immature to the most advanced.

The second type of model, on the other hand, combines maturity definitions with the use of questionnaires, for example, Likert-type questionnaires. The questionnaires should assess the organization's performance against the best practices described at each level. Thus, besides the described activities and dimensions, as in the Maturity Grids, the hybrid models evaluate these through questionnaires and classify the organization into a maturity level [17, 18].

Finally, the last type refers to models that have a specific and formal architecture, inspired by the CMM. In these cases, to address the required objectives (the maturity), common resources and organizational practices are organized by process areas. The CMM family models may not contain individual descriptions for each activity. Paula et al. [18] indicate the main difference between this type of model and the Maturity Grids: although both stratify the processes of interest into knowledge areas for maturity level classification, capability models go beyond, defining objectives and practices to be performed at each level, for each knowledge area.

2.1 Lean Maturity Models

According to the systematic review by Lameijer et al. [19], the academic and professional literature offers maturity models that guide the implementation of the Lean Six Sigma (LSS) approach. The study signals the latent need for scientific knowledge in the process of implementing approaches such as LSS, as the overall assessment of the quality and usefulness of existing models is negative. The review assesses how MM are grounded in relation to Organizational Development Theory. Nesensohn et al. [20], meanwhile, in their review on the application of MM to the Lean Construction (LC) approach, reveal there is no agreement on what LC maturity means. However, the authors point out that there is general agreement indicating that the concept of maturity is suitable to reflect the evolutionary path of the approach in organizations.

In contrast to these reviews, this study does not undertake to evaluate MM, but rather classifies its typology, identifies the attributes used in its development, and its relevance to the implementation process. Furthermore, this paper presents a synthesis of the models applied to lean approaches in general, not limiting itself to a single strand. Thus, it is expected to obtain a comprehensive perspective on the development and application of LMM in different contexts.

3 Research Methodology

To explore the extent of literature in the specific domain of MM in lean approaches, a Scoping Review was conducted. so that, even if without describing the findings in detail, this methodology identifies the appropriate parameters of a systematic review [8, 9]. Arksey and O'Malley [9] argue that this type of review helps identify existing research gaps in the literature in a focused manner. Therefore, in this case, the Scoping Review is conducted to investigate existing maturity models for implementing lean approaches. The methodological steps used in this paper follow the ones suggested by Armstrong et al. [8]: identification of the research question;

identification of relevant studies; study selection; data mapping; and collating, reviewing, and reporting the results.

Arksey and O'Malley [9] suggest that the central question of a Scoping Review should be broadly defined. Therefore, the main question of this one is what is known about the existing literature on maturity models developed to aid the implementation of lean approaches? From this general question, the entire scope of the research was defined to answer the following items: Q1. What are the existing models in the literature? Q2. Which are the most frequent types, according to the typology proposed by Fraser et al. [17]? Q3. In which productive sectors have these models been applied? Q4. In which countries were the studies conducted?

According to Armstrong et al. [8], some limitations can be considered in scope reviews, such as the databases used (according to their accessibility), the period of publication and the language of the articles. Scopus and Web of Science were chosen to cover interdisciplinary literature from different types of journals. The choice was made because they are popular bases in operations management [21–23] and because of their complementarity [24]. The search was performed in the titles, abstracts, and keywords fields. The search combined the terms "lean six sigma" or "lean" with the possible MM-related terms: "maturity model", "maturity grid" and "maturity level". Only articles in English published until 2021 were selected for analysis. The entire article identification and selection process is described and summarized in the PRISMA flow chart (see Fig. 1). Therefore, 46 articles were found in the WoS database and 58 articles in the Scopus database. From those 104 documents, 30 were duplicates and one was not a scientific article, leaving 73 articles for investigation.

For the selection of the studies, content analysis was used [25]. In a first analysis, after reading the abstract, 15 articles that did not develop or used MM were disregarded; 14 articles that addressed topics adverse to this research; and 5 articles that, due to lack of access, were not found in full.

From this first selection, 39 articles followed in the search to be read in full. In this step, the following exclusion criteria were used: MM that address continuous improvement, but were not applied to any aspect of the lean approach (9 articles); MM whose maturity measurement was directed to industry 4.0 (5 articles) or to supply chains (2 articles); literature review articles (3 articles); MM not used in the implementation phase (4 articles).

At the end of the selection phase, 16 articles remained. Following the proposed methodology, with the articles of interest selected, a data mapping was performed, enabling a specific data structure development according to the objectives of this study. Thus, to answer the research question, descriptive information was extracted from each selected article, such as where the study was conducted, the application sector, year of publication, and journal. In addition, the type of maturity model was analyzed – following the typology proposed by Fraser et al. [17].

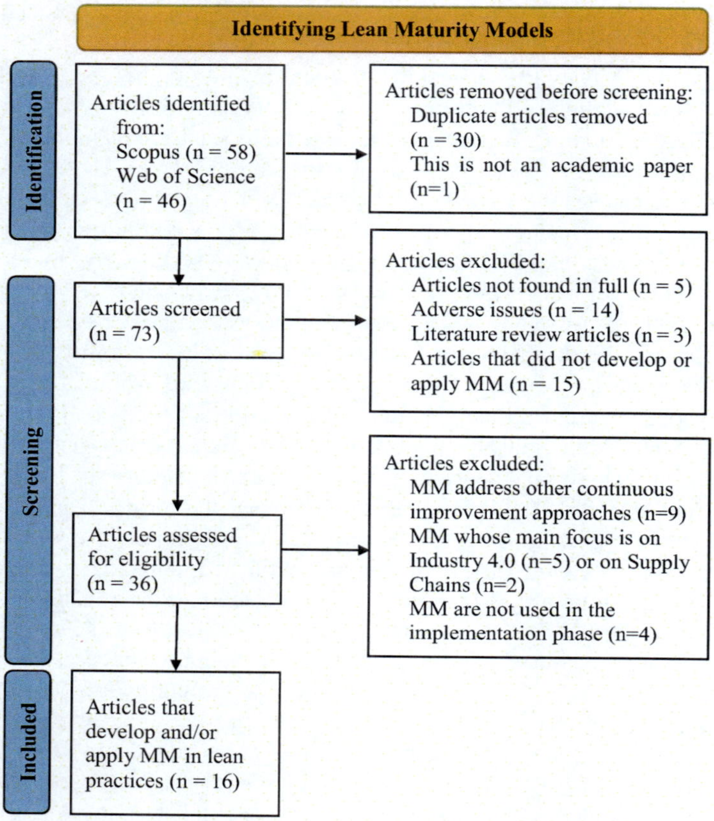

Fig. 1 PRISMA flowchart – Lean Maturity Model selection

4 Results and Discussions

This section aims to summarize, summarize, and explore the behavior of the data obtained by the Scoping Review, whose object of analysis was the LMM. From the 16 selected articles, 12 distinct LMMs were identified.

4.1 Profile of the Selected Articles

Among the selected articles, Bessant et al. [26] was the first to suggest different levels of development of continuous improvement skills [26]. However, this research considers its version adapted to lean manufacturing [27]. Looking specifically at the MM used in implementing lean approaches, the first model was developed and applied by Lemieux et al. [28] and Lemieux et al. [29] – a Maturity Grid adaptable to organizations in the luxury goods industry. After these one-off publications, in 2016, three new studies on the topic emerged. Whereas, between 2018 and 2021, every year had at least 2 publications. Although the number of publications rose to 4 in 2020, this is still not enough to state something about the temporal evolution of publications on the topic.

The main publication media for research on the topic are the journals Lean Six Sigma International, Total Quality Management & Business Excellence and Technical Journal, which correspond to 37.50% of the analyzed studies. The remaining publications occurred in 10 different journals in diverse areas (in particular operations management), which is justified by the diverse nature of the subjects related to lean approaches [30].

Seeking to answer question Q3, this research reveals that 25% of the studies did not take into consideration the industry in which the lean approach was being implemented. The Manufacturing and Construction sectors stand out, comprising 25% and 18.75% of the cases, respectively. As for Q4, this review indicates that MM is being structured as a guide for implementing lean approaches in several places around the world. Therefore, there is great divergence among countries that have developed or applied these artifacts for this purpose. However, a large portion of the analyzed articles (43.75%) does not even mention the country or locality in which the MM was validated. Table 1 illustrates the answers to Q3 and Q4. This lack of specificity in the development of these models may indicate possible gaps. For, without considering the sector and the place of application, it is understood that contextual aspects of the organizations, such as culture, size, complexity, planning parameters and production processes, are not taken into consideration.

4.2 Maturity Models as a Guideline to Lean Approach Implementations

Table 2 answers Q1 by presenting the 12 LMM found and a brief description – they are sorted by the publication date of the respective article. Considering the typology proposed by Fraser et al. [17] to answer Q2, half of the maturity models developed or applied in the selected articles are of the hybrid type, followed by Maturity Grid type models (33.33%) and, finally, by Capability Models (CMM) with 12.50%.

In the process of developing the LMM, the authors seek to characterize best practices and identify the evaluation parameters for successful implementation. In these cases, field research gains space and appears in many models to seek data through the experience and expertise of specialists in the theme. In this search for modeling that is more adherent to the reality of each production system, most authors bet on parameters such as Critical Success Factors (CSF), performance indicators, or Lean Principles in the composition of the model. The comparison

Table 1 Applicability of the LMM by sector and location

Article	Sector	Location
Lemieux et al. [28]	Luxury Sector	Generalized
Lemieux et al. [29]	Luxury Sector	Generalized
Nesensohn et al. [20]	Construction	Generalized
Maasouman and Demirli [31]	Generalized	Generalized
Verrier et al. [32]	Generalized	Generalized
Stålberg and Fundin [27]	General Manufacturing	Generalized
dos Santos Bento and Tontini [33]	General Manufacturing	Brazil
Reis et al. [34]	Coffee Industry	Colombia
dos Santos Bento and Tontini [35]	General Manufacturing	Brazil
Moya et al. [36]	General Manufacturing	Chile and France
Omotayo et al. [37]	Construction	Nigeria
Zanon et al. [38]	Chemical Industry	Generalized
Maier et al. [39]	Generalized	Austria
Rodegheri and Serra [13]	Construction	Brazil
Anthony and Antony [6]	Academic Institutions	United Kingdom
Maier et al. [40]	Generalized	Austria

Source: Elaborated by the authors

of the two cases below illustrates, and justifies, the diversity in the parameters used in the development of these artifacts.

The study by Nesensohn et al. [20] combines the CMMI framework with Lean Construction principles to develop the Lean Construction Maturity Model (LCMM). The model developed by Moya et al. [36], on the other hand, uses the CSFs to assess the level of readiness of SMEs to implement Lean Six Sigma. That said, it is noted that the first study has a specific focus on companies in the construction sector. To this end, the steps of the model development – from initial research on lean principles to the participation of experts – were carried out considering the construction sector, specifically. In the second study, which specifically covers the Lean Six Sigma approach, the authors limit themselves to manufacturing production processes in SMEs. In other words, every LMM should be developed based on its context: The lean approach, the sector or area of application, the size of the company, etc.

The Maturity Grids aim to guide managers in identifying improvements and consolidating best practices. Their application in LMM stands out for their replication in future work. Such as Lemieux et al. [29], who applied the maturity matrix developed by Lemieux et al. [28] to assist the implementation of the Leanaglie approach. Bessant's model [26], in its turn, was also used by Stålberg and Fundin [27] to analyze how the implementation of lean manufacturing is affected by external conditions. Although the authors of the analyzed articles have conducted a literature review on the topic, only half of the LMM used or adapted was models previously developed.

Table 2 Selected Lean Maturity Models

Authors	LMMs short description	Typology
Bessant et al. [26] and Stålberg and Fundin [27]	MM for organizations to increase their ability to implement continuous improvement. The framework is based on the concept of routine development and the evolution of organizational capability.	Maturity grid
Lemieux et al. [28, 29]	MM was developed to guide organizations to identify appropriate leagile improvement initiatives based on performance goals. The proposed framework is a causal matrix based on maturity.	Maturity grid
Verrier et al. [32]	Model based on CMMI framework aims to promote Lean and Green implementation synergistically. However, the framework does not provide maturity assessment methods.	Maturity grid
Maasouman and Demirli [31]	The model was developed by relating tools and processes of the Lean Manufacturing approach to performance measures of activities required at the operations level.	Hybrid
Rodegheri and Serra [13] and Nesensohn [20]	MM was developed to guide the implementation of Lean Construction by providing assessment, support and guidance in the maturation process. The model was developed using key Lean attributes.	CMM
Reis et al. [34]	MM integrated Lean and Green systems by formulating a conceptual framework. The framework measures maturity through metrics capable of evaluating key aspects of the practices.	Hybrid
dos Santos Bento and Tontini [33, 35]	MM was developed based on Toyota's management principles. The dimensionality and convergent validity of the tool are assessed through confirmatory factor analysis.	CMM
Moya et al. [36]	The model assesses the level of readiness of SMEs to implement Lean Six Sigma through an indicator calculated from Critical Success Factors (CSF). The indicator is inspired by the AIDA model.	Hybrid
Omotayo et al. [37]	The model was developed to implement lean practices in construction SMEs. The framework was built from the identification of essential variables and systems thinking approach.	Hybrid
Anthony et al. [6]	MM helps the implementation of Lean Six Sigma through action plans in academic institutions The framework was developed through literature review, questionnaires and case studies.	Maturity grid
Zanon et al. [38]	MM integrated Lean practices and Performance Measurement System (PMS) to leverage improvement efforts. The maturity levels of both approaches are surveyed and combined.	Hybrid
Maier et al. [39, 40]	MM aims to guide organizations in implementing the Lean Smart Maintenance approach. The model combines traditional aspects of lean practices with innovative aspects of Industry 4.0.	Hybrid

Source: Elaborated by the authors

Only two articles presented models whose classification was CMM. Coincidentally, both are about the same model, the Lean Construction Maturity Model (LCMM) [13, 20]. In general, the LCMM is a combination of CMMI definitions with Lean Construction principles and has the main objective of indicating the current state of maturity.

The differential of this model is that besides verifying the Lean maturity, it supports the company in the Lean maturity escalation, providing a tool to align and measure the continuous improvements. Therefore, the LCMM deserves to be highlighted as the only model that describes the maturity levels, assesses the current state of an organization, and also develops and controls improvement actions that drive the implementation of the practice. Although only two models are considered CMM-type, half of the models analyzed in this work use the CMM maturity levels in their entirety or as inspiration, reinforcing the relevance of the framework as the most traditional MM.

Besides supporting the implementation process of lean approaches, MM emerged as an artifact capable of providing synergy between different approaches. This is the case of Lean Smart Maintenance [39], a model that combines traditional aspects of lean practices with innovative aspects of Industry 4.0. According to Maier et al. [39], there is a latent need to adjust lean practices to digital transformations. Among the articles analyzed, the model developed by Zanon et al. [38] probably best exemplifies this capability of integration. In this case, the framework integrated Lean Manufacturing with Performance Measurement Systems (PMS) by surveying the maturity levels of both approaches in the literature and then correlating them. The results illustrate how Lean and PMS can evolve synergistically through continuous improvement efforts. Finally, another example of integration is the models that jointly address Lean and Green systems [32, 34]. In these cases, the studies managed to formulate a conceptual framework capable of providing synergy between the principles of the two approaches.

Most of the selected LMM were validated from case studies. In all of these, the artifacts achieved positive results and, in fact, contributed to successful implementations. When analyzing the LMM, one realizes that they deliver: a synthesis of lean best practices (separated by dimensions, processes, and maturity levels); a customized tool for measuring lean maturity; and improvement plans to leverage the lean maturation process. Although only the CMM models directly support improvements to raise an organization's maturity level, all models can indirectly assist in lean maturation. In several case studies, just accessing the maturity matrix or the maturity level assessment result was enough for managers of the organizations to create improvement plans on their own to advance the implementation successfully [6, 36].

5 Conclusion

The results of this Scoping Review reveal the diversity of LMM developed to support the implementation of lean approaches. For the most part, these are of a hybrid type and highly generalizable. The main constituent parameters of these frameworks are the Critical Success Factors (CSF) in lean approaches, performance KPIs, and Lean Principles. This study shows that as approaches adapt to new production systems, these parameters must be updated, giving rise to new models.

Most of the models analyzed are considered reproducible by their authors. The models by Bessant et al. [26] and Nesensohn et al. [41] were reproduced and adapted by Stålberg and Fundin [27] and Rodegheri and Serra [13], respectively, illustrating the relevance of developing a generalizable model for the literature. However, there are some practical limitations to this adaptation, the main one being the lean approach aspect. This becomes clear, for example, when we think that it makes no sense for an academic institution to apply a model developed from Lean Construction principles. Following this line, we conclude that how the models are developed may limit their reproduction in organizations of different sizes, nationalities, or types of productive systems.

In general, the MM analysis indicates to the organizations their Lean maturity level and the points to be developed. Although some MMs also help with improvement solutions, several case studies reveal that only the maturity matrix or the result of the maturity level assessment is enough for the managers of organizations to develop continuous improvement actions on their own to increase the lean maturity [6, 36].

The studies explored in this paper show how lean practices and principles are adapting to production systems and joining new approaches, such as Industry 4.0 and Green Systems [42, 43]. This movement of adaptation and integration brings many research opportunities, raising the need for specific innovative models to meet these new arrangements.

The current state of the art, regarding the application of LMM as a guide in implementation, exposed in this paper is expected to contribute to practitioners and academics. From the synthesis of LMMs raised, future research can use an existing model (or its constituent parameters) as a basis for the development of new models. Finally, through the synthesis exposed, practitioners can find a suitable framework for their business and use it as a guide to achieve success in the process of implementing lean approaches.

References

1. Womack, J.P., Jones, D.T.: Lean thinking – banish waste and create wealth in your corporation. J. Oper. Res. Soc. **48**, 1148 (1997). https://doi.org/10.1057/jors.2010.172
2. Nascimento, D., Caiado, R., Tortorella, G., Ivson, P., Meiriño, M.: Digital Obeya Room: exploring the synergies between BIM and lean for visual construction management. Innov. Infrastruct. Solut. **3** (2018)

3. Shah, R., Ward, P.T.: Lean manufacturing: context, practice bundles, and performance. J. Oper. Manag. **21**, 129–149 (2003). https://doi.org/10.1016/S0272-6963(02)00108-0
4. Antony, J., Ronald Snee, R.H.: Lean Six Sigma: yesterday, today, and tomorrow. Int. J. Qual. Reliab. Manag. **34**, 1073–1093 (2017)
5. Yadav, G., Desai, T.N.: Lean Six Sigma: a categorized review of the literature. Int. J. Lean Six Sigma. **7**, 2–24 (2016). https://doi.org/10.1108/IJLSS-05-2015-0015
6. Anthony, S.G., Antony, J.: A leadership maturity model for implementing Six Sigma in academic institutions – using case studies to build theory. Int. J. Lean Six Sigma. **12**, 675–692 (2021). https://doi.org/10.1108/IJLSS-01-2020-0011
7. Jørgensen, F., Matthiesen, R., Nielsen, J., Johansen, J.: Lean maturity, lean sustainability. IFIP Int. Fed. Inf. Process. **246**, 371–378 (2007). https://doi.org/10.1007/978-0-387-74157-4_44
8. Armstrong, R., Hall, B.J., Doyle, J., Waters, E.: "Scoping the scope" of a cochrane review. J. Public Health (Bangkok). **33**, 147–150 (2011). https://doi.org/10.1093/pubmed/fdr015
9. Arksey, H., O'Malley, L.: Scoping studies: towards a methodological framework. Int. J. Soc. Res. Methodol. Theory Pract. **8**, 19–32 (2005). https://doi.org/10.1080/1364557032000119616
10. Estampe, D., Lamouri, S., Paris, J.L., Brahim-Djelloul, S.: A framework for analysing supply chain performance evaluation models. Int. J. Prod. Econ. **142**, 247–258 (2013). https://doi.org/10.1016/j.ijpe.2010.11.024
11. Mettler, T., Rohner, P., Winter, R.: Management of the Interconnected World. In: Management of the Interconnected World (2010)
12. Jasper, V., Warse, K.: Applicability of the capability maturity model for engineer-to-order firms. Int. J. Technol. Manag. **48**, 219–239 (2009)
13. Rodegheri, P.M., Serra, S.M.B.: Maturity models to evaluate lean construction in Brazilian projects. Brazilian J. Oper. Prod. Manag. **17**(1), 20 (2020). https://doi.org/10.14488/bjopm.2020.016
14. Kosieradzka, A., Ciechańska, O.: Impact of enterprise maturity on the implementation of six sigma concept. Manag. Prod. Eng. Rev. **9**, 59–70 (2018). https://doi.org/10.24425/119535
15. Dooley, K., Subra, A., Anderson, J.: Maturity and its impact on new product development project performance. Res. Eng. Des. Theory, Appl. Concurr. Eng. **13**, 23–29 (2001). https://doi.org/10.1007/s001630100003
16. CMMI Product Team: CMMI for Development, Version 1.3
17. Fraser, P., Moultrie, J., Gregory, M.: The use of maturity models/grids as a tool in assessing product development capability. IEEE Int. Eng. Manag. Conf. **1**, 244–249 (2002). https://doi.org/10.1109/iemc.2002.1038431
18. De Paula, I.C., Fogliatto, F.S., Cristofari, C.A.: Method for assessing the maturity of product development management: a proposal. African J. Bus. Manag. **5**, 10285–10302 (2012). https://doi.org/10.5897/AJBM10.1553
19. Lameijer, B.A., De Mast, J., Does, R.J.M.M.: Lean six sigma deployment and maturity models: a critical review. Qual. Manag. J. **24**, 6–20 (2017). https://doi.org/10.1080/10686967.2017.12088376
20. Nesensohn, C., Bryde, D., Pasquire, C.: A measurement model for Lean Construction maturity. Lean Constr. J. **2016**, 1–9 (2016)
21. Caiado, R.G.G., Scavarda, L.F., Azevedo, B.D., Nascimento, D.L.D.M., Quelhas, O.L.G.: Challenges and benefits of sustainable industry 4.0 for operations and supply chain management—a framework headed toward the 2030 agenda. Sustain. **14** (2022). https://doi.org/10.3390/su14020830
22. Kreuter, T., Scavarda, L.F., Thomé, A.M.T., Hellingrath, B., Seeling, M.X.: Empirical and Theoretical Perspectives in Sales and Operations Planning. Springer, Berlin Heidelberg (2022)
23. Machado, E., Scavarda, L.F., Caiado, R.G.G., Thomé, A.M.T.: Barriers and enablers for the integration of industry 4.0 and sustainability in supply chains of msmes. Sustain. **13** (2021). https://doi.org/10.3390/su132111664
24. Mongeon, P., Paul-Hus, A.: The journal coverage of Web of Science and Scopus: a comparative analysis. Scientometrics. **106**, 213–228 (2016). https://doi.org/10.1007/s11192-015-1765-5

25. Ceryno, P.S., Scavarda, L.F., Klingebiel, K., Yüzgülec, G.: Supply chain risk management: a content analysis approach. Int. J. Ind. Eng. Manag. **4**, 141–150 (2013)
26. Bessant, J., Caffyn, S., Gallagher, M.: Evolutionary model of continuous improvement behaviour. Technovation. **21**, 67–77 (2001). https://doi.org/10.1016/S0166-4972(00)00023-7
27. Stålberg, L., Fundin, A.: Lean production integration adaptable to dynamic conditions. J. Manuf. Technol. Manag. **29**, 1358–1375 (2018). https://doi.org/10.1108/JMTM-02-2018-0055
28. Lemieux, A.A., Pellerin, R., Lamouri, S.: A mixed performance and adoption alignment framework for guiding leanness and agility improvement initiatives in product development. J. Enterp. Transform. **3**, 161–186 (2013). https://doi.org/10.1080/19488289.2013.818597
29. Lemieux, A.A., Lamouri, S., Pellerin, R., Tamayo, S.: Development of a leagile transformation methodology for product development. Bus. Process. Manag. J. **21**, 791–819 (2015). https://doi.org/10.1108/BPMJ-02-2014-0009
30. Salah, S., Rahim, A., Carretero, J.A.: The Integration of Six Sigma and Lean Management. (2010)
31. Maasouman, M.A., Demirli, K.: Development of a lean maturity model for operational level planning. Int. J. Adv. Manuf. Technol. **83**, 1171–1188 (2016). https://doi.org/10.1007/s00170-015-7513-4
32. Verrier, B., Rose, B., Caillaud, E.: Lean and Green strategy: the Lean and Green House and maturity deployment model. J. Clean. Prod. **116**, 150–156 (2016). https://doi.org/10.1016/j.jclepro.2015.12.022
33. dos Santos Bento, G., Tontini, G.: Developing an instrument to measure lean manufacturing maturity and its relationship with operational performance. Total Qual. Manag. Bus. Excell. **29**, 977–995 (2018). https://doi.org/10.1080/14783363.2018.1486537
34. Reis, L.V., Kipper, L.M., Giraldo Velásquez, F.D., Hofmann, N., Frozza, R., Ocampo, S.A., Taborda Hernandez, C.A.: A model for Lean and Green integration and monitoring for the coffee sector. Comput. Electron. Agric. **150**, 62–73 (2018). https://doi.org/10.1016/j.compag.2018.03.034
35. Bento, G.D.S., Tontini, G.: Maturity of lean practices in Brazilian manufacturing companies. Total Qual. Manag. Bus. Excell. **30**, S114–S128 (2019). https://doi.org/10.1080/14783363.2019.1665827
36. Moya, C.A., Galvez, D., Muller, L., Camargo, M.: A new framework to support Lean Six Sigma deployment in SMEs deployment. Int. J. Lean Six Sigma. (2018). https://doi.org/10.1108/IJLSS-01-2018-0001
37. Omotayo, T.S., Boateng, P., Osobajo, O., Oke, A., Obi, L.I.: Systems thinking and CMM for continuous improvement in the construction industry. Int. J. Product. Perform. Manag. **69**, 271–296 (2020). https://doi.org/10.1108/IJPPM-11-2018-0417
38. Zanon, L.G., Ulhoa, T.F., Esposto, K.F.: Performance measurement and lean maturity: congruence for improvement. Prod. Plan. Control. **32**, 760–774 (2021). https://doi.org/10.1080/09537287.2020.1762136
39. Maier, H.T., Schmiedbauer, O., Biedermann, H.: Validation of a lean smart maintenance maturity model. Teh. Glas. **14**, 296–302 (2020). https://doi.org/10.31803/tg-20200706131623
40. Maier, H.T., Schmiedbauer, O., Biedermann, H.: Development of a generic framework to assess asset management maturity within organisations. Teh. Glas. **15**, 235–241 (2021). https://doi.org/10.31803/tg-20210510161621
41. Nesensohn, C., Bryde, D., Ochieng, E., Fearon, D.: Maturity and maturity models in lean construction. Australas. J. Constr. Econ. Build. **14**, 45–59 (2014). https://doi.org/10.5130/ajceb.v14i1.3641
42. Caiado, R.G.G., Quelhas, O.L.G., Nascimento, D.L.M., Anholon, R., Leal Filho, W.: Measurement of sustainability performance in Brazilian organizations. Int. J. Sustain. Dev. World Ecol. **25** (2018). https://doi.org/10.1080/13504509.2017.1406875
43. Caiado, R.G.G., Lima, G.B.A., Gavião, L., Quelhas, O.L.G., Paschoalino, F.F.: Sustainability analysis in electrical energy companies by similarity technique to ideal solution. IEEE Lat. Am. Trans. **15** (2017). https://doi.org/10.1109/TLA.2017.7896394

Evaluation of Radio Frequency Identification in Hospitals Operations

Poline Andrade Balbi, Priscilla Cristina Cabral Ribeiro, Carlos Francisco Simões Gomes, and Marcelo Maciel Monteiro

Abstract The health system is quite complex because its process includes medicines, supplies, but patients either. Information about them involves privacy, registration, storage and analysis. Besides, equipment, medicines and supplies need a storage organization because its absence leads to high costs and waste of resources. With the intention of reducing these problems, some information technologies (IT), such as Radio Frequency Identification (RFID), improve material traceability, information recording, and increased accuracy. Information technologies (IT) demand evaluation, considering that they can be underestimated in a company, due to their high costs. Some stakeholders just consider the IT cost before the decision of implementing them. Hence, the absence of a complete evaluation leads the company to implement these technologies without any analysis. However, this evaluation is not simple and fast because it requires a structure supported by some attributes. In this context, the article aims to evaluate RFID in two hospitals, focusing on major operations, by a set of attributes. For this purpose, there was a literature review with an initial survey of articles in the main database. Then, there was the literature review, the structuring of the questionnaire, semi-structured interviews with two case studies, ending with a counterpoint between theory and field findings. After the case studies, the authors noticed that RFID increased profitability and inventory accuracy. On the other hand, there are factors to improve in the studied sample, such as system complexity, lack of technology knowledge, interference, return on investment, deployment cost, and employee training.

Keywords RFID · Information technology · Health · Hospitals

P. A. Balbi · P. C. C. Ribeiro (✉) · C. F. S. Gomes · M. M. Monteiro
Universidade Federal Fluminense, Niterói, RJ, Brazil
e-mail: priscillaribeiro@id.uff.br

V. M. López Sánchez et al. (eds.), *Industrial Engineering and Operations Management*, Springer Proceedings in Mathematics & Statistics 400,
https://doi.org/10.1007/978-3-031-14763-0_12

1 Introduction

To achieve improvements in hospital management, Radio Frequency Identification (RFID) technology can help reduce costs, increase control of adverse situations, improve patient satisfaction, make operations more agile, make decisions more accurate, improve the quality of customer service, balance the institution's economic and financial situation [1–4]. However, research and development processes on RFID in the healthcare sector have been slower than in other sectors [5]. For this, simpler processes can support the structure of these institutions and make them more agile [4].

It is unquestionable the need for IT in the Strategic Business Planning (PEE) of a company to collect, store, retrieve and disseminate information in the organization [6]. However, the IT must be evaluated before its implementation, because as stated by Grover et al. [7] it can be underestimated by users and/or senior management executives, resulting in budget cuts and hiring low-skilled IT executives. Currently, other authors continue to carry out assessments of IT to analyze the visibility of their implementation [8–12].

The objective of this paper is to evaluate RFID in two hospitals, focusing on their main operations, by a set of attributes. In this paper, the evaluation attributes will be divided into the following aspects: environment (procedure's rooms, surgeries' room and inventory), technical and economic.

The achievement of the objective occurs through a proposal of a set of attributes for the evaluation of RFID in operations: inventory control of assets (equipment), patients, and hospital supplies. Its contribution is to fill the gap in hospital management, aiming to improve the control of patients' information and resources. Furthermore, this evaluation could offer to the society a health care service focused on the patient, maintaining the objective of the institution and the well-being of its employees. The RFID evaluation will allow the hospital to better understand its processes, its costs, its environment, and its use of technology. Thus, senior management will be able to observe an increase in the quality of information, more accurate decision making, an increase in employee productivity, and a reduction in costs due to the optimal use of the resources [4].

According to the discussion above, the main research question is: How to evaluate RFID through evaluation attributes in two hospitals in their operations? Therefore, to support the research question, two secondary questions are formulated in this paper:

- How is RFID used in two large hospitals in Brazil?
- What is the assessment of hospital managers regarding the adoption of RFID in the hospital units where they work?

2 Literature Review

2.1 RFID in Preventing Inventory Losses and Their Impact in Supply Chain

According to Oliveira et al. [13], IT capabilities can significantly impact process performance, specifically in production processes and operations, improvements in products/services, and customer relationships, impacting profit and market share in a positive way. For greater integration between trading partners, IT can be used, such as some IoT devices with the RFID system, constituting a network formed by attaching the tag to each item in the stock. Thus, the administrator uses RFID readers to scan the objects. These portable readers can be wireless and intelligent, with processing capabilities and connected to middleware, establishing communication [14].

De Oliveira et al. [15] pointed out that, due to the dedication of researchers in the study of RFID, technicians have invested their time in developing low-cost RFID systems with a longer life cycle, using artificial intelligence systems, decision-making tools, and the inclusion of a variety of sensors. Thus, it is a fact that has increased the interest in IT, which is confirmed by the same authors. Besides, for Assis and Sagawa [16], RFID has advantages over barcodes, such as the information restrictions elimination, reading mistakes, duplicate readings, and others.

In companies with high complexity as hospitals, RFID provides inventory adjusted to demand, replenishment cycles aligned with the company's schedule, traceability of material flow, and adequate consumption per patient, which results in increased patient safety [17, 18]. The implementation of RFID on the shelves that contain the items to be stored allows to carry out the inventory at the time of storage, control the goods, update the information about the replacement of the product, reduce the lack of stocks and the increasing cost of inaccuracy (including the expiry date of perishable products). Thus, process efficiency, traceability, visibility, security, and trust among supply chain stakeholders are increased [19–21]. The integration of RFID and sensors in the health sector supports real-time data transmissions, such as vital signs (pulse rate), location (patient, medicines, and supplies), and activity [22].

Pérez et al. [9] state that, to increase the quality of patient care, the control of medicines (location, batch number, expiration date) is a priority, especially those which have a higher value. Technology plays an important role in this context, as the integration of RFID with management software contributes to a better development of tasks. Consequently, the physical contact between the nurse and the patient is eliminated, using this time to verify and answer internal phone calls, making the process more efficient. In addition, it allows data on adverse events, which lead to death, to be accessed for future improvement in the health care system in Brazil and outside world [23].

2.2 Information System and RFID Evaluation

The information system (IS) assessment discussed by DeLone and McLean (1992) [24] still influences the attributes adopted to assess IS and/or IT. The authors published an IS success assessment model based on the analysis of about 180 authors who had addressed the same topic and who identified the contributing factors to the success of the IS. The authors considered that the evaluation of IS to measure their success is fundamental in many companies' strategies. After 10 years of the published evaluation model, DeLone and McLean [25] updated it. Ribeiro et al. [8, 26] presented an evaluation method of RFID, and its organization inspired this article. According to these methods, this one organizes the attributes in three aspects: environment, technical and economic. The summary of the literature is in Table 1.

Table 1 Attributes for RFID evaluation

Aspects	Attributes	Authors
Environment	Wasted material (water and metal)	Miller [27]
	Noise	Miller [27] and Cassel et al. [11]
	Sensitivity	Martínez-Castro et al. [10]
	Scenario's characteristics	Pérez et al. [9], Colella et al. [28], and Assis and Sagawa [16]
Technical	RFID tags Quality	Miller [27], Colella et al. [28], and Assis and Sagawa [16]
	Reader configuration	Miller [27]
	Number of tags	Wei et al. [29] and Pérez et al. [9]
	Location	Pérez et al. [9], Farris et al. [30], and Güttler et al. [31]
	Frequency	Pérez et al. [9] and Colella et al. [28]
	Responsiveness	Miller [27] and Matbouly et al. [32]
	System performance	Colella et al. [28], Srichavengsup [33], and Golsorkhtabaramiri et al. [34]
	Ease of use	Venkatesh et al. [35] and Maydana et al. [36]
	Energy Efficiency	Golsorkhtabaramiri et al. [34]
	Security	Venkatesh et al. [35], Fan et al. [12] and Jeddi et al. [37]
	Storage capacity	Pérez et al. [9], Cai et al. [38], and Tiryakioglu et al. [39]
	Conformity with IoT	Farris et al. [30]
Economic	Cost (RFID)	Pérez et al. [9]
	Profitability	Pérez et al. [9] and Dovere et al. [40]

3 Method

To present a contribution to the stream of the research in IT/RFID, articles on RFID studies were searched. The main subjects searched were the control of patients, equipment and drugs. In addition, related to operations, there are possible future research: improvements in hospital processes and increasing in patient satisfaction, assisting in decision-making to invest, improvement in employee efficiency [4, 5, 18]; reduction of adverse events and consequent deaths [23]; qualitative research on hand hygiene control in hospitals, the usefulness of RFID to prevent the transmission of infections to patients and the risks associated with electromagnetic waves emitted by tags [41]; patient location on hospital premises (not only inside it), providing exact location [42]; inventory management in hospitals using RFID [18]; data generation for institutions interested in comparing their performance metrics with other hospitals [43]; internal rules in hospitals to control blood storage, transport and distribution conditions [44]. These gaps are important, as the development of research in these different areas of knowledge on the use of RFID in hospitals allows: achieving objectives in different sectors of hospitals; improvement of the processes that occur from the arrival of the patient to his/her hospital discharge; control of equipment, supplies, patients and even blood for transfusion; and evaluation and use of RFID.

Based on the information above, this article contributes to hospital management and the proper use of IT (through its evaluation with attributes and sub-attributes), considering what IT offers to hospital operations (inventory control of assets/equipment, control of patients and hospital supplies) and the improvement of patient safety.

The method was organized in stages to answer the main research questions, resulting from the search of papers and the literature review, presenting the factors (called attributes and sub-attributes), which must be considered to understand the phenomenon investigated in the article. After presenting the sub-attributes, they were grouped into three categories, namely: technical, environmental (from the place in where the material and their tags are located) and economic. Regardless of the different categories, these attributes are a cohesive set that makes RFID evaluation possible. This selection of attributes answers the research question and helps the hospitals to control the use of IT and its results in their business, considering the environmental conditions (water resistance, interference from equipment and metals), cost and profitability.

As reported in the previous section, the article by DeLone and McLean [24] is a work that has been the basis for the development of other articles on the topic. However, given the scope of his research, for certain IT, it needs to have

Table 2 Secondary research questions and actions to answer them

Secondary research questions	Actions
1. How does the RFID is used in two large hospitals in Brazil?	Field research: data collection and result analysis (open questions)
2. What is the evaluation from the managers related to the RFID adoption in hospitals?	Field research: data collection and result analysis (closed questions)

additions and subtractions. In the case of this article, as RFID is a hardware, it cannot have sub-attributes to evaluate it related to software, as it has different technical characteristics from other IT (not only because of it is hardware, but also because of its data storage capacity and generation of a large volume of data); its performance impacted environmental agents as already mentioned (water, equipment, constructions, metal, dust, etc.); and its cost of purchasing labels, implementation (layout changings, training and maintenance of the system that allows its use). Thus, it is not possible to generalize the authors' model, as these elements of the attribute categories (technical, environmental and economic) limit the boundaries of their use, evaluation and, therefore, the scope of the theory and its application to specific contexts. Table 2 presents the initiatives undertaken to obtain the answers to the secondary research questions.

The articles found in the search support the set of attributes of IT and RFID evaluation. They were collected in the Scopus and Web of Science databases, using the steps of bibliometric study that supports [15] and was the base of them [45]. These two papers, in different ways, guided the selection of the papers to compose the literature review. The research divided the attributes into three sets, such as environment, technical and economic. In the set of environmental attributes, there was a subdivision in the attribute 'characteristics of the scenario'. Its scores are in Table 4.

The main authors who are developing the research in IT evaluation are in Table 1, and their contributions and suitability for the research were analyzed. To organize contributions into attributes and sub-attributes the researchers applied Ribeiro's method [8, 26]. The answers to the main question are in Table 1 (attributes and sub-attributes), while the answers to the secondary questions are in Tables 3, 4 and 5. The answers to the close questions are in Tables 3 and 4 and Table 5 presents the answers to open questions. The scale of equal intervals between the points was: (1) very low; (2) low; (3) average; (4) high; (5) too high.

The researchers selected two healthcare organizations (hospital A and hospital B), which have in their operations the identification and traceability of patients, assets (equipment), and materials (medicines, vaccines, gloves, masks, etc.) by RFID. Therefore, the large size of the hospitals also influenced the choice, as well as previous contacts with the interviewees. Hospital A has two operations: operation 1 (OP1), identification and traceability of patients, and operation 2 (OP2), control of assets (equipment). Hospital B has operation 3 (OP3), the material control. The respondent 1 (R1) from hospital A is the IT manager, who has been in the position for 6 years; and the respondent 2 (R2) from hospital B, is a business analyst, who had started the activities shortly before the research (2 months).

4 Results and Discussion

4.1 Results

Table 3 presents the answers from the closed questions (scores) in order to evaluate RFID in the operations of the two hospitals. They are related to operation 1 (OP1) (patient identification and traceability) and operation 2 (OP2) (assets/equipment control), from hospital A, and the scores of operation 3 (OP3), referring to the material control, from hospital B. These scores answer to question 2, which is in Table 2:

Table 3 RFID evaluation by scores

Aspects	Attributes	R1 – OP1	R1 – OP2	R2 – OP3
Environment	Wasted material (water and metal)	3	2	4
	Noise	3	2	4
	Sensitivity	4	4	3
	Scenario's characteristics	Table 4	Table 4	NR
Technical	RFID tags Quality	5	5	NR
	Reader configuration	5	5	NR
	Number of tags	4	5	NR
	Location	4	4	NR
	Frequency	4	4	NR
	Responsiveness	5	4	NR
	System performance	5	5	NR
	Ease of use	4	4	4
	Energy Efficiency	3	3	3
	Security	5	4	4
	Storage capacity	5	5	4
	Conformity with IoT	4	4	NR
Economic	Cost (RFID)	5	5	NR
	Profitability	5	4	NR

NR no response

According to the organization of the environment aspect answers, separating the scenario attribute, Table 4 presents the scores of the sub-attributes:

4.2 Analysis and Discussion of the Results

The answers from the open questions obtained in interviews related to the two operations of hospital A and hospital B are analyzed below. Some questions were not answered by R2.

After analyzing separately, the OP1 and 2 of the hospital A and the OP3 of the hospital B, it was performed a comparative analysis between these operations. Concerning the environment aspect, in the OP 1 and 2 the tags presented high

Table 4 Scores of the sub-attributes of 'scenario's characteristics'

Operations	Scenario's characteristics	Scores
Operation 1	The wall material, which can reduce frequency signals	1
	The elevators that damage the location of objects	5
	The rooms types (open or semi-open)	3
	The object material to be controlled	2
Operation 2	The wall material, which can reduce frequency signals	1
	The elevators that damage the location of objects	5
	The rooms types (open or semi-open)	3
	The object material to be controlled	2

Table 5 Summary table of research results

Attributes	Aspects	Authors	Results
Environment	Wasted Materials and Noise interfere in tags reading	Miller [27] and Cassel et al. [11]	R1 – Average interference in the newborns' sector
	The sensitivity of the tag returns the proper signal	Martínez-Castro et al. [10]	R1 – The tag can read the data properly R2 – Average interference in metal and wet surfaces; the distance between the tag and the reader
	Scenario's characteristics	Miller [27] and Colella et al. [28]	R1 – Low reading influenced by wall material; elevators with not interference with the signal

(continued)

Table 5 (continued)

Technical	RFID tags Quality	Miller [27] and Colella et al. [28]	R1 – The system has high quality
	Reader configuration	Miller [27]	R1 – The reader has high performance in the reader configuration
	Number of tags	Pérez et al. [9] and Wei et al. [29]	R1 – Low tags interference when more than one tries to communicate with the system
	Location	Pérez et al. [9], Colella et al. [28], Farris et al. [30], and Güttler et al. [31]	R1 – The location is good throughout the facility generating low tag collisions
	Frequency	Pérez et al. [9] and Colella et al. [28]	R1 – Tags frequency increases the accuracy and responsiveness, and decreases the interference
	Responsiveness	Miller [27] and Matbouly et al. [32]	R1 – The factors that would influence in a communication delay between the reader and the tag have low influence in this case
	System performance	Colella et al. [28], Srichavengsup [33], and Golsorkhtabaramiri et al. [34]	R1 – The performance is on average R2 – The high score is related to the system's capacity of storage tags data
	Ease of use	Venkatesh et al. [35] and Maydana et al. [36]	R1 – The use of RFID with other tags has good acceptance R2 – Proper training prevents use constrains
	Energy Efficiency	Golsorkhtabaramiri et al. [34]	R1 – Average
Attributes	Aspects	Authors	Results
	Security	Venkatesh et al. [35], Fan et al. [12], and Jeddi et al. [37]	R1 – Cryptography ensures data security R2 – Information privacy is guaranteed
	Storage capacity	Pérez et al. [9], Cai et al. [38], and Tiryakioglu et al. [39]	R1 – RFID stands out comparing to other technologies
	Conformity with IoT	Farris et al. [30]	R1 – There is a virtual communication channel for data exchange
Economics	Cost (RFID)	Pérez et al. [9]	R1 – Very high
	Profitability	Pérez et al. [9] and Dovere et al. [40]	R1 – High

sensitivity, while OP 3 has a medium sensitivity. Hence, hospital B has a lower ability to return the signal properly compared to hospital A. Concerning the wasted materials and noise, there is a high reading of the tags in the control of the hospital material, while the patient's data is on average, and in the asset tracking is low, due to the interference of these attributes, confirming the literature [27]. Hospital A must obtain a proper tag for OP1, since it deals directly to the patients. In hospital B, the OP3 has a better reading, since their products do not have contact to the water.

In the technical aspect it can be noticed that the three operations had a high score in the 'ease of use' attribute, compared to other ITs and an average score in relation to energy efficiency at the moment of the tag reading, confirming the theory [34]. Although the scores of the storage capacity confirm the advantage of RFID to have a high capacity of storage, they show that the hospital A has tags that comprise more information in their operations and that also have a greater frequency. Finally, concerning the safety attribute, the difference between the scores can be justified due to the fact that they have different operations. While the OP1 treats the patients, the other operations deal with equipment and products, the safety attribute is more important, but not as the OP1.

5 Conclusion

It was noticed that, despite the positive performance of the RFID system in hospitals, there are disadvantages such as the complexity of the system; lack of knowledge about the technology; user training; deployment cost; return on investment; interference, which impairs the reading of the tags and the circulation of information. There are advantages, such as the agility of operations; reduction of errors, waste, and working hours; and speed and accuracy in the exchange of information.

For improvements in operations, it is suggested to provide better communication among the team and training of users during the RFID implementation. Hence, the technology would bring benefits to the hospital, and patient safety and satisfaction. Limitations can be overcome with coordinated and collaborative work between chain members (RFID suppliers, hospitals, customers, and employees).

This study can contribute to confirming that RFID has a high capacity to bring improvements to hospital operations. The health sector is looking for innovation and RFID adoption can provide benefits and conditions to serve society.

The conclusions obtained in the research are restricted only to the analyzed sample, so they cannot be generalized. As a suggestion for future work, it is recommended to apply the questionnaires to other hospitals in Brazil which have the RFID adopted in their operations. These applications in operations are not studied yet and could motivate decision-makers from other hospitals to adopt the technology.

References

1. Pinochet, L.H.C., Lopes, A.S., Silva, J.S.: Inovações e Tendências Aplicadas nas Tecnologias de Informação e Comunicação na Gestão da Saúde. Revista de Gestão em Sistemas de Saúde. **3**(2), 11–29 (2014)
2. Robinson, J., Starr, J., Vass, J.: Implementation of Radio-Frequency Identification in Hospitals. The University of Tennessee, Knoxville (2016)
3. Oda, O.: RFID na gestão da cadeia logística. CIO From IDG (2015). https://cio.com.br/rfid-na-gestao-da-cadeia-logistica/. Last acessed 02 Oct 2019
4. Deryahanoglu, O., Kocaoglu, B.: The effect of technology investments on hospital budget and patient safety: a simulation for emergency department. Int. J. Eng. Adv. Technol. **8**(6), 4650–4656 (2019)
5. Deryahanoglu, O., Kocaoglu, B.: Applications of RFID systems in healthcare management: a simulation for emergency department. Int. J. Innov. Technol. Explor. Eng. **8**(10), 786–792 (2019)
6. Falsarella, O.M., Jannuzzi, C.A.S.C.: Planejamento estratégico empresarial e planejamento de tecnologia de informação e comunicação: uma abordagem utilizando projetos. Revista Gestão e Produção. **24**(3), 610–621 (2017)
7. Grover, V., Jeong, S.R., Segars, A.H.: Information systems effectiveness: the construct space and patterns of application. Inf. Manag. **31**(4), 177–191 (1996)
8. Ribeiro, P.C.C., Scavarda, A.J., Batalha, M.O.: The application of RFID in brazilian harvest facilities: two case studies. Int. J. Eng.Bus. Manag. **3**(1), 57–63 (2011)
9. Pérez, M.M., González, G.V., Dafonte, C.: Evaluation of a tracking system for patients and mixed intravenous medication based on RFID technology. Sensors. **16**(12), 1–15 (2016)
10. Martínez-Castro, R.E., Jang, S., Nicholas, J., Bansal, R.: Experimental assessment of an RFID-based crack sensor for steel structures. Smart Mater. Struct. **26**(8), 1–13 (2017)
11. Cassel, M., Dépret, T., Piégay, H.: Assessment of a new solution for tracking pebbles in rivers based on active RFID. Earth Surf. Process. Landf. **42**(13), 1938–1951 (2017)
12. Fan, K., Jiang, W., Li, W., Yang, Y.: Lightweight RFID protocol for medical privacy protection in IoT. IEEE Trans. Ind. Informatics. **99**, 1–11 (2018)
13. Oliveira, D.D.L., Maçada, A.C.G.: IT capabilities' business value: analysis of multi-level performance in Brazilian organizations. Gest. Prod. **24**(2), 295–309 (2017)
14. Alwadi, A., Gawanmeh, A., Parvin, S., Al-Karaki, J.N.: Smart solutions for RFID based inventory management systems: a survey. Scalable Comput. **18**(4), 347–360 (2017)
15. De Oliveira, A.O., Oliveira, H.L.S., Gomes, C.F.S., Ribeiro, P.C.C.: Quantitative analysis of RFID' publications from 2006 to 2016. Int. J. Inf. Manag. **48**, 185–192 (2019)
16. Assis, R.D., Sagawa, J.K.: Assessment of the implementation of a Warehouse Management System in a multinational company of industrial gears and drives. Gest. Prod. **25**(2), 370–383 (2018)
17. Silva, M.Z., Francisco, C.F.: The influence of contingencies factors strategy and structure in the enterprise risk management in a hospital. Gest. Prod. **26**(1), e2315 (2019)
18. León-Araujo, M.D.C., Gómez-Inhiesto, E., Acaiturri-Ayesta, M.T.: Implementation and evaluation of a RFID smart cabinet to improve traceability and the efficient consumption of high cost medical supplies in a large hospital. J. Med. Syst. **43**(6), 178 (2019)
19. Chamekh, M., El Asmi, S., Hami, M.: A new architecture for supply-chain management. In: IEEE Annual Consumer Communications & Networking Conference, pp. 1–5, Las Vegas (2017)
20. Gaukler, G., Ketzenberg, M., Salin, V.: Establishing dynamic expiration dates for perishables: an application of RFID and sensor technology. Int. J. Prod. Econ. **193**, 617–632 (2017)
21. Qin, W., Zhong, R.Y., Dai, H.Y., Zhuang, Z.L.: An assessment model for RFID impacts on prevention and visibility of inventory inaccuracy presence. Adv. Eng. Inform. **34**, 70–79 (2017)
22. Gaynor, M., Waterman, J.: Design framework for sensors and RFID tags with healthcare applications. Health Policy Technol. **5**(4), 357–369 (2016)

23. Couto, R.C., Pedrosa, T.M.G. Roberto, B.A.D., Daibert, P.B., Abreu, A.C.C., Leão, M.L.: II Anuário da segurança assistencial hospitalar no Brasil – propondo as prioridades nacionais (2018). https://www.iess.org.br/cms/rep/Anuario2018.pdf. Last accessed 19 Dec 2019
24. Delone, W.H., McLean, E.R.: Information systems success: the quest for the dependent variable. Inf. Syst. Res. **3**(1), 60–95 (1992)
25. Delone, W.H., McLean, E.R.: The DeLone and McLean model of information systems success: a ten-year update. J. Manag. Inf. Syst. **19**(4), 9–30 (2003)
26. Ribeiro, P.C.C., Batalha, M.O., Scavarda, A.J.: Application of RFID in Brazilian harvest facilities: two case studies. In: 14th World Multi-Conference on Systemics, Cybernetics and Informatics Proceedings, vol. 1, pp. 199–203, Orlando, FL (2010)
27. Miller, J.: Criteria for evaluating RFID solutions for records and information. Inf. Manag. J. **35**(2), 31 (2007)
28. Colella, R., Catarinucci, L., Tarricone, L.: Measurement system for over-the-air evaluation of UHF RFID tags quality. Wirel. Power Transf. **4**(1), 33–41 (2017)
29. Wei, L., Yang, C., Chen, R.: RFID based stent graft dynamic assessment for health cares. IET Networks. **6**(1), 1–9 (2017)
30. Farris, I., Militano, L., Iera, A., Spinella, S.C.: Assessing the performance of a novel tag-based reader-to-reader communication paradigm under noisy channel conditions. IEEE Trans. Wirel. Commun. **15**(7), 4813–4825 (2016)
31. Güttler, F., Heinrich, A., Krauß, P., Guntermann, J., Bucourt, M., Teichgräber, U.: RFID-based real-time navigation for interventional magnetic resonance imaging: development and evaluation of a novel tracking system. J. Med. Dev. **11**(3), 1–19 (2017)
32. Matbouly, H., Zannas, K., Duroc, Y., Tedjini, S.: Analysis and assessments of time delay constrains for passive RFID tag-sensor communication link: application for rotation speed sensing. IEEE Sensors J. **17**(7), 2174–2181 (2017)
33. Srichavengsup, W.: Performance evaluation of anti-collision algorithms for RFID system with different delay requirements. Int. J. Adv. Comput. Sci. Appl. **8**(4), 523–532 (2017)
34. Golsorkhtabaramiri, M., Issazadehkojidi, N., Pouresfehani, N., Mohammadialamoti, M., Hosseinzadehsadati, S.M.: Comparison of energy consumption for reader anti-collision protocols in dense RFID networks. Wirel. Netw. **24**, 1–14 (2018)
35. Venkatesh, V., Aloysius, J.A., Hoehle, H., Burton, S.: Design and evaluation of auto-id enabled shopping assistance artifacts in customers' mobile phones: two retail store laboratory experiments. MIS Q. **41**(1), 83–113 (2017)
36. Maydana, T., Giraldo, L., González, Z., Schachner, B., Mayan, J., Luna, D., Benítez, S.: BarCode medication administration in ICU: learning from our nurses. Stud. Health Technol. Inform. **245**, 1095–1098 (2017)
37. Jeddi, Z., Khattab, A., Amini, E., Bayoumi, M.: Redundant bit security in RFIDs: architecture design and security performance evaluation. J. Circuits Syst. Comput. **26**(9), 1–25 (2017)
38. Cai, Y., Li, X., Li, M., Chen, X., Hu, H., Ni, J., Wang, Y.: Traceability and quality control in traditional chinese medicine: from chemical fingerprint to two-dimensional barcode. Evid. Based Complement. Alternat. Med. **251304**, 1–6 (2015)
39. Tiryakioglu, B., Kayakutlu, G., Duzdar, I.: Medical device tracking via QR code and efficiency analyze. In: Portland International Conference on Management of Engineering and Technology, pp. 3115–3128, Honolulu (2016)
40. Dovere, E., Cavalieri, S., Ierace, S.: RFID systems for moveable asset management: an assessment model. Int. J. Prod. Res. **55**(5), 1336–1349 (2016)
41. Tarantini, C., Brouqui, P., Wilson, R., Griffiths, K., Patouraux, P., Peretti-Watel, P.: Healthcare workers' attitudes towards hand-hygiene monitoring technology. J. Hosp. Infect. **102**, 413–418 (2019)
42. Chhetri, B.T., Alsadoon, A., Prasad, P.W.C., Venkata, H.S., Elchouemi, A.: Enhanced weighted centroid localization in RFID technology: patient movement tracking in hospital. In: International Conference on Advanced Computing & Communication Systems, pp. 910–915, Coimbatore (2019)

43. Sang, A.X., Tisdale, R.L., Nielsen, D., Loica-Mersa, S., Miller, T., Chong, I., Shieh, L.: How much time are physicians and nurses spending together at the patient bedside? J. Hosp. Med. **14**, E1–E6 (2019)
44. Dusseljee-Peute, L.W., Van der Togt, R., Jansen, B., Jaspers, M.W.: The value of radio frequency identification in quality management of the blood transfusion chain in an academic hospital setting. JMIR Med. Inform. (3), 7, e9510, 1–14 (2019)
45. Gomes, C.F.S., Ribeiro, P.C.C., Freire, K.A.M.: Bibliometric research in warehouse management system from 2006 to 2016. In: 22nd World Multi-Conference on Systemics, Cybernetics and Informatics, pp. 200–204, Orlando (2018)

Opportunities for Strengthening Brazilian Plastic Waste Management Adopting Structural Equation Modelling

Lucas Menezes Pereira and Francisco G. M. Freires

Abstract This article aims to provide a qualitative-quantitative theoretical basis for the elaboration of an assertive Supply Chain Management (SCM) project that maximizes the performance of a Brazilian system operating Reverse Logistics that collects and separates waste from urban centers and processes it through recycling and/or thermochemical conversion. For this purpose, exploratory research of bibliographic content was conducted on data from the generation, destination, processing, and recovery of plastic Municipal Solid Waste (MSW), in addition to the physical characteristics of this type of waste, plant infrastructure, socioeconomic profile of the urban population and urban development policies that manage materials. The quali-quantitative information of consolidated literature helped in the preparation of the survey – to be answered by professionals of this market segment – whose answers will be complementary to the literature on the mathematical approach by Structural Equation Modelling using the Partial Least Square method in the design of high-precision actions for the Brazilian need for plastic waste management. All this in a business model that generates employment and income, and produces electricity, among other benefits, in the conception of sustainability promoted by the Circular Economy. To complement the methodology used in this work, successful cases in the recovery of materials in addition to future trends in the generation of plastic waste, are also presented as some of the research results.

Keywords Plastic waste · Reverse Logistics · Supply Chain Management.

L. M. Pereira (✉) · F. G. M. Freires
Department of Mechanical Engineering, Federal University of Bahia. Rua Professor Aristides Novis, Salvador, Brazil
e-mail: lucasmpereira@ufba.br; francisco.gaudencio@ufba.br

V. M. López Sánchez et al. (eds.), *Industrial Engineering and Operations Management*, Springer Proceedings in Mathematics & Statistics 400,
https://doi.org/10.1007/978-3-031-14763-0_13

1 Introduction

Brazil generates 1.52 million tons of waste weekly, as stated in the 2020 Panorama of Solid Waste in Brazil, conducted by the Brazilian Association of Special Waste and Public Cleaning Companies – ABRELPE. According to the same association, despite the PNRS (NSWP-National Solid Waste Policy) aim of closing all dumps by 2014, 3001 municipalities discharged 40.5% of MSW in controlled dumps or landfills in an improper manner [1].

Behind the United States, China and India, Brazil is the fourth country in the world that most produces plastic waste [2], although the plastic recycling rate of this South American nation remains extremely low (1.2%) in comparison to the USA (34.60%) and China (21.92%) [3]. This problem is supposed to be caused by the insufficiency of facilities and collection vehicles [4], poor waste legislation [5], the population's lack of awareness about recycling programs [6], and the fact that recycled materials are taxed twice, once as raw materials and again as recycled [7].

Brazil's recycling market is kept extremely informal [8], and data from the countless autonomous waste pickers are scarce. Unfortunately, many of these people are homeless and treated as invisible, apart from society. Their work is done informally, which yields low incomes and inadequate social protection [9].

Recycling programs are uncommon in Brazil. According to the Business Commission on Recycling (CEMPRE), from the data of 2018, 1227 municipalities did the selective collection, which accounts for only 22% of Brazilian cities [10].

Whether they work alone or as members of cooperatives, waste pickers activity plays a great role, approximately 90% of recycling collection in Brazil [11], which means 1.7 million tons/year [12], with a significant portion of tailings, which turns out approximately 1.05 million tons/year the amount effectively recovered [13].

Brazil suffers from poor recycling numbers. According to ABREN (Brazilian Association of Energy Waste Recovery), a member of the International Solid Waste Association (ISWA), it is lost around 2.8 billion dollars annually in the treatment of diseases caused by the contact of Brazilians with MSW improperly discharged [14].

2 Methodology

This work on plastic waste is the result of a study started months ago involving professor and student in scientific research on Supply Chain Management in Reverse Logistics. As the first source for learning this subject, the student read the teacher's doctoral thesis about a qualitative discussion of how the physical characteristics of scrap tires could interfere with the efficiency and effectiveness of this material supply chain for recycling or incineration.

After reading, which served to give the student a greater insight into guidelines for assertive management of Municipal Solid Waste (MSW), the student had to choose what type of residue he would like to approach in the coming months of the research, in a more complete and conclusive study on other aspects besides the physical ones that could interfere in the performance (efficiency with effectiveness) of a logistics system that collects, separates, recycles and/or incinerates dry solid waste.

The student recalled that he could approach plastics, both because of the wide variety they have of chemical components – recycling rates are available from number 1 (simpler recycling) to 7 (more difficult to recycle) – as well as because it is a matter of urgent attention of Brazil. That is the increase in the production of plastic waste that accumulates in inadequate places of municipal territories, like open dumps.

The student then sought works in recent literature, in the range from 5 years ago to the current year, that brought information on the management of plastic waste in other countries – success management and trends of governments and private companies in the short and medium-term – that could inspire a great improvement for the current Brazilian situation.

After finding the results and cataloging them, the student started the reading phase of the mathematical theory involved in the Partial Least Square Structural Equation Modelling (PLS-SEM) and consulted an illustrative problem in the software SmartPLS 3 [15]. With the knowledge acquired, the student proposed in consultations with the guiding teacher, a formative structure model involving indicators and constructs (latent variables). According to the theory in [16], the determination of the structural equation with its coefficients should be after the application of a survey whose answers obtained from the respondents are used as samples for the calculation of the statistical significance of the elements themselves – and between them – of the proposed structural model, detection of possible problem of collinearity, among other quality aspects.

3 Plastic Waste Management - Recent Data, Success Cases, and Trends

3.1 MSW Data in the World

Regarding the production, use, and post-use destination of synthetic plastics, it is necessary to emphasize that, in any year, the fraction of each type of residue presents a difference in comparison to the percentage of each type of plastic produced. This is explained by the fact that different sectors of economic activity consume different polymeric mixtures with their respective service lives. To exemplify: in 2017, 71 million tons – 16% of all global plastic production (resin types, fibers, or additives) – were used by construction activity, but only 4% (about fourteen million tons) of global plastic waste was generated by it [17].

It is concluded that the components of the construction have a long service life. On the other hand, packaging was 158 million tons (36% of the total) of the plastic produced in that same year, but they constituted 46% (approximately 152 million tons) of the total waste, proof of the short service lives of most of these products.

Also in 2017, from the global total of 380 million tons of plastic waste generated, about 18.4% were recycled, 26.3% incinerated, and 55.3% discarded [18].

3.2 Success Cases

Successful cases in the management policy of MSW are already seen in countries of the European community, part of the world population that has historically been the most concerned about the sustainable aspect of economic activities on the planet. Thus, inspiration on the political and scientific aspects can be conducted in Brazil based on the progress reported in the following cases.

Regarding the production and recycling of plastic packaging by the UK industry, there has been considerable progress since 2000: the consumption of raw materials reduced by one third, also a drop of 85% in waste landfills, and a 70% reduction in the carbon emitted; all three in comparison to the 1990s [19]. Also in the UK, in 2016, the recycling rate of plastic bottles was 58% (343.33 million tons); the recycling rate of plastic pots, vats, and trays reached 32% (169.145 million tons), according to [20].

Also in 2016, but in the European Union (EU), the percentage of recycled plastic packaging was close to 41% of the total packages, much above the 22.5% demanded by its commission, according to [21].

The Deposit Return Scheme (DRS) consists of machines that charge consumers of bottled beverages a cash deposit when they are purchased. The deposit is retrieved when they return the bottle. This scheme in European countries has been successful, with values even exceeding 95% for the recycling rate, when home recycling is also included. In the UK, plastic bottles themselves have a recycling rate of 70%. About four million units (the other 30%) end up in destinations other than recovery [18].

3.3 Future Trends

Despite advances in the management of MSW such as those reported above, much more action is needed to achieve the goals set in climate change meetings, where treaties such as the Paris Agreement have been signed. Part of the measures, which appear in [18], are goals of governments and companies that outline sustainable trends in the management of MSWs for the planet's future:

- European industry forecasts a drop in the rate of waste destined to landfills, being this rate 10% at most, by 2035;
- The EU also targets the recycling rate of 65% of plastic packaging by 2025 and 70% by 2030;
- The French government has plans to achieve a 100% in plastic recycling rate within 7 years.

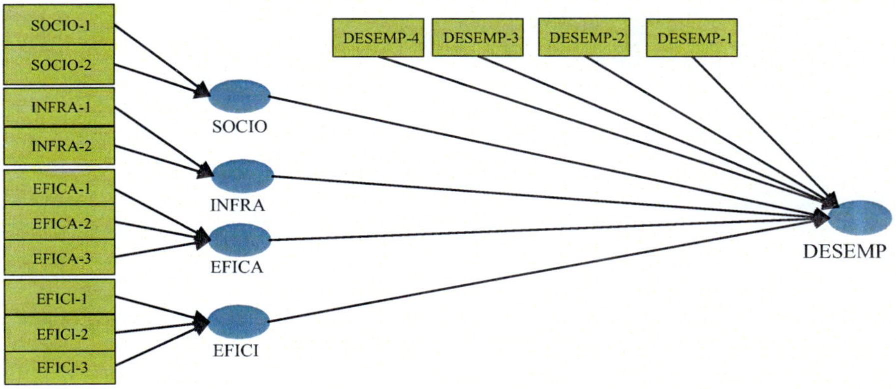

Fig. 1 Formative structural scheme produced in SmartPLS 3 software showing each indicator (in yellow) with its respective construct (in blue)

4 Mathematical Modeling by Structural Equations for MSW Management in Brazil

The modeling – by student and professor – of a high-performance Reverse Logistics system that collects, separates, processes, and recovers materials from plastic waste in Brazil has achieved considerable progress in SmartPLS 3 software, which uses the Partial Least Square-PLS method. Such progress is confirmed by model quality indices calculated when the simulation had been run.

The model in a formative scheme – from indicator to construct (see Fig. 1) – so far, has 14 indicators and 5 constructs. The Performance or Effectivity (DESEMP) construct is what one wants to assess the dependence concerning the other constructs and indicators. The other four constructs, in sequence, are the Reverse Logistics System Efficiency (EFICI), RLS Effectiveness (EFICA), Municipality Socioeconomic Aspects (SOCIO), and Municipality Infrastructure (INFRA).

Six out of the fourteen indicators were suggested by the doctoral thesis of the guiding professor [22]: EFICI1-complexity of waste, EFICI2-variety of waste (types of plastic: PET, HDPE, LDPE, PP, PS, PVC, or PUR...), EFICI3-variability of waste, EFICA1-market maturity, EFICA2-value of waste, EFICA3-volume processing. All the indicators are shown in Table 1.

5 Simulation

Until the preparation of this paper, the research team was able to simulate the system performance using randomly generated answers (see Tables 2 and 3) – through the

Table 1 Constructs and indicators for the structural model

Constructs (Latent variables)	Indicators
Reverse Logistics System (RLS) Efficiency	EFICI1-complexity of waste
	EFICI2-variety of waste (types of plastic: PET, HDPE, LDPE, PP, PS, PVC, or PUR...)
	EFICI3-variability of waste
RLS Effectiveness	EFICA1-market maturity
	EFICA2-value of waste
	EFICA3-volume processing
Performance (Effectivity)	DESEMP1-recycling rate
	DESEMP2-thermochemical conversion rate
	DESEMP3-business profitability
	DESEMP4-degree of plastics separation technologies
Municipality socioeconomic aspects	SOCIO1-socioeconomic profile of the municipality
	SOCIO2-population density of the municipality
Municipality infrastructure	INFRA1-degree of the presence of selective collection in the municipality
	INFRA2-presence of Deposit Return systems in the municipality

Table 2 Statistics on the randomly generated answers for each survey item. The abbreviation "Kurt." stands for Kurtosis and "Skewn." for Skewness

	No	Mean	Median	Min	Max	Standard Deviat.	Excess Kurt.	Skewn.
EFICI-1	1	3.083	3.000	1.000	5.000	1.308	−1.175	0.163
EFICI-2	2	2.967	3.000	1.000	5.000	0.001	−1.458	−0.008
EFICI-3	3	3.050	3.000	1.000	5.000	1.465	−1.470	−0.056
EFICA-1	4	2.933	3.000	1.000	5.000	1.424	−1.356	0.049
EFICA-2	5	3.067	3.000	1.000	5.000	0.001	−1.094	−0.079
EFICA-3	6	3.033	3.000	1.000	5.000	1.390	−1.232	−0.252
DESEMP-1	7	3.117	3.000	1.000	5.000	1.473	−1.431	−0.046
DESEMP-2	8	3.083	3.000	1.000	5.000	1.394	−1.236	−0.001
DESEMP-3	9	2.900	3.000	1.000	5.000	1.411	−1.290	−0.074
DESEMP-4	10	3.050	3.000	1.000	5.000	1.465	−1.376	−0.056
SOCIO-1	11	2.617	3.000	1.000	5.000	1.392	−1.337	0.228
SOCIO-2	12	2.967	3.000	1.000	5.000	0.001	−1.400	0.059
INFRA-1	13	3.017	3.000	1.000	5.000	1.443	−1.328	−0.098
INFRA-2	14	3.167	3.000	1.000	5.000	1.529	−1.491	−0.174

use of the Stat Trek website [23] – within the 5-point Likert Scale [24] of alternatives 1 up to 5 as to the degree of agreement that the evaluated item (indicator) contributes positively to the performance of the collection, separation, processing, and recovery of plastic waste.

Table 3 Random answers given to the questionnaire by 60 simulated respondents. Only answers from 10 people are shown for brevity

EFICI-1	EFICI-2	EFICI-3	EFICA-1	EFICA-2	EFICA-3	DESEMP-1	DESEMP-2	DESEMP-3	DESEMP-4	SOCIO-1	SOCIO-2	INFRA-1	INFRA-2
4	4	1	4	5	5	5	3	2	5	1	5	5	3
4	1	2	5	2	4	5	2	1	3	4	1	1	5
2	4	5	5	2	5	5	5	4	2	4	5	5	2
3	4	3	4	4	3	3	2	2	5	4	3	4	2
5	2	1	1	4	4	5	3	2	4	3	4	1	5
5	5	4	4	1	4	3	2	4	3	4	1	5	1
2	1	1	3	4	5	5	5	3	4	1	4	3	3
5	5	1	1	5	4	3	1	1	2	1	4	5	4
4	5	5	1	3	5	2	4	4	2	1	1	3	5
2	1	1	2	5	1	5	2	1	3	1	3	5	5

Sixty was the sample size chosen for the simulation of the answers to the questionnaire, which was chosen with support from the "10-times rule", which states that the minimum sample size should be equal to ten times the number of construct indicators with the most indicators [16]. It would then be forty the minimum sample size, but it has opted for sixty because it is a closer size that is intended to work on the real delivery of the questionnaire. Also, the following standard simulation parameters were maintained: Initial Weights: 1.0; Maximum number of iterations: 300; Stop criterion (10^-X): 7; not use Lohmoeller settings; Weighting scheme: Path.

It is shown in Table 3 the answers from the first 10 respondents. The 5-point Likert Scale contains the answer alternatives: 1 - Strongly disagree, 2 - Disagree, 3 - Neither agree nor disagree, 4 - Agree, 5 - Strongly agree.

6 Results

In the above conditions, path coefficients were calculated (see Fig. 2). The results show the good preliminary quality of the model in this initial study with random answers.

Almost all outer weights showed positive values (see Table 4), which is already a good start as only randomly generated answers are being used, which will be replaced by real answers from the interviewees in the next step of this study, in September (with final results to be documented later). Table 4 reveals which indicators are most relevant in this case of simulated answers.

The values for Effect size (see Table 5) show that all exogenous (predictor) constructs have a relevant effect – because they are higher than 0.020 – in explaining the results of the endogenous one, that is, how much an exogenous construct

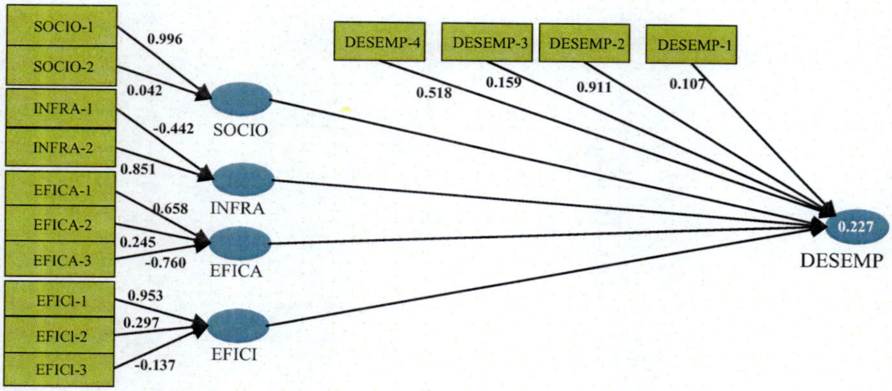

Fig. 2 Values obtained post-simulation that relate each indicator to its respective construct

Table 4 Outer weights obtained after simulation

	DESEMP	EFICA	EFICI	INFRA	SOCIO
DESEMP-1	0.107				
DESEMP-2	0.911				
DESEMP-3	0.159				
DESEMP-4	0.518				
EFICA-1		0.658			
EFICA-2		0.245			
EFICA-3		−0.760			
EFICI-1			0.953		
EFICI-2			0.297		
EFICI-3			−0.137		
INFRA-1				−0.442	
INFRA-2				0.851	
SOCIO-1					0.996
SOCIO-2					0.042

Table 5 Effect size values obtained after simulation

Effect size (f^2)	
	DESEMP
EFICA	0.048
EFICI	0.024
INFRA	0.026
SOCIO	0.121

contributes to the R^2 value of the target construct Performance or Effectivity (DESEMP). It is worth remembering that these values have an illustrative effect because the simulation was done, at this step of the work, using randomly generated answers following the 5-point Likert Scale.

The indices that detect potential collinearity between indicators are all adequate with values between 1.004 and 1.081 (see Table 6 for Outer VIF), which is another good finding for this step of random answers – only values under 1.000 or from 5.000 to above would indicate a problem, according to [25].

All Inner VIF values are also higher than 1 (Table 7), with values between 1.017 and 1.096.

R^2 (coefficient of determination) values of 0.20 are already considered high in studying consumer behavior [16] for example, because it involves many subjective and personal issues. Although the theme studied in this work allows more objective and direct assessment due to the presence of physical issues of the MSW and their recycling rates, it is still unclear the degree of strength with which market, socioeconomic issues of the population, and municipality infrastructure influence the performance of a system that collects, separates, processes and recovers plastic waste. The values displayed in Table 8 are not yet as high as desired, but it reflects the simulation with randomly generated answers.

Table 6 Outer VIF values obtained after simulation

Outer VIF (Variance Inflation Factor)	
DESEMP-1	1.050
DESEMP-2	1.081
DESEMP-3	1.044
DESEMP-4	1.053
EFICA-1	1.007
EFICA-2	1.042
EFICA-3	1.036
EFICI-1	1.020
EFICI-2	1.006
EFICI-3	1.024
INFRA-1	1.012
INFRA-2	1.012
SOCIO-1	1.004
SOCIO-2	1.004

Table 7 Inner VIF values obtained after simulation

Inner VIF	
	DESEMP
EFICA	1.088
EFICI	1.096
INFRA	1.017
SOCIO	1.047

Table 8 R^2 and R^2 adjusted

	R Square	R Square Adjusted
DESEMP	0.227	0.171

The proposed structural model is concise. According to [16], the more paths pointing toward a target (endogenous) construct, the higher its R^2 value; and for the evaluation of a structural model as good, it cannot be only used the values of R^2 and R^2 adjusted (that considers the far minimum structure necessary for the system model).

7 Conclusion

The progress of this study has already been noticed and documented. Once the final structural model and equation are found – proposing a high-performance Supply Chain Management – it is expected to be very significant: the reduction in the consumption of oil for virgin plastic production, the fall in greenhouse gas emissions, and the reduction of plastic waste inadequately destined in dumps – large methane emitters that accelerate global warming. The reduction of production costs is very likely to happen because no longer such high oil quantity will have to be used – a commodity that has had its price very high lately – to produce plastics, as recycling will be much more prominent than before.

The manufacturing of products never seen before is made possible by innovation combined with sustainability and can serve as a source of extra income. Patents can be generated and filed as recognition of the value of intellectual property in the search for solutions to people's needs on a business model advantageous to all parties.

Last but not least, it will be reached the generation of more jobs and the boosting of municipal, state, and national economies through higher recycling productivity, carefully observing the conditions of each company to achieve the best results: processing capacity, quality of manufactured products, proximity to the consumer, low production and delivery costs, competitive price in the market.

The model being developed may serve as a mirror to benefit other countries with still precarious MSW management infrastructure that presents similarities to the current situation in Brazil.

8 Future Developments

The next step in this study is to perform more random answer rounds simulating 60 respondents for each of the 14 questions. Try Bootstrapping method [16], which assesses the model reliability, significance, and quality in case of the set of answers does not behave as a normal distribution (bell shape). Also, Q^2 (Blindfolding Predictive and Relevance) [16] may be assessed. Then, observe the results from the software for a more reliable evaluation of the questionnaire consistency. From there, the questionnaire will be delivered to professionals in September (with final results to be documented later), and then the results will be examined definitively.

The structural model and equation found through data collected from respondents' answers will be used to develop case studies of improvement in companies with social impact, but it is also possible to use the model for the benefit of public or private management - collection, separation, processing, recovering - of solid waste, not only to the extent of a municipal approach but also expand it to state and national levels.

Prospecting and visits offered voluntarily by the research team for the case studies are thought to interact with managers and provide complementary knowledge on how to manage the Logistics and Supply Chain of their business to increase operational and financial performance.

It is very important to encourage the increase in research on MSW management not only in Brazil but all over the world. Many projects that boost the economy with sustainability, through the Circular Economy, can be developed. It is not ruled out the future expansion of this work to other MSWs materials – such as metals and glass – if there are stakeholders and mobilization by other researchers that can form partnerships with the research team author of this paper.

References

1. Brazilian Association of Special Waste and Public Cleaning Companies – ABRELPE.: Recommendations for solid waste management during the coronavirus pandemic (2020) https://abrelpe.org.br/download/3347. Last accessed 2022/03/30
2. De Aguiar, L.K., Manning, L.: In: Aktas, E., Bourlakis, M. (eds.) The Role of Informal and Semi-Formal Waste Recycling Activities in a Reverse Logistics Model of Alternative Food Networks BT - Food Supply Chains in Cities: Modern Tools for Circularity and Sustainability, pp. 145–169. Springer International Publishing, Cham (2020)
3. Kaza S, Yao L.C., Bhada-Tata P., Van Woerden F.: What a Waste 2.0: A Global Snapshot of Solid Waste Management to 2050 (2018). https://openknowledge.worldbank.org/bitstream/handle/10986/30317/9781464813290.pdf?sequence=13&isAllowed=y. Last accessed 2022/03/30
4. Alves, R., Ferreira, K.L.A., Lima, R., da Moraes, S.F.T.F.: An action research study for elaborating and implementing an electronic waste collection program in Brazil. Syst. Pract. Action Res. **34**, 91–108 (2021)
5. Bassin, I.D., Santana, F.L., Alves, I.R.F.S.B.J.P., Bassin, J.P.: Public policies and initiatives for waste minimization and recycling in Brazil. In: Singh, P., Milshina, Y., Tian, K., Borthakur, A., Verma, P., Kumar, A. (eds.) Waste Management Policies and Practices in BRICS Nations, 1st edn, (2021)
6. Ottoni, M., Dias, P., Xavier, L.H.: A circular approach to the e-waste valorization through urban mining in Rio de Janeiro, Brazil. J. Clean. Prod. **261**, 120990 (2020)
7. Demajorovic, J., Augusto, E.E.F., De Souza, M.T.S.: Reverse logistics of E-waste in developing countries: challenges and prospects for the Brazilian model. Ambient e Soc. **19**, 117–136 (2016)
8. Mancini, S.D., de Medeiros, G.A., Paes, M.X., et al.: Circular economy and solid waste management: challenges and opportunities in Brazil. Circ. Econ. Sustain. **1**, 261–282 (2021)
9. Oliveira, I.A.: Environmental justice and circular economy: Analyzing justice for waste pickers in upcoming circular economy in Fortaleza, Brazil. Circ. Econ. Sustain. **1**, 815–834 (2021)
10. Business Commitment to Recycling - CEMPRE.: Cempre Review: Selective collection profle, p. 24–25 (2019) https://cempre.org.br/wp-content/uploads/2020/11/CEMPRE-Review2019.pdf. Last accessed 2022/03/30
11. Lima, P.D.M., Colvero, D.A., Gomes, A.P., et al.: Environmental assessment of existing and alternative options for management of municipal solid waste in Brazil. Waste Manag. **78**, 857–870 (2018)
12. BRAZIL. Ministry of Regional Development. National Secretariat of Sanitation – SNS: National Sanitation Information System: diagnosis of municipal solid waste management - 2018. In: Brasília SNS/MDR, p. 247 (2019) http://www.snis.gov.br/downloads/diagnosticos/rs/2018/Diagnostico_RS2018.pdf, last accessed 2022/03/31
13. Penteado, C.S.G., de Castro, M.A.S.: Covid-19 efects on municipal solid waste management: what can efectively be done in the Brazilian scenario? Resour. Conserv. Recycl. **164**, 105152 (2021)
14. Brazilian Association for Energy Recovery of Waste – ABREN.: Institutional plan, March 2020. Brasília-DF, Brazil, 55:1–57 (2020). https://abren.org.br/wp-content/uploads/2020/06/Institutional-Plan-ABREN-2020-mar2020.pdf. Last accessed 2022/04/01
15. SmartPLS.: SmartPLS is here - Apply PLS-SEM Method in Minutes, With No Headaches! Available in: https://www.smartpls.com/. Last accessed 2022/03/12
16. Hair, J.F., et al.: A Primer on Partial Least Squares Structural Equation Modeling (PLS-SEM), 2nd edn. SAGE Publications, Thousand Oaks, CA (2017)
17. Rudolph, N., et al.: Understanding Plastics Recycling Economic, Ecological, and Technical Aspects of Plastic Waste Handling. Carl Hanser Verlag, Munich, Germany (2017)
18. Letcher, T.: Plastic Waste and Recycling: Environmental Impact, Societal Issues, Prevention, and Solutions. Academic, San Diego (2020)

19. HM Government.: Our waste, our resources: a strategy for England. pp. 1–146 (2018) Available in: https://assets.publishing.service.gov.uk/government/uploads/system/uploads/attachment_data/file/765914/resources-waste-strategy-dec-2018.pdf. Last accessed 2022/03/02
20. RECOUP points to need for investment in kerbside recycling. Letsrecycle.com, Oct. 12. 2017. Available in: https://www.letsrecycle.com/news/recoup-points-to-need-for-investment-in-kerbside-recycling/. Last accessed 2022/03/03
21. Plastics Europe.: Plastics-the Facts 2018: An analysis of European plastics production, demand and waste data. Available in: https://plasticseurope.org/knowledge-hub/plastics-the-facts-2018/ . Last accessed 2022/03/02
22. Freires, F.G.M.: The influence of physical components and cooperation between actors on the performance of reverse logistics systems: the case of scrap tires. Thesis (PhD in Management and Industrial Engineering), Faculty of Engineering. University of Porto, Porto, 253 p (2007)
23. STAT TREK.: Teach yourself statistics - Random Number Generator. Available in: https://stattrek.com/statistics/random-number-generator.aspx. Last accessed 2022/03/10
24. Likert Scale Response Options.: Mount Wachusett Community College Gardner, Massachusetts, USA. Available in: https://mwcc.edu/wp-content/uploads/2020/09/Likert-Scale-Response-Options_MWCC.pdf. Last accessed 2022/03/11
25. Hair, J.F., et al.: PLS-SEM: indeed a silver bullet. J. Mark. Theory Pract. **19**, 139–151 (2011)

Use of QFD to Prioritize Requirements Needed for Supplier Selection in an O&G Project

Sara Marques Oliveira de Araújo Souza, Cristiano Hora de Oliveira Fontes, and Francisco Gaudêncio Mendonça Freires

Abstract Due to the constant need to establish methods in the supplier selection process to avoid delays, reworks, cost increases and stakeholder discontent in projects in the Oil and Gas (O&G) sector, it is necessary to evaluate multi-criteria models and tools to prioritize both the industry regulations and strict technical specifications. This prioritization provides robustness in the construction of procurement contracts as well as effectiveness in the analysis of the technical proposal by engineering. The present study presents the use of the QFD (Quality Function Development) model in an O&G project in Bahia to build a validation system for high pressure fluid tests, simulating the pre-salt layer. The study aimed to prioritize the requirements to be met by suppliers of Service, Material, Equipment and Instruments from the design phase to the construction and delivery of the project. As a main result, the use of the model provided the integration between different disciplines in the group decision-making process, quantification of the team's judgments in numbers, understanding of the main risks between project requirements and establishing a technical proposal validation plan considering the requirements resulting priorities, enabling greater assertiveness for acquisitions and technical validations throughout the project life cycle.

Keywords QFD · Criteria · Supplier Selection.

1 Introduction

Due to the Oil usability in the world, in addition to being a fuel source for various chemicals such as solvents, plastics and medicines [1], the Oil and Gas (O&G) sector encompasses several systems necessary for exploration, such as the drilling and production process. Thus, the need to validate equipment of the Oil production

S. M. O. de Araújo Souza (✉) · C. H. de Oliveira Fontes · F. G. M. Freires
Polytechnic School, Federal University of Bahia, Salvador, BA, Brazil
e-mail: cfontes@ufba.br; francisco.gaudencio@ufba.br

V. M. López Sánchez et al. (eds.), *Industrial Engineering and Operations Management*, Springer Proceedings in Mathematics & Statistics 400,
https://doi.org/10.1007/978-3-031-14763-0_14

system, in order to avoid accidents and test the reliability of the system in an integrated way, is of great importance, as it will make it possible to recognize the productivity of the system in relation to infrastructure and interaction with the environment.

In this way, the supplier selection process becomes an important issue for decision making because the Oil and Gas industry requires compliance with norms and standards to ensure safety and operational quality. Another important factor is the technical requirements that must be met in any situation, simulating the pre-salt environment, with specific pressure, flow and temperature, requiring specialized instruments, equipment, materials and services. To ensure compliance with the requirements, the acquisition process needs to be assertive. How to guarantee that the project requirements will be met in the acquisition process, associating the team's experience, reasoned technical arguments and tradeoff analysis?

The selection of suppliers is one of the main activities in the sector, as it can directly impact the entire chain of product development and project operation. However, identifying which suppliers are suitable for each procurement requires prioritizing relevant criteria [2, 3].

Regarding methods aimed at supplier selection criteria, There is a growth in the classification methods, as Natural Language Processing [4, 5], Data Envelopment Analyses [6, 7] clustering techniques using C-means approaches [8, 9], Decision Tree [10–12], and Multi-Criteria Decision Analysis [13–20]. These methods can be applied in supplier chain classification involving projects associated with the Oil and Gas sector and other techniques such as Neural Networks, AHP (Analytic Hierarchy Process), KANO (Noriaki Kano) and QFD (Quality Function Development) can be applied together.

It was identified that in the last decade, issues related to the supply chain (Supply Chain Management, SCM) received priority in both business and academic environment. This fact is related to the need to assess the risks associated with the selection process of suppliers and not just the conventional cost criterion. This requires a multi-criteria decision-making (MCDM) approach since it involves conflicting and competing choices, or tradeoffs [13]. A supplier can provide a service or product of low quality but with a better delivery time, while another supplier may provide an unreliable delivery time, ensuring high quality of its delivery [21].

Some MCDM methods, applied to Oil and Gas Industry, have contributed to address the uncertainty of information using appropriate techniques such as fuzzy logic (among them QFD, Quality Function Development) [2, 22–24], Gray Systems Theory [25] and the combination of fuzzy and gray methods [26]. Assessing conflicting choices or tradeoffs has been shown to be of great importance and MCDM methods provide an efficient strategy for decision making [13, 16, 18, 19, 27–31].

This work applies QFD in the Oil and Gas sector since this technique is capable of considering the preferences of decision makers and the weights associated with the evaluation criteria in the supplier selection process [3, 13, 24, 32, 33]. In addition, QFD uses a planning tool that includes two interrelated matrices which are obtained by converting the absolute weight to a percentage of each criterion.

The motivation to use QFD in this study was mainly due it's applicability in the innovation project, considering the professional experience and engineering requirements. This work also required knowledge of the equipment, materials and services applicable to operate in extreme conditions of temperature, pressure and flow.

QFD has become a powerful planning tool, as it integrates analysis, control and optimization, being widely used by American companies [2, 34]. The main functions of QFD are aimed at in reducing design changes, shortening the innovation project development cycle, improving the project quality and promoting customer satisfaction. QFD can be applied either for new products, improvement of complex existing products or for service and process management such as optimizing and reducing the operating cost of producing a Toyota vehicle [2, 35].

This study presents an application of the QFD method for the selection and prioritization of criteria and requirements in the supplier selection process in a development and construction project associated with the Oil and Gas sector (Bahia, Brazil)., The application of the method focused on the planning stage of the acquisition of instruments, equipment, materials and services, called here in this study by "items". The paper is organized as follows: Sect. 2 presents the case study; Sect. 3 presents the results and discussions and Sect. 4 presents the conclusion and potential benefits.

2 Case Study

In order to apply the QFD in the process of identifying criteria for the selection of suppliers, a project for the construction and commissioning of a concrete structure for the validation of equipment and instruments was considered, simulating the real conditions of the pre-salt with about 300 meters of depth in a company in Bahia, Brazil.

As it is a project that covers specific factors such as pressure, flow and specific fluid circulation conditions, a multidisciplinary systemic approach was necessary to understand the needs to be met. To understand the project needs and identify which requirements were directly related, the use of QFD was proposed.

As already mentioned, the QFD performs the entire process of implementing the quality function through a series of matrices [7, 19] As shown in Fig. 1, the QFD seems like a house, as a row and column matrix filled with quantified values in the center. The left column represents the attributes, demands of the customer, requirement of project or the factors that must be attended and the requirement of the product (What and How). For example, how to design products or services to meet specific demands. The quantitative relationship between "What" and "How" is represented as a central element of the QFD process. Another parts: (A) represents customer needs; (B) represents engineering features or methods; (C) represents

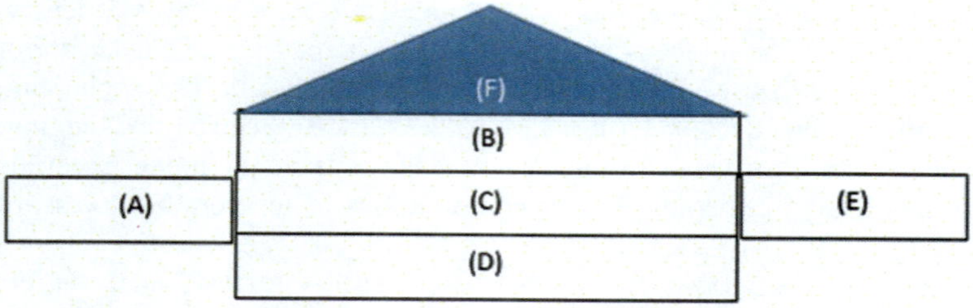

Fig. 1 QFD Features, Adapted Li, 2021

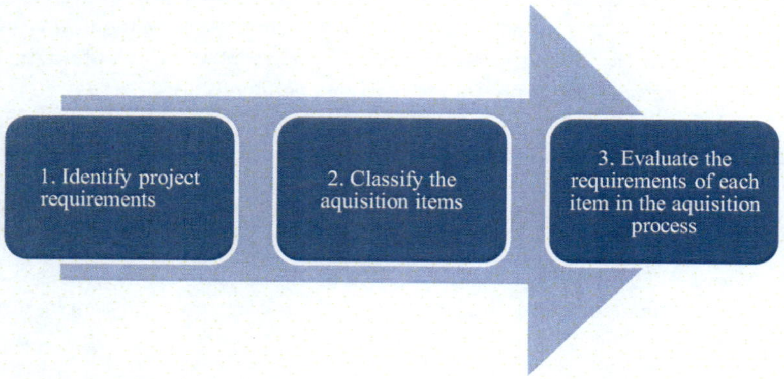

Fig. 2 Steps for using QFD in the real project - Case Study

the Matrix of relationship between "What" and "How"; (D) indicates the level of importance of expectations; (E) represents the market study about the needs presented; (F) represents the relationship between engineering characteristics and needs [4]. In this study, the part "E" was not considered.

To apply the QFD in the proposed project, the following process was considered, as shown in Fig. 2:

First step – Identitify Project requirements – was necessary understand with all the team all project requirements, defined by customers and stakeholders. All the requirement was listed and was the guide to the project scope;

Second step – Classify the aquisition items – Was important classify the acquisition in parts, called item, to understand the difficult level of purchase. For example, forged metal of heat exchangers need to attend specific standard due pression operation and the supplier must to be certified in specific standard. So, was classified the items in: equipment, instrument, service or material necessary to build and assembly the project;

Third step - Evaluate the requirements of each item in the acquisition process – was important found the factors that should be a negative influence on the process, for example, the material resistance or the fluid viscosity on the equipment.

Table 1 Description of weights according to item to be acquired in the Project

Item	Classification
Material	Low difficulty - 2
Service	High difficulty - 4
Equipment / Instrument	High difficulty - 4

Thus, having established the criteria and needs, the QFD was filled in the "What" field. Considering the structure of the QFD already described in Fig. 1, were assigned weights to items and notes to needs of this item.

The weight criterion had as its main factor the difficulty of acquisition, varying between 1 and 2, low difficulty, 3 as medium difficulty and 4 as high difficulty. The greater the specification of the item, for example, need to attend the specific standard or material, the greater the level of difficulty to be considered.

Item weights were assigned according to Table 1.

For this classification, the material, as it is already used in the national and international market, does not constitute a substantial difficulty for the project. However, the service in the O&G sector needs to be highly qualified in terms of operational safety, mainly. As well as the equipment and instruments, specific to meet high pressures, high temperatures and high flow level, it was considered with weight 4, high difficulty.

According to the classification of the items, the next step was to understand the needs to be met for each one and assign them a grade. The scores varied between 3, 6 and 9 considering the level of importance of the need in relation to the item.

Thus, the classification of each item in relation to weight and grade can be verified according to Table 2.

After the complete classification of the needs of each item, the requirements were assigned, that is, technical specifications of "how" it must be met and filled in the QFD worksheet. For each requirement listed, the trend (if higher or lower) and the unit of measure for each requirement were assigned, as shown in Fig. 3.

As with all the needs and requirements listed in the QFD, the degree of relevance between each need in relation to the requirement was evaluated. This step sought to classify which are the most important requirements for the selection of project suppliers, according to the scale:

- 9 (ou 6) - Strong relationship
- 6 (or 3) - Average ratio
- 1 (or 1) - Poor relationship
- Blank if not related.

The varieted scale makes it easier to highlight which principles will be most relevant. In Fig. 4, a clipping of the QFD can be seen containing the Needs, Requirements and grades assigned by the project's multidisciplinary team.

The multidisciplinary team was composed for 15 engineers specialists in industrial projects (mechanical, process and chemical engineer, electrical, automation, instrumentation, metrology and confiability). The team was consulted in group as

Table 2 Rating and Need Grade of each Item

Item	Need	Classification	Grade
Material	Regulated registration	2	9
	MSDS (Chemical Product Information Sheet)	2	6
	Durability	2	3
	Security Criteria	2	9
	reliability	2	9
Service	Skilled Labor	4	9
	Market experience	4	6
	Agility	4	9
	reliability	4	9
	Punctuality	4	9
	Flexibility	4	6
	Organization	4	6
	technical capacity	4	6
	Quality Programs	4	9
Equipment / Instrument	reliability	4	9
	Meeting technical requirements	4	9
	Functionality	4	9
	Modularity	4	3
	Easy Maintenance	4	6
	Lowest price	4	6
	Easy Operation and Assembly	4	3
	Compliance with legal requirements	4	9
	Compliance with security requirements	4	9

Regulation of the material/Equipment/Service	Nº Variety of parts	Fail-safe assembly (Poka Yoke)	Market components	Mass of each part	Dimensions of each part	Nº of tools for assembly	Nº of assembly operations	Mass of the set
quantity	quantity		quantity	Kg	M	quantity	quantity	Kg
↑	↓	↑	↑	↓	↓	↓	↑	↓

Fig. 3 Exemple of tendence and requirement unit

meeting. The requirements were collected by scope declaration approved by project customer. When divergent notes, the engineer must justify the argument to the team to keep the note.

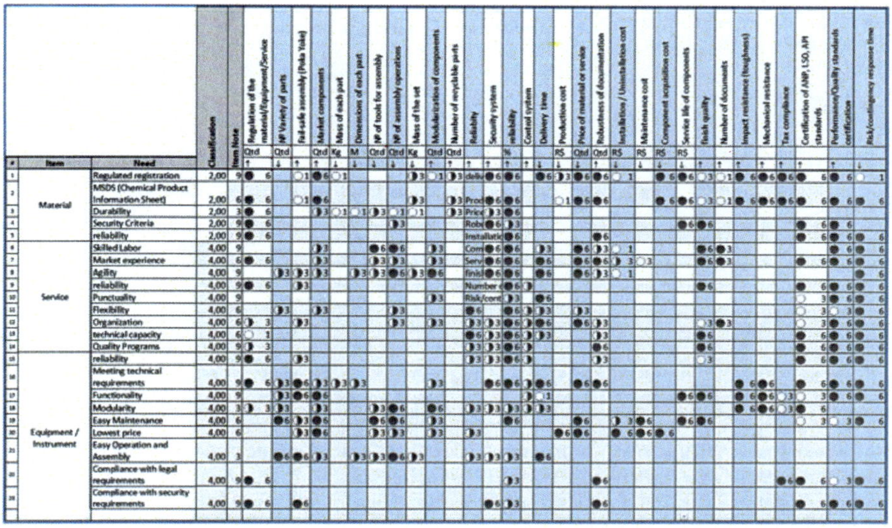

Fig. 4 QFD clipping containing the Needs and Requirements per Item to be acquired in the project

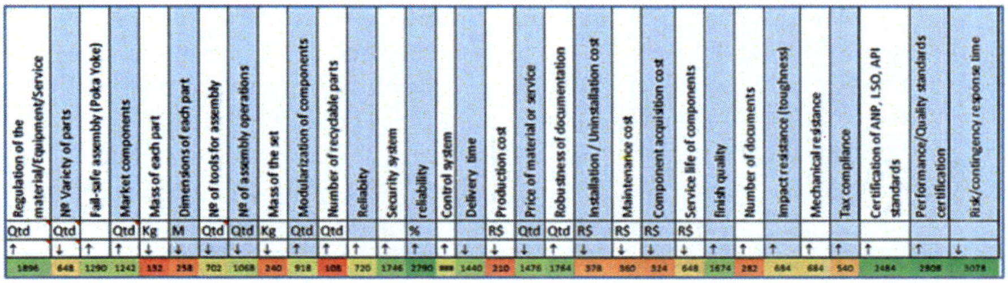

Fig. 5 Selection of requirements according to QFD

3 Results and Discussions

Thus, using the assigned value considering the grade and weight of each need, it was possible to understand which are the most important requirements that should be prioritized in the supplier selection process, as shown in Fig. 5.

Was considered first ten most important due great relevance between Need by Item as represented in the Table 3.

In this work, the QFD method was used to select the main criteria to be considered in suppliers of the O&G sector, in the process of acquiring necessary items in the construction and assembly process (instruments and equipment, services and materials).

Table 3 Ranking of the 10 requirements selected by the QFD

Ranking	Selected requirements
1	Risk Response Time
2	Performance / Quality
3	Certification of API, ANP and ISO standards
4	reliability
5	Material/Equipment/Service Regulation
6	Security system
7	Robustness of documentation
8	Finish quality
9	Price of material or service
10	Service response time

QFD was used to structure and translate item needs into requirements and standards (for example, American Petroleum Institute, Petroleum National Agency), as a key criterion in the supplier selection process.

Since this is a project with a high level of risk due to high pressure at a great depth of operation, whose objective is to validate the quality of fluid tests, the priority criteria resulting from the QFD were:

1. Risk Response Time configures the first criterion selected – considering the risks inherent to the operation, environment, safety of people and assets, suppliers need to be increasingly prepared with a contingency plan and risk mitigation. Thus, it is necessary to present the regulated procedure to the risk plan provider and verify the mitigation plan to certification action plans. To ensure adequate treatment of the risks, was necessary atribute responsibility to contract supplier in the purchase contract.
2. The second most important selection criterion listed is Performance and Quality in the operation historically, such as experience in projects in the sector and level of success in deliveries. This criteria is related to the level of operational quality of the equipment, instrument and service to be purchased. Its important understand the market and the quality requisition most applied to Oil and Gas projects.
3. Certification in API, ISO and other standards in the sector - the standards and norms guide from the best practices of reliability and use of appropriate materials, hydrostatic test conditions and other tests, ideal conditions for validation of the equipment considering high pressures and flows to the desired organizational structure for operation;
4. Reliability configures assertiveness in the use of instruments, equipment within the tolerance limit established by engineering. The reliability process must be present into the conceptual project, acquisition, fabrication, delivery and comissioning tests. The supplier must provide evidence each cicle lyfe part, sharing information and risk mitigation plan.
5. The Regulation establishes the standard compliance of the provision of services in a legal way. With internal political and operation procedures, the supplier

must attend the legal regulation and provide future plan to maintain the improvement process.
6. The Security System confirms the existing concern not only of the concessionaire of the oil and gas sector, but of all who participate in the supply chain, whether by service or providing inputs, equipment and instruments. To attend this requeriment, the supplier must report the safety operation status and actions to keep the operation and management safety;
7. The robustness in the documentation results in more structured and systemic processes necessary for the high standard of equipment and services in the sector. It is also associated with the regulatory standards required to be able to supply;
8. Quality of the finish is important to attend the techinical requirment considering the correct function, fabrication and integration system. The inspection process can propouse ways to keep the quality in the equipment, service or material.
9. The price of the material or service for a long time was one of the main factors in supplier selection. Nowadays it is realized that having a competitive price is not enough. In addition to having a competitive price, the supplier needs to meet several factors that configure the success of the acquisition in the project;
10. Service response time – speaking of projects, time management is an important factor, always associated with the quality and reliability of the service to be provided.

In addition to listing the most important requirements for each item to be purchased, the QFD "roof" assesses conflicts or tradeoffs between requirements. This analysis made it possible to understand the risks between the requirements, and later, an action plan was identified for the most critical conflicts, such as the relationship between lower price and safety, since to understand the safety requirements it will be necessary to purchase valves control for each system in the project.

4 Conclusion

Considering the importance of developing and commissioning projects in the O&G sector with quality to meet the requirements of the National Agency of Petroleum, Gas and Biofuels, such as meeting regulatory and safety requirements, establishing a method of prioritizing criteria for selecting suppliers is fundamental. In this study, the engineering team used the QFD method to prioritize the criteria related to the items to be acquired for the construction of the work, in the acquisition planning stage. The items were divided by applicability: Equipment and Instruments, Service and Material.

The advantage of this method lies in the ability to transform the decision makers' verbal assessments into numbers and ranking, which are more accurate. One of the main benefits of using QFD, among others, in the planning stage of project acquisitions, is the reduction of rework and changes in the project, as the engineering and Supply Chain team will be more assertive in the search for suppliers in the market; Reduction of customer complaints, cost and time reduction; Increase in customer satisfaction; Reduction of the risk of non-compliance with contracts or guarantee clauses. The use of the method also strengthened the integration between engineers and other project disciplines due to constant brainstorming to identify important factors, impacts of the selection process in the medium and long term and the possibility of using QFD in other projects with the same objective, helping decision making and supplier selection.

The risk response time was the most voted requirement, and this requirement applies to all items, because, due to the complexity of the project in question, the system needs to meet high pressure, temperature and flow. The Supplier must consider a contingency plan throughout the process, from layout and design validation to post-delivery tryout.

This study provided: (1) integration between different disciplines (mechanics, chemistry, instrumentation, automation, safety, environment) in the group decision-making process; (2) quantification of various judgments in numbers; (3) understood the risks between requirements through the tradeoffs between all requirements; (4) it gave weight to the criteria of items that will be acquired by the project, such as equipment and instruments, services and materials; (5) establish a technical proposal validation plan considering the priority requirements resulting from the QFD, enabling greater assertiveness in the choice of supplier.

The use of QFD provided more accurate results in qualitative and subjective data, reducing the imprecision of the multidisciplinary team's judgments.

A limitation of QFD in a project of great complexity according to this study, is the fact that the project is still in the development phase, so it will need adjustments and increments in the items evaluated. Future applications may consider new criteria and attributes and already analyze the supplier base available both in the national and international markets, selecting the groups of suppliers that best meet the priority requirements of this study, thus reducing process risks throughout the entire process in the entire process resulting in value creation in the supply chain, from item classification, supplier selection, procurement monitoring and project commissioning as a whole.

References

1. Anderson, R.N.: Petroleum analytics learning machine' for optimizing the Internet of Things of today's digital oil field-to-refinery petroleum system. In: Proceedings in 2017 IEEE International Conference on Big Data (Big Data), p. 4542e4545 (2017, Dec)
2. Li, F., et al.: Rational supplier selection based on two-phase deep analysis considering Fuzzy QFD and Game Theory. Mob. Inf. Syst. **2021**, 1–12 (2021). https://doi.org/10.1155/2021/9979017
3. Karsak, E.E., Dursun, M.: Taxonomy and review of non-deterministic analytical methods for supplier selection. Int. J. Comput. Integr. Manuf. **29**(3), 263–286 (2016). https://doi.org/10.1080/0951192X.2014.1003410

4. Bo, Y., Meifang, Y.: Construction of the knowledge service model of a port supply chain enterprise in a big data environment. Neural Comput. & Applic. **33**(5), 1699–1710 (2021). https://doi.org/10.1007/s00521-020-05044-w
5. J, J.K: Mathematical modeling of Indian Petroleum Supply Chain. Int. Res. J. Modern. Eng. Technol. Sci. **07**, 2582–5208 (2020) www.irjmets.com
6. Afroz, Z., Shafiullah, G.M., Urmee, T., Higgins, G.: Modeling techniques used in building HVAC control systems: a review. Renew. Sust. Energ. Rev. **83**(February), 64–84 (2018). https://doi.org/10.1016/j.rser.2017.10.044
7. Amalnick, M.S., Habibifar, N., Hamid, M., Bastan, M.: An intelligent algorithm for final product demand forecasting in pharmaceutical units. Int. J. Syst. Assur. Eng. Manag. **11**(2), 481–493 (2020). https://doi.org/10.1007/s13198-019-00879-6
8. Yin, X.F., Khoo, L.P., Chong, Y.T.: A fuzzy c-means based hybrid evolutionary approach to the clustering of supply chain. Comput. Ind. Eng. **66**(4), 768–780 (2013). https://doi.org/10.1016/j.cie.2013.09.025
9. Keskin, G.A.: Using integrated fuzzy DEMATEL and fuzzy C: means algorithm for supplier evaluation and selection. Int. J. Prod. Res. **53**(12), 3586–3602 (2015). https://doi.org/10.1080/00207543.2014.980461
10. Zhang, Z., Han, H., Cui, X., Fan, Y.: Novel application of multi-model ensemble learning for fault diagnosis in refrigeration systems. Appl. Therm. Eng. **164**(July 2019), 114516 (2020). https://doi.org/10.1016/j.applthermaleng.2019.114516
11. Ramakurthi, V.B., Manupati, V.K., Machado, J., Varela, L.: A hybrid multi-objective evolutionary algorithm-based semantic foundation for sustainable distributed manufacturing systems. Appl. Sci. (Switzerland). **11**(14) (2021). https://doi.org/10.3390/app11146314
12. Serapião, A.B.S., Mendes, J.R.P.: Classification of petroleum well drilling operations with a hybrid particle swarm/ant colony algorithm. Lecture Notes in Computer Science (Including Subseries Lecture Notes in Artificial Intelligence and Lecture Notes in Bioinformatics), 5579 LNAI(January 2016), 301–310 (2009) https://doi.org/10.1007/978-3-642-02568-6_31
13. Kaviani, M.A., Karbassi Yazdi, A., Ocampo, L., Kusi-Sarpong, S.: An integrated grey-based multi-criteria decision-making approach for supplier evaluation and selection in the oil and gas industry. Kybernetes. **49**(2), 406–441 (2020). https://doi.org/10.1108/K-05-2018-0265
14. Kar, A.K.: A hybrid group decision support system for supplier selection using analytic hierarchy process, fuzzy set theory and neural network. J. Comput. Sci. **6**, 23–33 (2015). https://doi.org/10.1016/j.jocs.2014.11.002
15. Wu, C., Lin, Y., Barnes, D.: An integrated decision-making approach for sustainable supplier selection in the chemical industry. Expert Syst. Appl. **184**(May), 115553 (2021). https://doi.org/10.1016/j.eswa.2021.115553
16. Wood, D.A.: Supplier selection for development of petroleum industry facilities, applying multi-criteria decision making techniques including fuzzy and intuitionistic fuzzy TOPSIS with flexible entropy weighting. J. Nat. Gas Sci. Eng. **28**, 594–612 (2016). https://doi.org/10.1016/j.jngse.2015.12.021
17. Fagundes, M.V.C., Hellingrath, B., Freires, F.G.M.: Supplier selection risk: a new computer-based decision-making system with Fuzzy extended AHP. Logistics. **5**(1), 13 (2021). https://doi.org/10.3390/logistics5010013
18. Ristono, A.P., Santoso, P.B., Tama, I.P.: A literature review of criteria selection in supplier. J. Indus. Eng. Manag. **11**(4), 680 (2018). https://doi.org/10.3926/jiem.2203
19. Wang, C.N., Tsai, H.T., Ho, T.P., Nguyen, V.T., Huang, Y.F.: Multi-criteria decision making (MCDM) model for supplier evaluation and selection for oil production projects in Vietnam. PRO. **8**(2) (2020). https://doi.org/10.3390/pr8020134
20. Akhtar, M., Ahmad, M.T.: A stochastic fuzzy multi-criteria group decision-making for sustainable vendor selection in Indian petroleum refining sector. Benchmarking. (2021). https://doi.org/10.1108/BIJ-09-2020-0500
21. Dweiri, F., Kumar, S., Khan, S.A., Jain, V.: Designing an integrated AHP based decision support system for supplier selection in automotive industry. Expert Syst. Appl. **62**, 273–283 (2016)

22. Dursun, M., Karsak, E.E.: A QFD-based fuzzy MCDM approach for supplier selection. Appl. Math. Model. **37**, 5864–5875 (2013)
23. Wu, C., Barnes, D.: A literature review of decision-making models and approaches for partner selection in agile supply chains. J. Purch. Supply Manag. **17**(4), 256–274 (2011). https://doi.org/10.1016/j.pursup.2011.09.002
24. Bevilacqua, M., Ciarapica, F.E., Giacchetta, G.: A Fuzzy-QFD approach to supplier selection. J. Purchas. Supply Manag. **12**, 14–27 (2006). https://doi.org/10.1016/j.pursup.2006.02.001
25. Çelikbilek, Y.: A grey analytic hierarchy process approach to project manager selection. J. Organ. Chang. Manag. **31**(3), 749–765 (2018)
26. Wang, H., Duanmu, L., Lahdelma, R., Li, X.: A fuzzy-grey multicriteria decision making model for district heating system. Appl. Therm. Eng. **128**, 1051–1106 (2018)
27. Liou, J.J.H., Chang, M.H., Lo, H.W., Hsu, M.H.: Application of an MCDM model with data mining techniques for green supplier evaluation and selection. Appl. Soft Comput. **109**, 107534 (2021). https://doi.org/10.1016/j.asoc.2021.107534
28. Chai, J., Liu, J.N.K., Ngai, E.W.T.: Application of decision-making techniques in supplier selection: a systematic review of literature. Expert Syst. Appl. **40**(10), 3872–3885 (2013). https://doi.org/10.1016/j.eswa.2012.12.040
29. Aouadni, S., Aouadni, I., Rebaï, A.: A systematic review on supplier selection and order allocation problems. J. Indus. Eng. Int. **15**(s1), 267–289 (2019). https://doi.org/10.1007/s40092-019-00334-y
30. Barata, J.F.F., Quelhas, O.L.G., Costa, H.G., Gutierrez, R.H., de Lameira, V.J., Meiriño, M.J.: Multi-criteria indicator for sustainability rating in suppliers of the oil and gas industries in Brazil. Sustainability (Switzerland). **6**(3), 1107–1128 (2014). https://doi.org/10.3390/su6031107
31. Tarei, P.K., Thakkar, J.J., Nag, B.: A hybrid approach for quantifying supply chain risk and prioritizing the risk drivers: a case of Indian petroleum supply chain. J. Manuf. Technol. Manag. **29**(3), 533–569 (2018). https://doi.org/10.1108/JMTM-10-2017-0218
32. Ho, W., Dey, P.K., Lockström, M.: Strategic sourcing: a combined QFD and AHP approach in manufacturing. Supply Chain Manag. Int. J. **16**(6), 446–461 (2011)
33. Alinezad, A., Seif, A., Esfandiari, N.: Supplier evaluation and selection with QFD and FAHP in a pharmaceutical company. Int. J. Adv. Manuf. Technol. **68**, 355–364 (2013). https://doi.org/10.1007/s00170-013-4733-3
34. Zairi, M., Ginn, D.: East meets West: a critical analysis on the evolution, growth and transfer of QFD from Japan to the West. Asian J. Quality. **4**(1), 1–19 (2003)
35. Kapur, A., Potters, L., Mallalieu, L.B.: WE-C-214-02: six sigma tools for a "No-Fly" patient safety oriented, quality checklist driven, paperless multi-center radiation medicine department. Med. Phys. **38**(6), 3807 (2011)

Consumer Medicines Disposal Behavior: Insights Towards a More Sustainable Chain

Nicole Cecchele Lago, Camila Kolling, Milene Schaiane Auler, Janine Fleith de Medeiros, and José Luis Duarte Ribeiro

Abstract The amount of waste related to the improper disposal of medicines by the end consumer increases propositionally to the pharmaceutical industry's growth. Although many studies propose public policies that encourage the environmentally proper disposal of medicines, the domestic waste remains the main form of disposal. In this context, there are opportunities for research to understand consumer behavior and, based on this, generate insights to make the medicines chain more sustainable. This study aims to investigate medicine disposal behavior of consumers in Southern Brazil. A quantitative research was conducted through the application of a survey with a non-probabilistic sample of 204 consumers. Data were analyzed using descriptive statistics through frequency analysis. The results show that most consumers discard end-of-use medicines in the household waste, sink or toilet. Consumers say they have not received information or guidelines for environmentally proper disposal. Based on the findings, insights are provided to stimulate pro-environmental consumer behavior and to the design of circular business models in the pharmaceutical supply chain.

Keywords Medicine disposal · Pro-environmental behavior · Circular business model.

N. C. Lago · C. Kolling (✉) · J. L. D. Ribeiro
Industrial Engineering Graduate Program, Universidade Federal do Rio Grande do Sul (UFRGS), Porto Alegre, Brazil

J. F. de Medeiros
Industrial Engineering Graduate Program, Universidade Federal do Rio Grande do Sul (UFRGS), Porto Alegre, Brazil

Graduate Program on Business Administration and Graduate Program on Environmental Sciences, Universidade de Passo Fundo (UPF), Passo Fundo, Brazil

M. S. Auler
Graduate Program on Business Administration and Graduate Program on Environmental Sciences, Universidade de Passo Fundo (UPF), Passo Fundo, Brazil

V. M. López Sánchez et al. (eds.), *Industrial Engineering and Operations Management*, Springer Proceedings in Mathematics & Statistics 400,
https://doi.org/10.1007/978-3-031-14763-0_15

1 Introduction

Medicine disposal has become a worldwide challenge, and it has drawn the attention of policymakers, health professionals, pharmaceutical companies, and the community in general [1]. The amount of waste related to the inappropriate disposal of medicine waste increases proportionally to the pharmaceutical industry numbers [2]. High environmental risks are related to the medicine's disposal stage [3, 4].

Especially important is the disposal decision of unused medicines by the end consumer, as these medicines can negatively affect the environment if not treated and disposed of properly [3]. In Brazil, there is a lack of regulations and programs to inform and educate consumers about environmentally proper medicine disposal [5], making the disposal practices adopted by end consumers an important problem.

Despite the recent growth in research on the product disposal stage of the consumer decision-making processes, the field lacks further investigations that enable a better understanding of the phenomenon and most suitable actions to guide consumers regarding environmentally appropriate medicine disposal practices [6]. In this context, it becomes important to advance the understanding of consumer behavior to minimize environmental impacts in the disposal phase.

Based on the above, this study aims to investigate the medicine disposal behavior of consumers in Southern Brazil. Based on this, insights are provided to stimulate pro-environmental consumer behavior, aiming to bring to light the need to design circular business models in the pharmaceutical supply chain.

This paper is organized as follows. After this introduction, Sect. 2 briefly reviews the literature about pro-environmental behavior, pro-environmental disposal behavior, and medicine disposal. Section 3 describes the methodological procedures, detailing the survey conducted with medicine consumers and the techniques used for data collection and analysis. Section 4 presents the analysis and discussion of the results. Section 5 presents the main conclusions, highlighting the study's theoretical and practical contributions, limitations, and directions for future research.

2 Theoretical Background

2.1 *Pro-Environmental Behavior: A Brief Overview*

Pro-environmental behavior (PEB) is defined as the conscious search for minimizing the negative impacts of individual actions in the natural or built environment [7]. Such behaviors can manifest themselves in several domains, as long as they are centered on actions that reduce the negative impact or generate a positive impact on the environment [8]. Recycling practices [9–11], reduction of waste production [12], minimization of greenhouse gas emissions [13], purchase of green products [14, 15], use of sustainable transport [16–18] and environmentally proper waste disposal [9, 12] are some examples.

Behavior change requires breaking old unsustainable habits to establish new ones [19, 20]. The aim is to establish the PEB as the standard, as little or no cognitive effort is required [21]. Specifically, a specific PEB can be stimulated through knowledge of the determinants that influence or inhibit this PEB [12, 20]. Such determinants can be classified into external factors (such as social norms, cost, and convenience) and individual factors (which include demographic and psychographic variables such as attitude and environmental awareness) [22].

2.1.1 Pro-Environmental Disposal Behavior

Product disposal is an important phase in the consumer's decision-making process because, in addition to being linked to environmental protection, it can open paths for economic and social development [10, 23]. In the environmental sphere, it is well understood that waste management is an actionable and effective means to reduce post-consumer environmental impact [23, 24]. In this sense, management solutions must involve the consumer since the disposal decision taken by consumers defines the course of unused products [12]. Despite the belief that individual disposal actions are not sufficient to contribute positively to the environment has prevailed for a long time [25], recent studies show that consumers can lead large-scale environmental change through their individual disposal choices [10, 23, 24].

Disposal choices require additional efforts from the consumer since the product's lifecycle has been completed for him [26]. Thus, in principle, there is no more interest or benefit to be obtained with the product that is in end-of-use. In this context, studies sustain that environmental awareness can direct consumers towards pro-environmental practices, even in the post-consumption phase [25, 27]. Conceptually, environmental awareness is related to selfish and socio-altruistic value orientations and beliefs about the consequences of environmental changes on valued objects [28]. Specifically, consumers with high awareness about environmental problems are more likely to behave pro-environmentally in disposal decisions [25]. Such consumers are responsible for the product's potential impact on the environment and feel personally empowered to provide environmentally appropriate end-of-life for the product to be discarded [9, 29].

However, some research advocates that environmental awareness cannot change behavior [12, 30]. Thus, market incentives emerged as strategies to promote PEBs [31]. These incentives offer rewards to consumers in exchange for their participation in the disposal process [10, 32]. Indeed, the promotion of PEBs appears to be most effective when the individual's well-being is maximized under the constraints he faces [7]. In this context, psychological, functional, and financial benefits need to be emphasized to boost consumer participation in products disposal programs [6]. Particularly, it is known that the provision of monetary incentives increases the recycling intentions of consumers with little concern for the environment [27]. Furthermore, monetary benefits have a positive effect in promoting short-term PEBs [30, 31]. Nonetheless, these authors found that such benefits do not offset the costs of long-term maintenance of the PEB.

Communication also acts as an incentive to boost changes in products disposal behavior [10, 29]. For example, persuasive communication strategies positively influence recycling behaviors [24]. In addition, knowledge about suitable places for disposal is a premise to guide consumers towards environmentally appropriate behavior. Specifically, information promotes intrinsic pro-environmental motivations in consumers because once well-informed, they know how to structure their actions [31]. However, the information applied alone is limited to altering consumers' disposal decisions [12, 26]. Therefore, although essential, it must be combined with other more effective approaches to produce positive effects in pro-environmental disposal behaviors.

2.2 *Medicines Disposal*

The incorrect disposal of medicines by the population is one of the main ways medicines reach the environment [33]. This problem is global and recent studies have been concerned with exploring alternatives to deal with the problem from the consumer's perspective [34–36]. This is because although efforts have already been dedicated to the development of medicines based on green chemistry [37–39] (already designed to reduce the environmental impact on the disposal stage), the closing of the lifecycle of medicines will continue depending on the end consumers' disposal action when medicines are expired or are no longer helpful [5].

For example, Brazil annually produces between 10 to 28 thousand tons of medicine waste, with half of this waste amount being disposed of by end consumers [40]. In this context, a survey conducted with 540 Brazilians found that 66% of respondents dispose of the unused or expired medicines in the garbage [33]. Furthermore, 71.9% of respondents say they have never received any information about the proper disposal of medicines, and 54.4% do not know if their cities have specific medicine collectors and where they are located [33]. These authors conclude these numbers are mainly due to the absence of a Brazilian reverse logistics system added to the low investment in environmental education to inform the population.

However, a similar reality is found in other economies. A study of 415 medicine consumers in Saudi Arabia found that 89% discard unused medicines in household waste [41]. These researchers found that women significantly more adopt this disposal option than men. In Malta, inappropriate medicine disposal is also standard practice [42]. The results of an applied survey with 422 respondents confirmed that only a little more than half of the respondents (54.7%) know that unused medicines should not be thrown into the bathroom sink or toilet; even so, they do [42].

In Hong Kong, 54% of respondents dispose of expired or unused medicines and household solid waste [43]. Despite this, 62.3% of respondents are aware that this practice causes significant damage to the environment. Based on these results, the authors propose a scheme for returning unused medicines to increase the convenience of disposal, combining this action with environmental education at the national level. In Serbia, another emerging economy, the researchers found that while most respondents believe that the most appropriate way to dispose of unused medication is to return it to the pharmacy (81.9%), the method of disposal most reported is to throw them in the trash (59.1%) [35].

In general, the practical implications of these studies are aimed at creating more efficient and effective laws and regulations to regulate the action of the final consumer, in addition to drawing attention to the responsibility of pharmacies and pharmaceutical industries. In this regard, it is known that in developed economies, such as France and the United States, the pharmacies are obligated by law to participate in the collection system of unused medicine [34, 36]. Indeed, there is evidence that this legal obligation has increased the volume of medicines disposed of at collection sites [34].

Despite this significant increase found in the United States, the results of this survey also revealed that only 14% of consumers interviewed know the closest place to carry out an environmentally proper disposal. An older study conducted in 2017 identified a similar result among students at an American university: only 24% of respondents reported knowledge about the National Prescription Drug Take-Back Day, a program existent in the country since 2010 [44]. Furthermore, only 4% reported using the service to dispose of unused medicines [44].

This brief review of the global medicines disposal scenario highlights the urgent need to advance the understanding of medicines consumers' disposal behavior, mainly in emerging economies.

3 Method

Descriptive research with a quantitative approach was carried out. The survey was conducted with consumers of medicines through questionnaires developed based on the parameters reported in the previous subsection (Pro-environmental Disposal Behavior). This study used a non-probabilistic convenience sampling method. The data collection instrument was sent to a sample of 240 medicine consumers in drugstores and public health clinics of a town that is representative of the interior of Brazil. 36 inconsistent responses were discarded from the data analysis. Thus, the final sample of valid and complete responses is composed of 204 consumers.

The questionnaire comprised three main sections. The first one addressed behavioral aspects related to medicines. The frequency of consumption, medicine and packaging disposal practices, and the motivations for proper disposal were investigated. The second section comprised psychographic variables, addressing consumers' perception regarding the appropriate medicine destination, their level of environmental awareness, and information regarding the appropriate ways of disposal. The last section investigated consumers' demographic variables (gender and age). Overall, the questions had dichotomous response options. A 5-point Likert scale was only used to assess the influence of financial and communication incentives on the motivation to dispose of medicines at specific collection sites and the set of variables that addressed environmental awareness. Finally, for the data analysis, a general description of frequency was performed.

4 Results and Discussion

Table 1 summarizes the descriptive statistics of the demographic, psychographic and behavioral variables of the consumers in the sample.

In general, the sample comprises more women than men. Most consumers are 39 years old or younger and take medicines frequently. A large part of the respondents did not receive information on medicine disposal and the harmful effects on the environment. Most consumers discard unused medicines and their packaging in inappropriate places, such as household waste, sinks or toilets. Despite this, most consumers revealed a high level of environmental awareness. Specifically, although environmental awareness may influence the decision of what is considered the safest means of disposal, real behavior is not always equivalent to awareness [45]. In this sense, to match consumers' behavior and attitude, it is necessary to make the consequences of consumers' inappropriate practices more tangible [45].

In this case, a strategy that may work is to increase access to a broader and more tangible set of information about the consequences of inappropriate

Table 1 Results of the frequency analysis

Demographic and psychographic variables	Frequency n (%)	Behavior and psychographic variables	Frequency n (%)
Male	49 (24.02)	Medicines disposal in the domestic waste, toilet or sink	140 (68.62%)
Female	155 (75.98)	Medicines disposal in collectors suitable for this purpose	(31.38%)
≤ 39 years	131 (64.22)	Packaging disposal in the domestic waste	176 (86.28%)
≥ 40 years	73 (35.78)	Packaging disposal collectors suitable for this purpose	28 (13.72%)
Low consumption frequency	7 (3.44)	Low environmental awareness	14 (6.86%)
Intermediate consumption frequency	50 (24.50)	Intermediate environmental awareness	35 (17.15%)
High consumption frequency	147 (72.06)	High environmental awareness	155 (75.99%)
Did not receive disposal information	184 (90.20)	Would discard it correctly if there were communication campaigns	142 (69.61%)
Received disposal information	20 (9.80)	Would discard correctly if there were financial incentives	139 (68.14%)

medicine disposal, as most consumers stated that they would dispose of medicines in appropriate places if they were subjected to communication campaigns. However, communication campaigns on environmental issues based only on information provision have stopped influencing consumer attitudes and it is being replaced by the more aggressive development of social and environmental marketing [6]. Furthermore, there is evidence that communication is only effective when it sensitizes and makes social and environmental risks tangible to consumers [41]. This can be emphasized mainly through the risks to the health of citizens [3] and to the environment in which they live.

The results also showed that financial incentives could help consumers to dispose of medicines correctly. Thus, pro-environmental medicine disposal behavior can also be stimulated by providing financial incentives for consumers to dispose of medicines at specific collection sites through a reward system that provides discounts for future purchases. Retailers and manufacturers of pharmaceutical products can manage this system. Consequently, their images would be associated with environmental sustainability and incentive strategies. Therefore, they would achieve a win-win status in the relationship with consumers, industry, and the environment [37].

Interestingly, most consumers say they do not know the correct way to dispose of unused medicines because they have not received information. In particular, this finding reveals the relevance of this study because the information and involvement with the final consumer is the starting point of any business model that aims at circularity [3]. In this regard, communicating the best practices for medicine disposal on medicines labels has the potential to clearly spread information [34]. Thus, just as labelling strategies influence food consumers' purchasing decisions [46], medicine packaging labels are expected to influence consumer disposal decisions positively. Additionally, for those consumers who dispose of unused medicines at appropriate points, the strategy is to increase and keep the convenience of these points. Probably, such consumers do not need communication or financial incentives. Thus, providing accessible disposal points for end-of-use medicines is critical to keep these individuals engaged with proper disposal and to initiate the design of circular medicines chains.

Based on findings associated with inappropriate medicines disposal and mainly misinformation related to environmentally proper disposal means, it is possible to infer little attention has been devoted to this. Looking closely at the disposal stage is essential for reducing negative environmental impact because this predominantly occurs after consumption. In this scenario, it is understood that regulatory actions can be effective. Thus, this study is in line with previous studies that have already proposed creating more efficient and effective laws and regulations for the Brazilian pharmaceutical sector [33]. Specifically, it is suggested that Brazilian legislation hold retailers and pharmaceutical industries accountable for implementing and managing disposal programs. This seems to work in France, the United States, and Serbia, where the pharmacies are legally obligated to participate in the collection system of unused medicine [34–36]. Therefore, public authorities should be responsible for encouraging, regulating, and inspecting medicine waste management, as they are considered intermediary actors in this value network [47]. Engaging the different actors in the medicines chain is essential for solving the problem of inappropriate disposal and, above all, for the design of circular business models.

5 Conclusions

Literature lacks research that understand consumer behavior and propose practices to encourage environmentally proper disposal. This study investigated medicines disposal behavior of consumers in Southern Brazil. The results show that most consumers discard unused medicines in domestic waste instead of disposing of them at appropriate collection points. More specifically, consumers say they have not received information or guidelines for environmentally proper disposal. It is possible to conclude that although consumers have revealed a high level of environmental concern, the initiatives developed and communication strategies employed have not been sufficient to encourage the proper disposal of medicines. Based on the results, it was possible to shed light on some insights to provoke medicines chain stakeholders to reflect on this emerging issue to provide guidelines to stimulate pro-environmental consumer behavior regarding disposal.

The research offers relevant contributions to the disposal behavior literature and, particularly, to the practical context of medicine disposal to solve the environmental problem associated with inappropriate medicine disposal by the end consumer. Theoretically, the study advanced in research on the proper disposal of medicines from a behavioral perspective, complementing previous studies carried out on technical solutions. Furthermore, it was provided helpful starting points for designing circular business models in the pharmaceutical supply chain. Insights provided can also help in structuring the medicine reverse logistics processes and developing the green supply chain in this sector.

Despite the theoretical and managerial contributions of the study, some limitations need to be acknowledged. First, consumers answered psychographic and behavioral variables by themselves, which may not reflect their real actions. Thus, considering that the disposal of medicines is not frequent and can hardly be witnessed in loco since it can occur in domestic environments, projective research techniques could be employed to confirm the results found. In addition, the study used only descriptive statistics to start looking at the medicines disposal problem. Since it was found that most consumers do not dispose of properly (but also that there are incentive opportunities to reverse this behavior), future studies could be conducted to understand the factors and variables that impact the decision to dispose of medicines in appropriate places. Similarly, factors that represent barriers to environmentally proper disposal could be investigated. Similar studies could be conducted in other disposal situations since this decision process stage is little explored. Another limitation is that the present study generates embryonic insights to structure circular business models in the medicines chain based on consumer engagement. Therefore, future research could integrate all stakeholders to generate more effective and viable solutions, involving all actors in the chain.

Acknowledgments The authors thank the Coordination for the Improvement of Higher Education Personnel in Brazil (CAPES - Coordenação de Aperfeiçoamento de Pessoal de Nível Superior) and the Brazilian National Council for Scientific and Technological Development (CNPq - Conselho Nacional de Desenvolvimento Científico e Tecnológico) for the financial support received to conduct this research.

References

1. Ayele, Y., Mamu, M.: Assessment of knowledge, attitude and practice towards disposal of unused and expired pharmaceuticals among community in Harar city, Eastern Ethiopia. J. Pharm. Policy Pract. **11**, 1–7 (2018). https://doi.org/10.1186/s40545-018-0155-9
2. Yoada, R.M., Chirawurah, D., Adongo, P.B.: Domestic waste disposal practice and perceptions of private sector waste management in urban Accra. BMC Public Health. **14**, 1–10 (2014). https://doi.org/10.1186/1471-2458-14-697
3. de Campos, E.A.R., ten Caten, C.S., de Paula, I.C.: End-of-use and end-of-life medicines—insights from pharmaceutical care process into waste medicines management. Environ. Sci. Pollut. Res. (2021). https://doi.org/10.1007/s11356-021-14661-4
4. Siegert, M.W., Lehmann, A., Emara, Y., Finkbeiner, M.: Addressing the use and end-of-life phase of pharmaceutical products in life cycle assessment. Int. J. Life Cycle Assess. **25**, 1436–1454 (2020). https://doi.org/10.1007/s11367-019-01722-7
5. Pereira, A.L., de Vasconcelos Barros, R.T., Pereira, S.R.: Pharmacopollution and Household Waste Medicine (HWM): how reverse logistics is environmentally important to Brazil. Environ. Sci. Pollut. Res. **24**, 24061–24075 (2017). https://doi.org/10.1007/s11356-017-0097-9
6. Cruz-Cárdenas, J., Arévalo-Chávez, P.: Consumer behavior in the disposal of products: forty years of research. J. Promot. Manag. **24**, 617–636 (2018). https://doi.org/10.1080/10496491.2018.1405514
7. Kollmuss, A., Agyeman, J.: Mind the gap: why do people act environmentally and what are the barriers to pro-environmental behavior? Environ. Educ. Res. **8**, 239–260 (2002). https://doi.org/10.1080/13504620220145401
8. Nordlund, A.M., Garvill, J.: Value structures behind proenvironmental behavior. Environ. Behav. **34**, 740–756 (2002). https://doi.org/10.1177/001391602237244
9. Trudel, R., Argo, J.J., Meng, M.D.: Trash or recycle? How product distortion leads to categorization error during disposal. Environ. Behav. **48**, 966–985 (2016). https://doi.org/10.1177/0013916515577635
10. Trudel, R., Arg, J.J., Meng, M.D.: The recycled self: consumers' disposal decisions of identity-linked products. J. Consum. Res. **43**, 246–264 (2016). https://doi.org/10.1093/jcr/ucw014
11. Lu, B., Wang, J.: How can residents be motivated to participate in waste recycling? An analysis based on two survey experiments in China. Waste Manag. **143**, 206–214 (2022). https://doi.org/10.1016/j.wasman.2022.02.034
12. Barr, S.: Factors influencing environmental attitudes and behaviors: a U.K. case study of household waste management. Environ. Behav. **39**(4), 435–473 (2007). https://doi.org/10.1177/0013916505283421
13. Brick, C., Lewis, G.J.: Unearthing the "Green" personality: Core traits predict environmentally friendly behavior. Environ. Behav. **48**, 635–658 (2016). https://doi.org/10.1177/0013916514554695

14. De Canio, F., Martinelli, E., Endrighi, E.: Enhancing consumers' pro-environmental purchase intentions: the moderating role of environmental concern. Int. J. Retail Distrib. Manag. (2021). https://doi.org/10.1108/IJRDM-08-2020-0301
15. Lago, N.C., Marcon, A., Ribeiro, J.L.D., de Medeiros, J.F., Brião, V.B., Antoni, V.L.: Determinant attributes and the compensatory judgement rules applied by young consumers to purchase environmentally sustainable food products. Sustain. Prod. Consum. **23**, 256–273 (2020). https://doi.org/10.1016/j.spc.2020.06.003
16. Manca, S., Fornara, F.: Attitude toward sustainable transport as a function of source and argument reliability and anticipated emotions. Sustain. **11** (2019). https://doi.org/10.3390/su10023288
17. Kowalska-Pyzalska, A., Kott, M., Kott, J.: How much polish consumers know about alternative fuel vehicles? Impact of knowledge on the willingness to buy. Energies. **14**, 1–19 (2021). https://doi.org/10.3390/en14051438
18. Kormos, C., Gifford, R., Brown, E.: The influence of descriptive social norm information on sustainable transportation behavior: a field experiment. Environ. Behav. **47**, 479–501 (2015). https://doi.org/10.1177/0013916513520416
19. Ertz, M., Sarigöllü, E.: The behavior-attitude relationship and satisfaction in Proenvironmental behavior. Environ. Behav. **51**, 1106–1132 (2019). https://doi.org/10.1177/0013916518783241
20. Stern, P.C.: Toward a coherent theory of environmentally significant behavior. J. Soc. Issues. **56**, 407–424 (2000). https://doi.org/10.1111/0022-4537.00175
21. Langenbach, B.P., Berger, S., Baumgartner, T., Knoch, D.: Cognitive resources moderate the relationship between pro-environmental attitudes and green behavior. Environ. Behav. **52**, 979–995 (2020). https://doi.org/10.1177/0013916519843127
22. Li, D., Zhao, L., Ma, S., Shao, S., Zhang, L.: What influences an individual's pro-environmental behavior? A literature review. Resour. Conserv. Recycl. **146**, 28–34 (2019). https://doi.org/10.1016/j.resconrec.2019.03.024
23. Sun, M., Trudel, R.: The effect of recycling versus trashing on consumption: theory and experimental evidence. J. Mark. Res. **54**, 293–305 (2017). https://doi.org/10.1509/jmr.15.0574
24. White, K., Macdonnell, R., Dahl, D.W.: It's the mind-set that matters: the role of construal level and message framing in influencing consumer efficacy and conservation behaviors. J. Mark. Res. **48**, 472–485 (2011). https://doi.org/10.1509/jmkr.48.3.472
25. Van Birgelen, M., Semeijn, J., Keicher, M.: Proenvironmental consumption behavior investigating purchase and disposal. Environ. Behav. **41**, 125–146 (2009)
26. Abbas, H., Farooquie, J.A.: Return and disposal of unused medicines: a customer perspective of reverse logistics. Int. J. Bus. Manag. Invent. ISSN., 59–66 (2013)
27. Shultz, P.W., Oskamp, S.: Effort as a moderator of the attitude-behavior relationship : general environmental concern and recycling. Soc. Psychol. **59**(4), 375–383 (1996). https://doi.org/10.2307/2787078
28. Stern, P.C., Dietz, T.: The value basis of environmental psychology. J. Soc. Issues. **50**, 65–84 (1994). https://doi.org/10.1111/j.1540-4560.1994.tb02420.x
29. Jacobs, S., Sioen, I., De Henauw, S., Rosseel, Y., Calis, T., Tediosi, A., Nadal, M., Marques, A., Verbeke, W.: Marine environmental contamination: public awareness, concern and perceived effectiveness in five European countries. Environ. Res. **143**, 4–10 (2015). https://doi.org/10.1016/j.envres.2015.08.009
30. Uren, H.V., Roberts, L.D., Dzidic, P.L., Leviston, Z.: High-status pro-environmental Behaviors: costly, effortful, and visible. Environ. Behav. **53**, 455–484 (2021). https://doi.org/10.1177/0013916519882773
31. Steinhorst, J., Klöckner, C.A.: Effects of monetary versus environmental information framing: implications for long-term pro-environmental behavior and intrinsic motivation. Environ. Behav. **50**, 997–1031 (2018). https://doi.org/10.1177/0013916517725371
32. Sangwan, K.S.: Key activities, decision variables and performance indicators of reverse logistics. Procedia CIRP. **61**, 257–262 (2017). https://doi.org/10.1016/j.procir.2016.11.185
33. Quadra, G.R., Silva, P.S.A., Paranaíba, J.R., Josué, I.I.P., Souza, H., Costa, R., Fernandez, M., Vilas-Boas, J., Roland, F.: Investigation of medicines consumption and disposal in Brazil: a

study case in a developing country. Sci. Total Environ. **671**, 505–509 (2019). https://doi.org/10.1016/j.scitotenv.2019.03.334
34. Ehrhart, A.L., Granek, E.F., Nielsen-Pincus, M., Horn, D.A.: Leftover drug disposal: customer behavior, pharmacist recommendations, and obstacles to drug take-back box implementation. Waste Manag. **118**, 416–425 (2020). https://doi.org/10.1016/j.wasman.2020.08.038
35. Paut Kusturica, M., Golocorbin-Kon, S., Ostojic, T., Kresoja, M., Milovic, M., Horvat, O., Dugandzija, T., Davidovac, N., Vasic, A., Tomas, A.: Consumer willingness to pay for a pharmaceutical disposal program in Serbia: a double hurdle modeling approach. Waste Manag. **104**, 246–253 (2020). https://doi.org/10.1016/j.wasman.2020.01.029
36. Lima, M.L., Luís, S., Poggio, L., Aragonés, J.I., Courtier, A., Roig, B., Calas-Blanchard, C.: The importance of household pharmaceutical products disposal and its risk management: example from Southwestern Europe. Waste Manag. **104**, 139–147 (2020). https://doi.org/10.1016/j.wasman.2020.01.008
37. Hua, M., Lai, I.K.W., Tang, H.: Analysis of advertising and a points-exchange incentive in a reverse supply chain for unwanted medications in households based on Game Theory. Int. J. Prod. Econ. **217**, 259–268 (2019). https://doi.org/10.1016/j.ijpe.2019.02.004
38. Khan, H.K., Rehman, M.Y.A., Malik, R.N.: Fate and toxicity of pharmaceuticals in water environment: an insight on their occurrence in South Asia. J. Environ. Manag. **271**, 111030 (2020). https://doi.org/10.1016/j.jenvman.2020.111030
39. Viegas, C.V., Bond, A., Vaz, C.R., Bertolo, R.J.: Reverse flows within the pharmaceutical supply chain: a classificatory review from the perspective of end-of-use and end-of-life medicines. J. Clean. Prod. **238** (2019). https://doi.org/10.1016/j.jclepro.2019.117719
40. Torres, A.C.G.: Pertinência da normativa estadual e distrital sobre logística reversa aplicada ao setor de medicamentos no Brasil. Cad. Ibero-Americanos Direito Sanitário. **5**, 41 (2016) https://doi.org/10.17566/ciads.v5i1.238
41. Alhomoud, F.K., Alsadiq, Y., Alghalawin, L., Alhifany, A., Alhomoud, F.: Pharmacy students' knowledge and practices concerning the storing and disposal of household medication in Saudi Arabia. Curr. Pharm. Teach. Learn. **13**, 5–13 (2021). https://doi.org/10.1016/j.cptl.2020.08.004
42. West, L.M., Stewart, D., Cordina, M.: Mixed-methods approach to determine adherence, knowledge and behavioral determinants associated with medication wastage. Res. Soc. Adm. Pharm. **16**, 654–662 (2020). https://doi.org/10.1016/j.sapharm.2019.08.003
43. Chung, S., shan, Brooks, B.W.: Identifying household pharmaceutical waste characteristics and population behaviors in one of the most densely populated global cities. Resour. Conserv. Recycl. **140**, 267–277 (2019). https://doi.org/10.1016/j.resconrec.2018.09.024
44. Vatovec, C., Van Wagoner, E., Evans, C.: Investigating sources of pharmaceutical pollution: survey of over-the-counter and prescription medication purchasing, use, and disposal practices among university students. J. Environ. Manag. **198**, 348–352 (2017). https://doi.org/10.1016/j.jenvman.2017.04.101
45. Griskevicius, V., Cantú, S.M., Van Vugt, M.: The evolutionary bases for sustainable behavior: implications for marketing, policy, and social entrepreneurship. J. Public Policy Mark. **31**, 115–128 (2012). https://doi.org/10.1509/jppm.11.040
46. Silayoi, P., Speece, M.: Packaging and purchase decisions: an exploratory study on the impact of involvement level and time pressure. Br. Food J. **106**, 607–628 (2004). https://doi.org/10.1108/00070700410553602
47. de Campos, E.A.R., de Paula, I.C., Pagani, R.N., Guarnieri, P.: Reverse logistics for the end-of-life and end-of-use products in the pharmaceutical industry: a systematic literature review. Supply Chain Manag. **22**, 375–392 (2017). https://doi.org/10.1108/SCM-01-2017-0040

A Model to Estimate Operators' Performance in Accomplishing Assembly Tasks

Francesco Facchini, Daniela Cavallo, and Giovanni Mummolo

Abstract In the fourth industrial revolution, the digital network is the basis of smart manufacturing systems. In work environments 4.0, the operators' role is drastically changed. There is increasing utilisation of innovative devices, and new technologies have changed work activities into more cognitive than physical tasks. According to scientific studies in the new industrial era, the operators' skill to process the information related to a single task plays a crucial role in improving the manufacturing systems' effectiveness. The methods available in scientific literature to assess the operator's performance are mainly focused on the cognitive and physical efforts required by the task. In other words, they depend on tasks complexity and neglect human behaviour over time and the workers' abilities. Therefore, an evaluation including the skills and properness of a specific operator to perform an assigned task needs more investigation. Consistent with this research gap, the paper aims to develop an information-based theoretical model allowing to estimate an operator's performance index to accomplish an assembly job by evaluating both the tasks' complexity and the operator's skill.

The model is applied to an automotive company to test and evaluate the potential applications of the methodology that go beyond the case study developed. The results proved the effectiveness of the model in estimating the operators' performance, providing a job schedule based on task complexity and workers' abilities.

Keywords Human Performance · Assembly Tasks · Operator's Performance Index

F. Facchini (✉) · D. Cavallo
Polytechnic University of Bari, Bari, Italy
e-mail: francesco.facchini@poliba.it

G. Mummolo
University of Bari, Bari, Italy

V. M. López Sánchez et al. (eds.), *Industrial Engineering and Operations Management*, Springer Proceedings in Mathematics & Statistics 400,
https://doi.org/10.1007/978-3-031-14763-0_16

1 Introduction

The human factor is considered a cost-effective alternative to expensive automated solutions and an easily interchangeable high flexible resource [1]. The increasing adoption of innovative devices (i.e. sensors, actuators, control processing units, and communication devices) [2] due to the fourth industrial revolution is changing the smart factory's tasks [3]. The "smart-operator" has to perform tasks within a manufacturing process, learn from the activity, predict specific situations, identify inefficiency, and make decisions [4]. The decision-making tasks are defined as conditions in which the decision-maker must evaluate the options and select the best one. The complexity of these tasks depends on the hardness to identify the optimal configuration. In smart manufacturing systems, the decision-making tasks are needed to ensure the flexibility of assembly lines, increase the variety of the products and reduce the operative costs [5]. If, on the one hand, the complexity of the modern assembly line allowed an increase in the customisation of the products [6], on the other hand, the growth of product variants led to a rise in the amount of information to be processed by each operator [7]. According to [8, 9], the complexity of assembly tasks is strictly related to the amount and the content of information. The assembly tasks of discrete parts represent the core of the manufacturing process. Consistent with this point of view, most large international companies consider assembly tasks as the main important value-added activities. In this regard, Mital et al. [10] proved that a proper design of a part to be assembled could significative reduce the task complexity and the corresponding assembly time. In this regard, it is very interesting to note that the tasks required by assembly jobs include both cognitive and physical efforts. Each assembly job can be more 'physical' than 'cognitive' oriented according to physical efforts prevailing over the cognitive ones, or vice versa. In no case, the manual assembly job can require only cognitive or physical effort.

According to Bi and Salvendy [11], the evaluation methods to assess the operator's performance in assembly jobs are mainly focused on task analysis. Generally, they do not consider human behaviour over time and the worker's ability. From this perspective, a holistic study has been conducted to develop an information-based theoretical model allowing to estimate an operator's performance index to accomplish an assembly job by evaluating both the tasks' complexity and the operator's dexterity skill. Consistent with this aim, the amount and the content of information of an assembly task have been related to the corresponding time needed by operators with different dexterity skills in accomplishing it.

The remainder of the paper is organised as in the following. In Sect. 2, the theoretical background of task complexity, with reference to assembly jobs, is investigated. A model for estimating operators' performance considering the assembly jobs' complexity and the dexterity phenomenon is presented in Sect. 3. Then, the model is applied to a full case study from the automotive industry (Sect. 4). Finally, in Sects. 5 and 6, discussions and conclusions are presented, respectively.

2 The Theoretical Background of Task Complexity

In the industry 4.0 context, the concept of task complexity plays a crucial role to design and optimising manufacturing systems. The task complexity assessment is one of the most important challenges faced by manufacturing companies. Similarly, the strategies to reduce the complexity of the tasks represent one of the main impactful approaches to minimise costs and increase production performances [12]. Currently, the scientific literature does not provide a unique definition of "task complexity" [13]. In 1993, March and Simon identified the task complexity by evaluating three objective qualities: uncertainty in the possible alternatives, inexact or unknown data, and hardness to divide the task into independent sub-task [14]. According to Latham and Yukl, complex tasks are characterised by performance identified with multiple quantitative and nonquantitative dimensions. In particular, the latter led to a significant increase in complexity [15]. At the aggregate level, complexity is defined as the difficulty degree in predicting the system proprieties (e.g., assembled product) when are given the properties of the system's parts (e.g., components) [16]. Fast-Berglund et al. provide two definitions of complexity under an objective and subjective perspective [17]:

- Objective complexity depends on the nature of products, hierarchical structures, processes, variety, and strength of interactions.
- Subjective complexity depends on different individuals' factors such as skills, competence, and experience.

An evaluation method to estimate the complexity degree of semifinished parts allowed for developing a guideline to design finished-product adopting ease assembled parts and considering the systems boundaries condition (i.e., work cycle, operative sequences equipment, and system layouts) [18]. Boothroyd et al. introduced the Design for Assembly (DFA) method to assess the assembly's complexity, analysing different cycle times collected in a sample of empirical observations [19]. Similarly, the assembly's complexity is related to the quality of the available data required to assemble the components [20].

ElMaraghy and Urbanic developed an information-based model to identify the product and process complexity in some manufacturing work environments. The model depends on the total quantity of information, diversity of information, and information content [21]. Braha and Maimon proved that the total assembly time is a linear function of the information content [22]. A new entropy measure is introduced to quantify the assembly's complexity related to a possible number of alternatives needed to complete the task [23]. Similarly, Busogi et al. evaluate the manufacturing choice complexity by considering the number of possible alternatives and the part similarity [24]. In most cases, the current studies faced the performance of an assembly line, neglecting the losses due to aspects related to both human performance and task complexity. Generally, they are considered adopting different and independent assessments methods. The model proposed contributes to overcome this limit by jointly estimating the task complexity in terms of the amount of information to be processed (expressed in bits) and the operators' performance as a function of their dexterity skill.

3 The Model Description

Human performance in a manufacturing system is difficult to evaluate since it depends on multiple quantitative and nonquantitative (e.g., emotional) variables requiring complex analysis [25]. In the case of repetitive tasks, cognitive experts stated that the human factor is strongly related to the worker's learning capacity, leading to reduced cycle time during the work shift. [26]. In 2009, an analytical model proved that, in the case of repetitive tasks, the cycle time decreases with the number of task repetitions (n), according to learning (λ) and tiredness (τ) phenomena depending on human behaviour [27]. Assuming a dexterity phenomenon (ε) given in Eq. 1, the time required to assemble one unit (MT), in the case of the repetitive manual task, can be evaluated according to Eq. 2.

$$\varepsilon = \lambda - \tau \tag{1}$$

$$MT(n) = MT(1) - \varepsilon \ln(n) \, [sec] \tag{2}$$

Where $MT(n)$ is the time required to perform the task at n repetition, $MT(1)$ is the time required to perform the task at the first repetition, and n identifies the number of repetitions. Similarly, the Fitts law (Eq. 3) proved a linear dependency between MT and the corresponding amount of information to be processed (I_c), depending on β-constant estimated on empirical observations [28]:

$$MT = \beta \cdot I_c \tag{3}$$

Samy and Elmaraghy estimate the amount of information required to accomplish an assembly manual task (I_c) as shown below (Eq. 4) [18].

$$I_c = \left[\frac{n_p}{N_p} + CI_{product}\right] \cdot \left[\log_2\left(N_p + 1\right)\right] + \frac{n_s}{N_s} \cdot \log_2\left(N_s + 1\right) \; [bit] \tag{4}$$

Where, N_p and N_s are the total numbers of parts and fasteners, respectively, n_p and n_s are the number of unique parts and fasteners, respectively, and $CI_{product}$ is the product assembly complexity index depending on the physical characteristic of the part (i.e., handling and insertion attributes).

In the present work, the operator's performance index (IP), defined as the performance of the operator to accomplish a manual assembly task, is evaluated by the rate between the information required to accomplish an assembly manual task (I_c) and the time required to perform the task at n repetition ($MT(n)$), as shown in Eq. 5.

$$IP(n) = \frac{I_c}{MT(n)} \left[\frac{bit}{sec}\right] \tag{5}$$

IP is a positive value increasing with the operator's dexterity skill to perform a specific manual assembly task.

4 Case Study

The proposed model is tested on an assembly line of a large automotive company located in the south part of Italy. A campaign to collect the time required to perform each task's repetition (*MT*) has been planned in a manual workstation (*WS*) of an assembly line of pumps of a high-pressure diesel injection system. The assembly line consists of 15 *WS*s grouped in 6 highly automated (*WA*), 4 completely manual (*WM*), and 5 with small degrees of automation (*WSA*). The configuration of the flowline is shown in Fig. 1.

In the *WM* assessed, identified with code 150, the body of the pump is manually assembled with a polygonal ring, an eccentric shaft and a flange. The first task consists of checking the presence of the o-ring and of a bushing in the flange; this inspection is out of the evaluation since it is not considered an assembly task. The second task consists of assembly flange (4) with eccentric shaft (3); the polygonal ring (2) is assembled with (3) and (4), and the body of pump (1) is assembled with (4), (3), and (2). Finally, the operator inserts six screws (5) on the flange. The screws

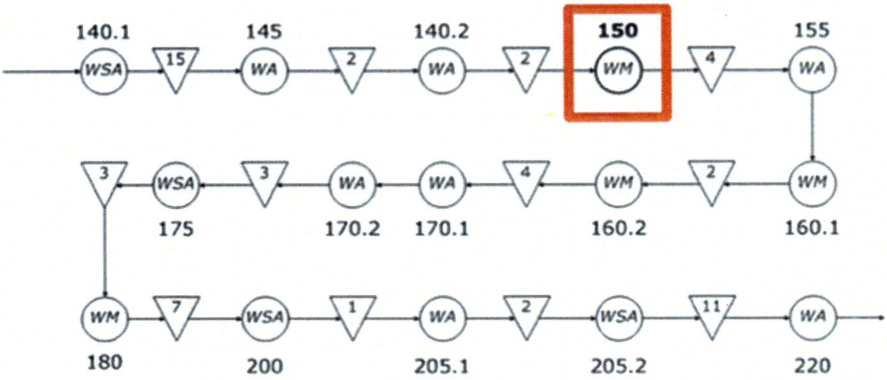

Fig. 1 Flowline of the assembly line of pumps

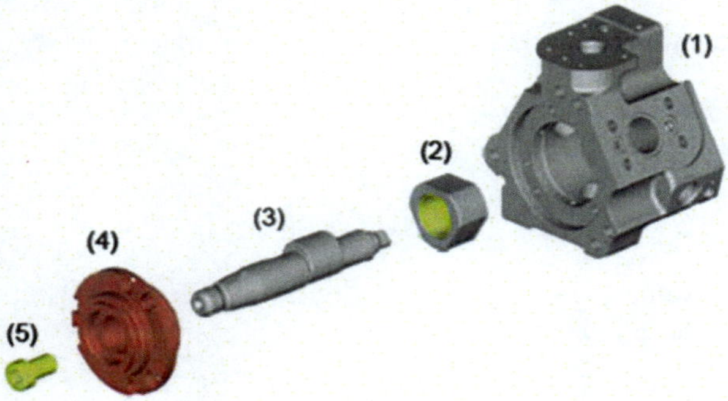

Fig. 2 Body and components to be assembled. Pump body (1); polygonal ring (2); eccentric shaft (3); flange (4); screw (5)

will be screwed up in the next workstation. In Fig. 2, the body and the components assembled in the WS are shown.

The $CI_{product}$ is evaluated according to [18], and it is given by the individual assembly complexity indices of all parts (CI_{part}). For this purpose, the weighted average values of the part assembly complexity factors (C_{part}) are estimated assuming the average difficulty factors for handling (C_h) and insertion (C_i). The "complexity matrix" (Tables 2 and 3) allows identifying C_h and C_i, is built-up assuming the type of components parts (as a row) and the corresponding assembly attributes obtained by Table 1 (as a column). For this purpose, the different handling and insertion factors are examined separately.

With J and K the numbers of handling and insertion attributes of each part, respectively, the sum of the average difficulty factor (S) allows for evaluate C_h and C_i shown in Eqs. 6 and 7, respectively.

$$C_h = {}^{S}\!\big/_{J} \tag{6}$$

$$C_i = {}^{S}\!\big/_{K} \tag{7}$$

Then, C_{part} for each of the four components can be estimated according to Eq. 8.

$$C_{part} = \frac{[C_h \cdot (S \cdot C_h) + C_i \cdot (S \cdot C_i)]}{S \cdot C_h + S \cdot C_i} \tag{8}$$

CI_{part} and $CI_{product}$ are identified from Eqs. 9 and 10, considering the amount of the part to be assembled (N), respectively.

$$CI_{part} = x_p \cdot C_{part} \tag{9}$$

Table 1 Manual assembly attributes from Design for Assembly methodology

Group	Attribute	Symbol	Description	Average difficulty factor
Handling attributes	Symmetry ($\alpha + \beta$)	Sym	$\alpha + \beta < 360$	0.70
			$360 \leq \alpha + \beta < 540$	0.84
			$540 \leq \alpha + \beta < 720$	0.94
			$\alpha + \beta = 720$	1.00
	Size	Siz	>15 *mm*	0.74
			6 *mm* < *size* ≤ 15 *mm*	0.81
			<6 *mm*	1.00
	Thickness	Th	>2 *mm*	0.27
			0.25 *mm* < *size* ≤ 2*mm*	0.50
			≤0.25 *mm*	1.00
	Weight	W	<4.53 *kg*	0.50
			≥4.53 *kg*	1.00
	Grasping and manipulation	GM	Easy to grasp and manipulate	0.91
			Not easy to grasp and manipulate	1.00
	Assistance	Ass	Using one hand	0.34
			Using one hand with grasping aids	1.00
			Using two hands	0.75
			Using two hands with assistance	0.57
	Nesting and tangling	NT	Parts do not severely nest or tangle and are not flexible	0.58
			Parts severely nest or tangle or are flexible	1.00
	Optical magnification	Op	Not necessary	0.80
			Necessary	1.00
Insertion attributes	Holding down	Hd	Not required	0.54
			Required	1.00
	Alignment	Al	Easy to align or position	0.86
			Not easy to align or position	1.00
	Insertion resistance	In	No resistance	0.87
			Resistance to insertion	1.00
	Accessibility and vision	AV	No restrictions	0.57
			Obstructed access or restricted vision	0.81
			Obstructed access and restricted vision	1.00

(continued)

Table 1 (continued)

Group	Attribute	Symbol	Description	Average difficulty factor
	Mechanical fastening processes	Mfp	Bending	0.34
			Riveting	0.58
			Screw tightening	0.42
			Bulk plastic deformation	1.00
	Non-mechanical fastening processes	Nmfp	No additional material required	0.58
			Soldering processes	0.67
			Chemical processes	1.00
	Non-fastening processes	Nfp	Manipulation of parts or sub-assemblies (fitting or adjusting of parts)	0.75
			Other processes (liquid insertion)	1.00

Table 2 Handling complexity matrix

	N	Attributes										
Part name (i)	Sym	Siz	Th	W	GM	Ass	NT	Op	J	S	C_h	
Pump body	1	1.00	0.81	0.27	1.00	0.91	0.75	1.00	0.80	8	6.54	0.82
Polygonal ring	1	0.70	0.81	0.27	0.50	1.00	0.34	0.58	0.80	8	5.00	0.63
Eccentric shaft	1	0.84	0.81	0.27	0.50	1.00	0.34	0.58	0.80	8	5.14	0.64
Flange	1	0.70	0.81	0.27	1.00	1.00	0.75	0.58	0.80	8	5.91	0.74

Table 3 Insertion complexity matrix

	N	Attributes									
Part name (i)		Hd	Al	In	Av	Mfp	Nmfp	Nfp	K	S	C_i
Pump body	1	1.00	0.86	0.87	0.57				4	3.30	0.83
Polygonal ring	1	0.54	1.00	1.00	0.57				4	3.11	0.78
Eccentric shaft	1	0.54	1.00	0.87	0.57				4	2.98	0.75
Flange	1	0.54	0.86	0.87	0.57	0.42			5	3.26	0.65

$$CI_{product} = \sum CI_{part} \tag{10}$$

Where x_p in Eq. 9 is the percentage of the x-th dissimilar parts (i.e., N_i/N_{tot} according to the number of parts (N) shown in Tables 2 and 3).

The amount of information to be processed (I_c) referred to the case study considered, identified by Eq. 4, provides 4.25 [*bit*]. According to Eq. 5, the operator's performance index over time, $IP(n)$, depends on I_c and on time required to perform the task at n repetition, $MT(n)$. Starting from data collected on WM150 (three days, one work shift (8 hours) per day, same operator), the $MT(n)$ have been computed (Figs. 3, 4, and 5), and the corresponding $IP(n)$ have been estimated (Figs. 3, 4, and 5).

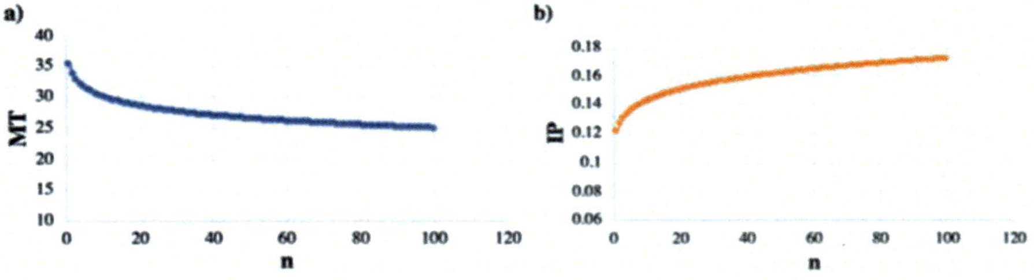

Fig. 3 Movement Time (**a**) and operator's performance index (**b**) estimated at day 1

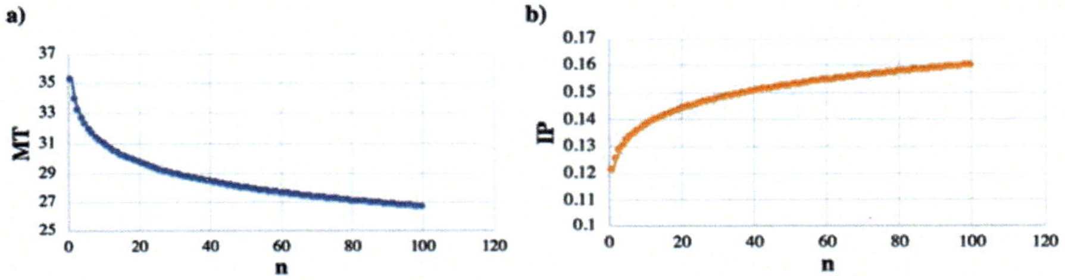

Fig. 4 Movement Time (**a**) and operator's performance index (**b**) estimated at day 2

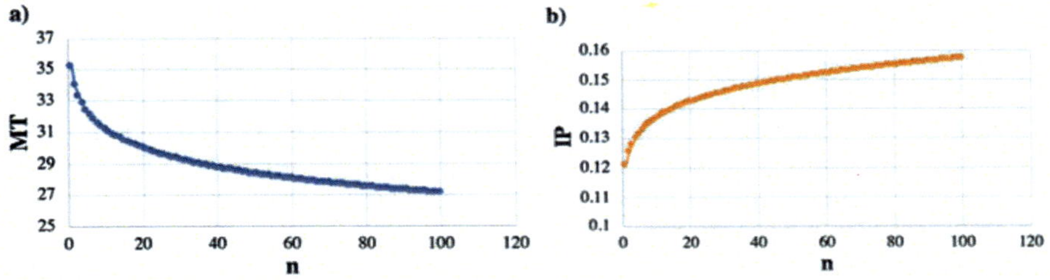

Fig. 5 Movement Time (**a**) and operator's performance index (**b**) estimated at day 3

In the case study considered, the assembly job at *WM* 150, was conducted by a woman of 45 years, highly experienced. The dexterity (ε) of this worker was evaluated per each work shift (Eq. 2). The daily average *IP* and ε are shown in Fig. 6.

5 Discussion

The information-based model developed allows estimating the operator's performance index, *IP*; it is evaluated from the time needed by operators at the first repetition, $MT(1)$, the dexterity of the worker, ε, depending on $MT(n)$, and the manual assembly task complexity, I_c. In the case study developed, the model

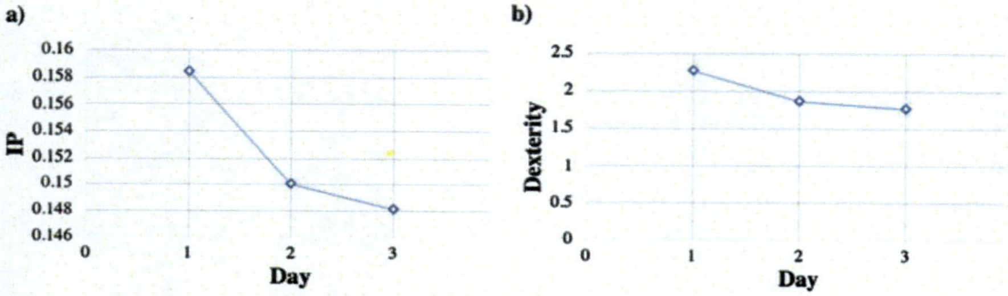

Fig. 6 Average IP (**a**) and dexterity (**b**) values per day

allowed identifying *IP* in accomplishing an assembly task of a pump of a high-pressure diesel injection system.

The results achieved proved that *IP* is strongly related to both the task features and to human behaviour, evaluated in terms of information content, I_c, and worker's dexterity, ε, respectively.

The outcomes shown in the previous section proved that the operator's performance index, *IP*, increases with decreasing in $MT(n)$, observed in the same work shift (Figs. 3, 4, and 5). This means that the operator's performance in the same work shift improves over time. In other words, the amount of information processed in the unit time of the same operator for the same task in the same work shift increases with the increase of the number of repetitions.

On the contrary, the average *IP*, evaluated on different work shifts slightly decreased over time (Fig. 6a). This means that if, on the one hand, the operator's performance improves in the same workday, on the other hand, the operator's performance slightly worsens on different workdays. This effect, consistent with the existing scientific research [29, 30], could be a consequence of the worker's alienation in the repetition of the same jobs. The model's capability to identify this phenomenon depends on the reduction of the effect due to the operator's dexterity (ε) over time (Fig. 6b), resulting from the impossibility to optimise the manual performance beyond a certain limit.

Interestingly, the *IP*, evaluated on different workdays, assumed very similar values (i.e., included between 0.145 and 0.16) for the same operator. This highlights the model's capability to identify a single performance index per operator. It emerged, indeed, that the changes in dexterity (ε) lead to minimal changes in the operator performance (Fig. 7). Consistent with this consideration, the case study proved that *IP* provides a joint indication, including human behaviour (related to dexterity) and task complexity (related to I_c). In other words, *IP* will assume a single value per operator to identify him/her performance in accomplishing an assigned task.

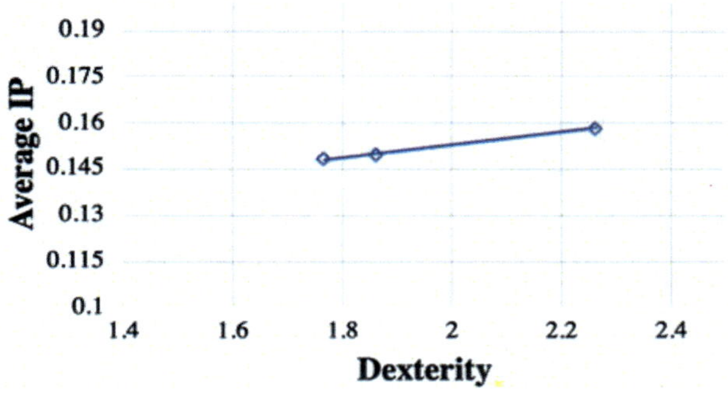

Fig. 7 Average IP-dexterity dependence

According to information theories background, *IP* can be defined as that parameter that allows estimating the processing speed of 1 bit for each operator when the amount of information and the time needed to perform the set of tasks are known.

6 Conclusion and Further Research

A model allowing to evaluate the operator's performance index in accomplishing a manual assembly task with prevalent physical efforts is developed. Consistent with the purpose of the work, the model allowed to estimate the amount of information processed in the unit time to accomplish an assembly job evaluating both the amount and the content of information of the task and the corresponding time needed by operators, with different dexterity skill, to accomplish it. From a managerial perspective, the operator's performance index can be used as an indicator to assign the operator to a specific task to improve the human-system performance. The developed model allows understanding the performance changes during the work shift and on the different workdays. The model paves the way for industrial applications having important managerial implications, especially in digital work environments where subjective aspects could affect productivity. Since the model application is based on only one operator and three observed days, further analysis on a more extensive and heterogeneous sample (e.g., man/woman, different level of experience, different age) and multiple days of observation, are required. Therefore, further research should investigate the model's applicability to a larger sample of participants and to different tasks to evaluate the variability of the operator's performance index.

References

1. Boenzi, F., Digiesi, S., Mossa, G., Mummolo, G., Romano, V.A.: Modelling workforce aging in job rotation problems. IFAC-PapersOnLine. **28**(3), 604–609 (2015)
2. Barreto, L., Amaral, A., Pereira, T.: Industry 4.0 implications in logistics: an overview. Proc. Manuf. **13**, 1245 (2017)
3. Wichmann, R.L., Eisenbart, B., Gericke, K.: The direction of industry: a literature review on industry 4.0. In: Proceedings of the International Conference on Engineering Design, ICED, vol. 2019-Augus, pp. 2129–2138 (2019)
4. Gazzaneo, L., Padovano, A., Umbrello, S.: Designing smart operator 4.0 for human values: a value sensitive design approach. Proc. Manuf. **42**, 219 (2020)
5. Zhu, X.: Modeling product variety induced manufacturing complexity for assembly system design, ProQuest Diss. Theses (2009)
6. Jovane, F., Koren, Y., Boër, C.R.: Present and future of flexible automation: towards new paradigms. CIRP Ann. - Manuf. Technol. **52**(2), 543 (2003)
7. Elmaraghy, W., Elmaraghy, H., Tomiyama, T., Monostori, L.: Complexity in engineering design and manufacturing. CIRP Ann. - Manuf. Technol. **61**(2), 793 (2012)
8. Urbanic, R.J., ElMaraghy, W.H.: Modeling of manufacturing process complexity. In: Advances in Design. Springer, London (2006)
9. ElMaraghy, W.H., Urbanic, R.J.: Assessment of manufacturing operational complexity. CIRP Ann. - Manuf. Technol. **53**(1), 401 (2004)
10. Mital, A., Desai, A., Subramanian, A., Mital, A.: Product Development: a Structured Approach to Consumer Product Development, Design, and Manufacture, 2nd edn. Elsevier, Amsterdam (2014)
11. Bi, S., Salvendy, G.: A proposed methodology for the prediction of mental workload, based on engineering system parameters. Work Stress. **8**, 355 (1994)
12. Stager, U., Amann, W., Maznevski, M.: Managing Complexity in Global Organizations. Wiley, Chichester (2012)
13. Blecker, T., Abdelkafi, N.: Modularity and delayed product differentiation in assemble-to-order systems: analysis and extensions from a complexity perspective. In: International Series in Operations Research and Management Science, vol. 87, (2006)
14. March, J.G., Simon, H.A.: Organizations. 1958. NY Wiley, New York (1993)
15. Latham, G.P., Yukl, G.A.: A review of research on the application of goal setting in organizations. Acad. Manag. J. **18**(4), 824 (1975)
16. Weaver, W.: Science and complexity. Am. Sci. **36**(4), 536 (1948)
17. Fast-Berglund, Å., Fässberg, T., Hellman, F., Davidsson, A., Stahre, J.: Relations between complexity, quality and cognitive automation in mixed-model assembly. J. Manuf. Syst. **32**(3), 449 (2013)
18. Samy, S.N., Elmaraghy, H.: A model for measuring products assembly complexity. Int. J. Comput. Integr. Manuf. **23**(11), 1015 (2010)
19. Boothroyd, G., Dewhurst, P., Knight, W.A.: Product design for manufacture and assembly. Comput. Aided Des. **26**, 505 (2010)
20. Morse, E.P.: On the complexity of mechanical assemblies. In: Proceedings of the ASME Design Engineering Technical Conference, vol. 3, (2003)
21. ElMaraghy, W.H., Urbanic, R.J.: Modelling of manufacturing systems complexity. CIRP Ann. - Manuf. Technol. **52**(1), 363 (2003)
22. Braha, D., Maimon, O.: The measurement of a design structural and functional complexity. IEEE Trans. Syst. Man, Cybern. Part A Syst. Humans. **28**(4), 527 (1998)
23. Zhu, X., Hu, S.J., Koren, Y., Marin, S.P.: Modeling of manufacturing complexity in mixed-model assembly lines. J. Manuf. Sci. Eng. Trans. ASME. **130**(5), 051013 (2008)
24. Busogi, M., Ransikarbum, K., Oh, Y.G., Kim, N.: Computational modelling of manufacturing choice complexity in a mixed-model assembly line. Int. J. Prod. Res. **55**(20), 5976–5990 (2017)

25. Chen, W.: Analysis of man-machine-environment system in industrial design and comprehensive evaluation of products man-machine relationship. IOP Conf. Ser. Mater. Sci. Eng. **746**(1), 012039 (2020)
26. Fan, G., Li, A., Zhao, Y., Moroni, G., Xu, L.: Human factors' complexity measurement of human-based station of assembly line. Hum. Factors Ergon. Manuf. **28**(6), 342 (2018)
27. Digiesi, S., Kock, A.A.A., Mummolo, G., Rooda, J.E.: The effect of dynamic worker behavior on flow line performance. Int. J. Prod. Econ. **120**(2), 368 (2009)
28. Fitts, P.M.: The information capacity of the human motor system in controlling the amplitude of movement, 1954. J. Exp. Psychol. **121**, 262 (1992)
29. Coburn, D.: Job alienation and well-being. In: Health and Work Under Capitalism: an International Perspective (2019)
30. Ward, K.: Human and alienating work: what sex worker advocates can teach Catholic social thought. J. Soc. Christ. Ethics. **41**(2), 261 (2021)

Efficiency Assessment of Public Transport Vehicles Using Machine Learning and Non-parametric Models

Bruna Cristina Siqueira Kaiser, Renan Silva Santos, Rodrigo Goyannes Gusmão Caiado, Luiz Felipe Scavarda, and Paulo Ivson Netto

Abstract The transportation sector is essential for today's global economy, as it tan-gents a wide range of issues such as mobility, urban planning, and economic development. Understanding the performance of vehicles is fundamental for the Brazilian economy since millions of passengers are carried by public transport every day, and this sector represents a significant share of the national GDP. Although in literature, there is a range of suitable approaches for efficiency analysis, the fourth industrial revolution has leveraged the way of acquiring data (e.g., via digital technologies), bringing the need for more advanced data analytics models to explore and process the data beforehand, as well as dealing with uncertainty. In this sense, this paper aims to provide a novel approach to assessing the efficiency of public transport vehicles by combining fuzzy clustering and Data Envelopment Analysis models in a real case study with data from embedded sensors in buses in Rio de Janeiro. A more robust integrated approach for evaluating operational efficiency can assist decision-makers and consumers in better comprehending the relationship between energy (fuel) consumption and bus efficiencies. This could enable the authorities and public transport management departments to develop appropriate policies and strategies and to reconstruct certain features of the inefficient routes, thereby increasing the operational efficiency of land transportation, reducing mobility costs, and even decreasing the carbon footprint.

Keywords Fuzzy clustering · Data envelopment analysis · Transportation sector.

B. C. S. Kaiser (✉) · R. G. G. Caiado · L. F. Scavarda · P. I. Netto
Pontifical Catholic University of Rio de Janeiro (PUC-Rio), Rio de Janeiro, Brazil

R. S. Santos
Pontifical Catholic University of Rio de Janeiro (PUC-Rio), Rio de Janeiro, Brazil

Federal Fluminense University (UFF), Niterói, Brazil

V. M. López Sánchez et al. (eds.), *Industrial Engineering and Operations Management*, Springer Proceedings in Mathematics & Statistics 400,
https://doi.org/10.1007/978-3-031-14763-0_17

1 Introduction

In the dimension of passenger displacement, the transportation sector tangents essential issues such as mobility, urban planning, and economic development. Because of this, this sector demands investments and synergies between the public and private sectors [1], because as private companies offer vehicles for passenger transportation, public institutions are responsible for investments in infrastructure in the road network. According to the National Confederation of Transport (CNT), in 2019, 293.9 million passengers were transported in 9 Brazilian capitals, Belo Horizonte-MG, Curitiba-PR, Fortaleza-CE, Goiânia-GO, Porto Alegre-RS, Recife-PE, Rio de Janeiro-RJ, Salvador-BA e São Paulo-SP, in the scope of regular and chartered transport, as classified by the National Land Transport Agency (ANTT) [2]. Additionally, in 2018, the gross domestic product (GDP) of the transport sector, in which the passenger transport sector is included, was R$ 256.08 billion, in front of the R$ 6.83 trillion of national GDP [3].

Evaluating the operational efficiency of land transportation is an essential practice for the sector, since it can generate ways to optimize the operation, reduce costs and decrease the carbon footprint [4]. There are also other factors that highlight the importance of efficiency evaluation in the industry mentioned above, such as rising fuel prices and changes in legislation regarding gas emissions [5]. In this context, understanding the performance of vehicles is fundamental and for this purpose several measures can be used, such as fuel/km, cost/km [6] and mean time between failures (MTBF) [7]. In order for the measurements to be genuinely representative, the reliability of the data sources must play an important role [8, 9], and the adoption of the Internet of Things (IoT) paradigm has stood out, since it supports several types of applications from the most diverse domains for monitoring, anomaly detection and data collection in real-time [10].

There is a range of suitable methodologies for assessing efficiency levels in the literature, but these use limited and subjective measures [11, 12]. As numerous factors impact the volumes of data collected, integrated approaches that use specific, advanced analysis techniques can complement already widely used models by addressing the uncertainties associated with approaches that enable performance comparisons [13] used an integrated approach, fuzzy clustering, and DEA, in their studies to generate an analysis more feasible with reality through efficiency analysis of more homogeneous groups by the fuzzy paradigm. [14] proposed a framework, from the integrated approach with fuzzy logic, clustering and DEA, to assist in identifying environmental components that influence the levels of efficiency of the analyzed units for efficiency calculation and ranking construction.

Although the relevance of the transportation sector is evident and studies on efficiency using the DEA methodology are directed to it [15, 16] a focus on the environmental theme [17, 18] is observed in front of only a few studies on operational transportation efficiency [19]. This paper aims to fill the gap that exists in studies addressing operational efficiency assessment of the transportation sector. To this end, a case study undertaken in a passenger transportation company in Brazil, from an integrated approach with Fuzzy K-means and DEA, was conducted. In this sense, this study brings contributions in practical and theoretical dimensions, since it brings elements of support for decision making for managers in the transportation sector, through applications of integrated methodologies, thus addressing issues of operational efficiency obtained through DEA, considering uncertainties (fuzzy).

This article is organized as follows: Sect. 2 is dedicated to the construction of the theoretical basis, in which the theme of operational efficiency and its measurement through DEA in the transportation sector is presented. Section 3 is reserved for the methodology aspects, where the focus is on the Fuzzy K-means and DEA themes. Section 4, in turn, delimits the case study, bringing the constitutive elements of the case and contains the results of the DEA model used to perform the analysis of vehicle efficiency performance, in addition to discussions about the study carried out. Finally, in Sect. 5 the conclusions and indications of theoretical and practical implications are made.

2 Background

Efficiency in the passenger transport sector can be understood from the relationship that is established between input and output variables, which can be associated with the operational dimensions (fuel consumption and use of mechanical parts) or social (quality of service offered to the population) [20]. There is a range of issues regarding the operational aspects that touch on them. In this sense, [21] highlighted that the combination of the engine and its respective type of transmission, are sources of model variation and can, therefore, provide significant operational differences. A very latent issue is fuel consumption and its consequent CO2 emissions [22], a recurring theme in sustainability studies (Richardson, 2005). Moreover, fuel consumption relates to the issue of technical (energy) efficiency, which can be measured considering two paradigms, that of maximizing outputs [20, 23, 24] and that of minimizing inputs [23, 24].

The transportation sector is highly dependent on fossil fuels, as a consequence, some indicators are widely used to relate important quantities. Such as fuel efficiency, which measures the amount of work derived from combustion and fuel consumption that expresses the relationship between fuel consumed per vehicle and distance traveled [25]. The literature has addressed the issue of technical efficiency from some models, such as DEA, which allows calculating degrees of efficiency, establishing efficiency frontiers, and ranking units based on their respective performances [26]. However, this approach has limitations if large volumes of data are used to assign performance measures. In order to overcome this problem, the Fuzzy K-means method presents itself as a feasible alternative for an initial treatment of the data for later efficiency analysis of specific groups that have similar degrees of pertinence. Therefore, this approach takes into account degrees of similarities and dissimilarities between the data allocated in specific clusters [27].

Traditionally, many studies presented in the literature evaluate issues related to efficiency performance in the transportation sector with the exclusive use of the DEA method. [28] applied the DEA method to perform a technical efficiency analysis of Indian state-owned public transport companies and as the main results brought the performance of the sector and practical suggestions for improvement to promote the increase of the overall efficiency level of the sector. [24] conducted a diagnostic analysis of the public transport sector from the bidding paradigm. To do so, she used the DEA method to estimate efficiency indices to define more appropriate criteria for selecting bids to generate benefits for the population and the government. [11] evaluated the eco-efficiency of an automotive industry from the perspective of sustainable development using the DEA method, whose main result was the proposition of a method to select vehicles according to their eco-efficiency score.

There is also a trend of productions considering hybrid approaches, considering fuzzy clustering to confer homogeneity to the data and subsequent efficiency analysis with DEA in several areas of knowledge, such as in the energy sector [29], in the health sector [13], in the banking sector [14], among others. [30], on the other hand, evaluated efficiency in the transportation sector. Therefore, they analyzed the efficiency of scheduled public transport routes, DEA was applied to measure the efficiency and fuzzy clustering approach to rank the routes to have a more understandable evaluation of the results. The main results indicated that fuzzy clustering tools are efficient for clustering high-dimensional data sets and present a practical utility of the model for a diagnostic evaluation of route performance. Table 1 provides an overview of efficiency analysis studies in the transportation sector with a breakdown of the respective inputs and outputs.

It is observed that the majority of efficiency analysis studies only adopt the DEA as an analysis method, thus evidencing its importance to evaluate entities from a technical, sectorial, sustainable point of view [31, 32]. The study conducted by [30] highlighted the importance of adopting mixed methods, producing results closer to real conditions, considering the use of fuzzy clustering associated with DEA.

3 Methods

This section presents the methods used to analyze vehicle data from the public transport company mentioned in Sect. 1. To avoid decreasing the homogeneity of the analyzed set, increasing the possibility of performance results being affected by external factors. To deal with data inaccuracy a fuzzy partition clustering model was used to divide the data set into k groups before analyzing efficiency through DEA models. The analysis relies on a combination of Fuzzy K-means and DEA, as detailed below. The variables used in the DEA model consist of one input (the bus fuel consumption) and one output (traveled distance), with the Decision-Making Units (DMUs) being the buses. As the buses travel in predefined routes, with a fixed number of travels a day, the model used is oriented to the minimization of the input

Table 1 Summary of efficiency analysis studies in the transportation sector

Authors	Application	Method	Inputs	Outputs
Agarwal et al. (2011)	Technical efficiency of the public transport sector	DEA	Fleet size, Total staff, Fuel consumption and Accident per lakh kilometer	Bus utilization, Passenger kilometers, and Load factor
Hanauerová (2019)	Technical efficiency of public procurements in the bus transportation sector	DEA	Consumption of both fuel and equipment, effective driving hours, the total number of seats, and the number of staff employed	Passenger and seat kilometers
Heymann et al. (2021)	Eco-efficiency of load transport vehicles	DEA	Load capacity	Fuel efficiency, energetic consumption and total emissions
Pham et al. (2020)	Efficiency of the public transport sector	Clustering, Fuzzy logic and DEA	Annual operation times, cancellation times, per-week sailing frequency, number of ships, gross vessel tonnage, voyage time and distance and distance from the ticket office to ferry ship ramp	Number of passengers and the proportions of general passengers and islanders

(fuel consumption), since from an operational point of view, it is more achievable than increasing traveled distance, given the routes fixed characteristics.

3.1 Fuzzy K-Means Clustering

Data analysis based on clustering approaches can enable more assertive interpretations since they consider degrees of similarity and dissimilarity, ultimately leading to grouping these data in subsets with common characteristics, the so-called clusters. In this sense, [27] highlight that the clustering approach makes it possible to analyze enormous volumes of data, since the subgroups are expectedly more homogeneous than the entire dataset, reducing the variability among them. Nevertheless, the literature has been pointing out some problems with this approach. One of the main criticisms stems from the fact that clusters are generalist, since in crisp clustering, it is usually considered that membership of all data in a given cluster is equal to 1, e.g., or the data is part of the cluster or not. However, this classification does not capture all the nuances of real applications. In this context, approaches such as Fuzzy K-means clustering can be used, since they have a more realistic approach, capturing the uncertainty that typically exists when dealing with real-world data [33].

Fuzzy K-means is a clustering technique based on the original K-means clustering algorithm [34], but incorporating the concepts of Fuzzy Logic [35], to provide a realistic classification of data in which each data point has a probability of pertaining to each of the clusters, instead of simply pertaining or not, which, as aforementioned, is inadequate when uncertainty needs to be considered. The basic idea of Fuzzy K-means is to use a numerical programming algorithm in order to minimize the following objective function:

$$J_q(U, V) = \sum_{j=1}^{N} \sum_{i=1}^{K} (u_{ij})^q d^2(X_j, V_i); \qquad K \leq N \tag{1}$$

Where U is a fuzzy K-partition of the data, V is a set of possibilities for K, q is any real number greater than 1, X_j is the j-th m-dimensional feature vector, V_i is the centroid of the i-th cluster, u_{ij} is the degree of membership of X_j in the i-th cluster, $d^2(X_j, V_i)$ is any chosen distance metric between X_j and V_i, N is the number of data points and K is the number of clusters [33]. The q parameter functions as the weighting exponent for u_{ij} and controls the "fuzziness" of the resulting clusters [36]. The minimization problem can be solved with algorithms like the one presented in [33] by Gath and Geva.

The results of Fuzzy K-means are the centroids of the clusters and a membership degree for each cluster for each data point, so that higher membership of a data point to a given cluster represents a higher probability that the data point belongs to the cluster.

When dealing with real data, fuzzy clustering techniques encompass some issues, such as the variability of the data, and the fact that the number of clusters and location of centroids is not known a priori. So the analysis must include cluster validation metrics, such as Dunn Index or the Silhouette Method [37].

3.2 DEA Models

DEA is a method suitable for comparison of performance or, in other words, efficiency measurements [38], that has been extensively used in various fields and applications, from companies to countries [39]. DEA evaluates the efficiency of production units, called DMU in terms of inputs and outputs.

It is crucial to notice that DEA also provides means to identify the least efficient DMUs, as well as a mean for benchmarking these because of the most efficient

DMUs, since the performance in DEA is always relative, e.g., it estimates the performance of a DMU with relation to the others, not absolutely [40].

DEA is based on the concept of efficiency presented in [41]. Linear Programming methods to solve the optimization problem, like the one in [38], are used to determine the so-called efficient frontier and the projection of the inefficient DMUs. The general form of the optimization problem for the DEA-CCR (named after Charnes, Cooper and Rhodes, who devised the solution in [38]) is presented as follows.

The objective is to minimize:

$$h_0 = \sum_{i=1}^{m} v_i x_{i0} \tag{2}$$

Subject to the restrictions:

$$\sum_{r=1}^{s} u_j y_{j0} = 1 \tag{3}$$

$$\sum_{j=1}^{s} u_j y_{jr} - \sum_{i=1}^{m} v_i x_{ir} \leq 0 \qquad r = 1, \ldots, s \tag{4}$$

$$u_j v_i \geq 0 \quad j = 1, \ldots, n \quad i = 1, \ldots, m \tag{5}$$

Where x_{ir} and y_{jr}, always positive, are the input i and output j of DMU r, s is the number of DMUs, v_i and u_i are the respective weights and the objective function h_0 represents the efficiency.

Some variations of the DEA-CCR, such as the DEA-BCC (named after Banker, Charnes and Cooper [42]) were developed over the years. The DEA-BCC is expressed as follows:

The objective is to minimize:

$$h_0 = \sum_{i=1}^{m} v_i x_{i0} + v_* \tag{6}$$

Subject to the restrictions:

$$\sum_{r=1}^{s} u_j y_{j0} = 1 \tag{7}$$

$$\sum_{j=1}^{s} u_j y_{jr} - \sum_{i=1}^{m} v_i x_{ir} - v_* \leq 0 \quad r = 1, \ldots, s \tag{8}$$

$$u_j v_i \geq 0 \quad j = 1, \ldots, n \quad i = 1, \ldots, m \tag{9}$$

$$v_* \in \mathbb{R} \tag{10}$$

In the DEA-BCC model, the parameter v_* can be interpreted as the scale factor, encompassing one of three scale returns situations: increasing, constant or negative [42, 43].

The efficiency obtained from the DEA-BCC is also called Technical Efficiency (TE), and the efficiency given by the DEA-CCR model is called Production Efficiency (PE). As DEA-CCR considers variable scale returns, a Scale Efficiency can be defined as [44]:

$$SE\ (x_k, y_k) = \frac{PE\ (x_k, y_k)}{TE\ (x_k, y_k)} \tag{11}$$

4 Case Study and Results

In this section, a case study dealing with the use of DEA to evaluate operational efficiency in a bus transport company is discussed. This application is a real case of passenger vehicle performance evaluation in the context of Industry 4.0. Data acquisition from the buses occurred through embedded IoT sensors that monitored real operational parameters of the vehicles to provide insights into operational benefits, such as reducing fuel consumption and applying predictive maintenance. To ensure greater homogeneity of the DMUs analyzed, only diesel-powered buses that performed transportation under the same conditions were considered.

The company under analysis is a bus transportation company that operates several bus lines in Rio de Janeiro. During the month of September of 2020, each bus was monitored to obtain the data of fuel consumption and traveled distance. Table 2 shows the first five entries of data used. The total number of buses under analysis is 241.

The complete dataset of 241 entries was then clustered into three groups using k-means clustering with the help of the R programming language. The results of the clustered data are shown in Fig. 1.

The number of clusters (k = 3) was chosen by analyzing measures such as the Dunn Index and mean Silhouette [37], with the respective obtained values of 1.8959 and 0.8386, corresponding to the best values when compared to other k in the range of 2–10 clusters for this dataset.

Table 2 Example data from buses

Vehicle ID	Total Fuel Consumption (l)	Total Traveled Distance (km)
41001	847	2335
41004	2469	6236
41005	568	1495
41009	3406	9118
41010	4758	10998

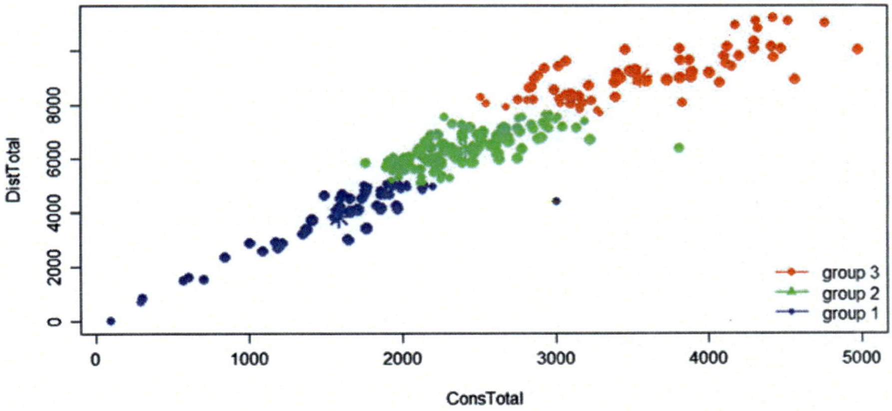

Fig. 1 Clusters were obtained after applying fuzzy k-means to bus consumption and distance data

Table 3 Summary of the clusters

Cluster	N° of elements	Min. Membership Degree	Max Membership Degree	Avg. Membership Degree
Cluster 1	57	0.49	1	0.78
Cluster 2	110	0.47	1	0.83
Cluster 3	74	0.49	1	0.83

The summary of the clusters is shown in Table 3. It can be noticed that, as expected given the validation metrics, the clusters are well defined (average membership higher than 0.75 for all the clusters). Also, cluster 2, which comprises elements that do not have low traveled distance and low fuel consumption nor travel a high distance with a high fuel consumption, is the largest one, with almost half of the elements.

As can be seen from Fig. 1, cluster 1 groups the buses which traveled less in that month, thusly consuming less fuel and cluster 3 represents the vehicles which traveled more and consequently used more fuel. This difference in the characteristics of the groups makes it adequate to analyze the groups separately. Also, cluster 2 is the cluster with the most elements, making it a good representative of the group. Thus, we used DEA cluster 2 to evaluate efficiency. The results are shown in Fig. 2.

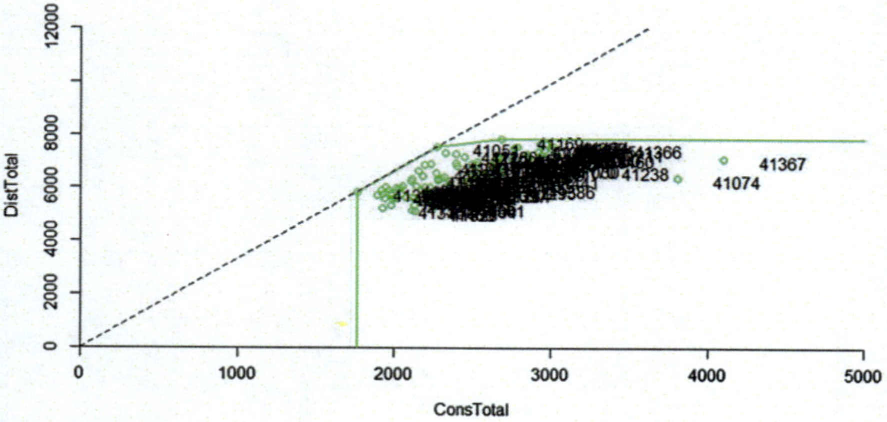

Fig. 2 DEA plot for cluster 2

Table 4 Most and least efficient DMUs for cluster 2

TE			PE			SE		
Ranking	ID	Eff.	Ranking	ID	Eff.	Ranking	ID	Eff.
1	41169	1	1	41051	1	1	41051	1
2	41051	1	2	41316	0,99690	2	41440	0.99998
3	41316	1	3	41226	0,94356	3	41366	0.999974
4	41226	0,94392	4	41293	0,94113	4	41215	0.99997
5	41293	0,94214	5	41261	0,93220	5	41088	0.99992
...	...	...	...	...	...	...	...	...
237	41295	0,68562	237	41295	0,683715	237	41164	0.89733
238	41386	0,65707	238	41386	0,655277	238	41315	0.88810
239	41238	0,62514	239	41238	0,624287	239	41169	0.87906
240	41367	0,52130	240	41367	0,520952	240	41466	0.87717
241	41074	0,50516	241	41074	0,504175	241	41320	0.87359

The five most and least efficient DMUs for cluster 2, ranked by Technical Efficiency (TE), Production Efficiency (PE) and Scale Efficiency (SE), oriented to the minimization of the input are shown in Table 4

It can be seen that, although units 41295, 41386, 41238, 41367 and 41074 can be considered inefficient both from technical and production points of view, they are not the most inefficient when returns to scale are evaluated. Actually, when considering returns to scale, even the most inefficient unit, 41320, is 87% as efficient as the most efficient, 41051. Unit 41051 is the only one with total efficiency in the three analyses. Units 41316, 41226, and 41293 can be considered efficient from the technical and production perspective, although they do not appear in the top-of-scale return ranking. This can be due to, as aforementioned, scale returns being high even at the bottom of the ranking, meaning a smaller variability in this measure of efficiency when compared to technical or production efficiency.

5 Conclusion and Future Research

5.1 Concluding Remarks and Implications

In this work, the energy efficiency of the vehicles of a public transport company was evaluated. The clustering + DEA approach showed that it is possible to evaluate a representative set of units or divide the whole set into smaller sets to compare vehicles with similar characteristics. The approach used also showed the value of data that can be collected from vehicles and leveraged to increase efficiency. This allows, combined with other kinds of data, to establish a benchmark among the vehicles, clearly showing the most efficient, which could drive decision-makers to prioritize the most efficient vehicles, which in turn increases operational efficiency and, at the same time, reduces energy (fuel) consumption.

Therefore, the conduction of this study brought contributions in mainly two dimensions, the practical and the theoretical. From a practical point of view, as it is a case study, the present study provided public transport fleet managers with information that enables the optimized allocation of vehicles, prioritizing as soon as the most economical vehicles are arranged on priority routes or at peak times, for example. In addition, this type of analysis can provide helpful information to detect points of improvement in route management [6] and contribute to providing a public service that meets the real needs of the population [30]. From a theoretical point of view, the present study contributes to minimizing the existing gap in the literature on operational efficiency analysis in the transport sector [19], using an innovative approach, exploring tools consolidated in the academy jointly, and achieving satisfactory results.

5.2 Limitations and Further Research

Although the present study has taken an integrated approach to analyzing the efficiencies of mass transit vehicles, the results obtained are limited to the analysis of vehicles' traveled distance and fuel consumption, due to the number of selected variables (available). In this context, the complete approach must involve data from other operational (or correlated) characteristics, such as vehicle age, driver characteristics, average number of passengers, type of fuel used, route, institutional characteristics and other possible outcomes, such as emissions of carbon dioxide. This situation allows considering, at the same time, the human, technological and organizational aspects, in a holistic perspective of efficiency analysis, which provides results that better describe the reality [45]. In addition, other exploratory data analysis approaches can be used, such as fuzzy k-medoids, whose selection of representative objects, called medoids, is done to minimize the overall difference within each cluster. However, when using a more significant number of variables, principal components analysis (PCA) is recommended since the principal components, calculated from inter-correlated dependent variables, highlight the most important information, thus allowing to visualize patterns of similarity between them.

Regarding the DEA models, future research should rely on these data and evolve upon present work using DDF (Directional Distance Function), which offer some advantages when including multiple inputs-multiple outputs problems with undesirable variables in efficiency calculations [40]. Moreover, non-radial approaches can be used, minimizing and/or maximizing the respective inputs and outputs, verifying the performance. The inverted frontier approach can also provide important performance comparison scenarios in DEA models [11]. Finally, regression models can be configured as a viable alternative for understanding the influence of external variables on vehicle performance.

Acknowledgments This work was supported by the Coordenação de Aperfeiçoamento de Pessoal de Nível Superior – Brasil - CAPES [Finance Code 001] & [Grant Number 88881.198822/2018-01]; Brazilian National Council for Scientific and Technological Development – CNPq [311757/2018-9]; Carlos Chagas Filho Foundation for Research Support of the State of Rio de Janeiro – FAPERJ [Grant number E-26/201.363/2021; E26/211.298/2021]

References

1. Russo, F., Pellicanò, D.S.: Planning and sustainable development of urban logistics: from international goals to regional realization. WIT Trans. Ecol. Environ. **238**, 59–72 (2019). https://doi.org/10.2495/SC190061
2. ANTT: Relatório Anual de Atividades. Agência Nac. Transp. Terr. **5**, 129 (2019)
3. CNT.: Anuario CNT do Transporte 2021. 1–25 (2021)
4. Wang, C.N., Le, T.Q., Yu, C.H., Ling, H.C., Dang, T.T.: Strategic environmental assessment of land transportation: an application of DEA with undesirable output approach. Sustain. **14** (2022). https://doi.org/10.3390/su14020972
5. Hofman, T., Dai, C.H.: Energy efficiency analysis and comparison of transmission technologies for an electric vehicle. 2010 IEEE Veh. Power Propuls. Conf. VPPC 2010. 31–36 (2010). https://doi.org/10.1109/VPPC.2010.5729082
6. Pina, V., Torres, L.: Analysis of the efficiency of local government services delivery. An application to urban public transport. Transp. Res. Part A Policy Pract. **35**, 929–944 (2001). https://doi.org/10.1016/S0965-8564(00)00033-1
7. Falcone, D., Silvestri, A., Duraccio, V.: MKBF, MKTR: new indexes for the maintenance management in transport. Int. Work. Harbour, Marit. Multimodal Logist. Model. Simul. (2003)
8. Nascimento, D., Caiado, R., Tortorella, G., Ivson, P., Meiriño, M.: Digital Obeya room: exploring the synergies between BIM and lean for visual construction management. Innov. Infrastruct. Solut. **3**, 1 (2018)
9. Muniz, M.V.P., Lima, G.B.A., Caiado, R.G.G., Quelhas, O.L.G.: Bow tie to improve risk management of natural gas pipelines. Process. Saf. Prog. **37** (2018). https://doi.org/10.1002/prs.11901
10. Du, X., Zhou, Z., Zhang, Y., Rahman, T.: Energy-efficient sensory data gathering based on compressed sensing in IoT networks. J. Cloud Comput. **9** (2020). https://doi.org/10.1186/s13677-020-00166-x
11. Heymann, M.C., Paschoalino, F.F., Caiado, R.G.G., Lima, G.B.A., Pereira, V.: Evaluating the eco-efficiency of loading transport vehicles: a Brazilian case study. Case Stud. Transp. Policy. **9**, 1688–1695 (2021). https://doi.org/10.1016/j.cstp.2021.06.018

12. Mahlberg, B., Luptacik, M.: Eco-efficiency and eco-productivity change over time in a multisectoral economic system. Eur. J. Oper. Res. **234**, 885–897 (2014). https://doi.org/10.1016/j.ejor.2013.11.017
13. Omrani, H., Shafaat, K., Emrouznejad, A.: An integrated fuzzy clustering cooperative game data envelopment analysis model with application in hospital efficiency. Expert Syst. Appl. **114**, 615–628 (2018). https://doi.org/10.1016/j.eswa.2018.07.074
14. Kazemi, S., Mavi, R.K., Emrouznejad, A., Mavi, N.K.: Fuzzy clustering of homogeneous decision making units with common weights in data envelopment analysis. J. Intell. Fuzzy Syst. **40**, 813–832 (2021). https://doi.org/10.3233/JIFS-200962
15. Güner, S., Coşkun, E.: Estimating the operational and service efficiency of bus transit routes using a non-radial DEA approach. EURO J. Transp. Logist. **8**, 249–268 (2019). https://doi.org/10.1007/s13676-018-0123-1
16. Li, Q., Bai, P.R., Chen, Y., Wei, X.: Efficiency evaluation of bus transport operations given exogenous environmental factors. J. Adv. Transp. **2020** (2020). https://doi.org/10.1155/2020/8899782
17. Caiado, R.G.G., Quelhas, O.L.G., Nascimento, D.L.M., Anholon, R., Leal Filho, W.: Measurement of sustainability performance in Brazilian organizations. Int. J. Sustain. Dev. World Ecol. **25** (2018). https://doi.org/10.1080/13504509.2017.1406875
18. Caiado, R.G.G., Lima, G.B.A., Gavião, L., Quelhas, O.L.G., Paschoalino, F.F.: Sustainability analysis in electrical energy companies by similarity technique to ideal solution. IEEE Lat. Am. Trans. **15** (2017). https://doi.org/10.1109/TLA.2017.7896394
19. Zhou, G., Chung, W., Zhang, Y.: Measuring energy efficiency performance of China's transport sector: a data envelopment analysis approach. Expert Syst. Appl. **41**, 709–722 (2014). https://doi.org/10.1016/j.eswa.2013.07.095
20. Mezghani, M.A., Boujelbene, Y.: The efficiency of public road transport in Tunisia: Validation by the DEA method. 1–12
21. Scavarda, L.F., Schaffer, J., Scavarda, A.J., da Cunha Reis, A., Schleich, H.: Product variety: an auto industry analysis and a benchmarking study. Benchmarking. **16**, 387–400 (2009). https://doi.org/10.1108/14635770910961399
22. Huang, Y., Surawski, N.C., Organ, B., Zhou, J.L., Tang, O.H.H., Chan, E.F.C.: Fuel consumption and emissions performance under real driving: comparison between hybrid and conventional vehicles. Sci. Total Environ. **659**, 275–282 (2019). https://doi.org/10.1016/j.scitotenv.2018.12.349
23. Society, R.S.: The measurement of productive efficiency author (s): M. J. Farrell source. J. Royal Stat. Soc. Ser A (General). **120**(3) (1957) Published by : Wiley for the Royal Statistical Society Stable: http://www.js. J. R. Stat. Soc. 120, 253–290 (2017)
24. Hanauerová, E.: Assessing the technical efficiency of public procurements in the bus transportation sector in the Czech Republic. Socio Econ. Plan. Sci. **66**, 105–111 (2019). https://doi.org/10.1016/j.seps.2018.07.010
25. National Academies Press: Chapter 5. In: Real prospects for energy efficiency in the United States. Natl. Res. Counc (2010)
26. Charnes, A., Cooper, W.W., Rhodes, E.: Measuring the efficiency of decision making units. Eur. J. Oper. Res. **2**, 429–444 (1978)
27. Valente de Oliveira J., Pedrycz, W.: Advances in Fuzzy Clustering and its Applications. (2007)
28. Agarwal, S., Yadav, S.P., Singh, S.P.: DEA based estimation of the technical efficiency of state transport undertakings in India. Opsearch. **47**, 216–230 (2010). https://doi.org/10.1007/s12597-011-0035-4
29. Azadeh, A., Ghaderi, S.F., Anvari, M., Saberi, M., Izadbakhsh, H.: An integrated artificial neural network and fuzzy clustering algorithm for performance assessment of decision making units. Appl. Math. Comput. **187**, 584–599 (2007). https://doi.org/10.1016/j.amc.2006.08.092
30. Pham, T.Q.M., Lee, G., Kim, H.: Toward sustainable ferry routes in Korea: analysis of operational efficiency considering passenger mobility burdens. Sustain. **12**, 1–22 (2020). https://doi.org/10.3390/su12218819

31. Machado, E., Scavarda, L.F., Caiado, R.G.G., Thomé, A.M.T.: Barriers and enablers for the integration of industry 4.0 and sustainability in supply chains of MSMEs. Sustain. **13**, 11664 (2021)
32. Caiado, R.G.G., Scavarda, L.F., Azevedo, B.D., de Nascimento, D.L.M., Quelhas, O.L.G.: Challenges and benefits of sustainable industry 4.0 for operations and supply chain management—a framework headed toward the 2030 agenda. Sustain. **14** (2022). https://doi.org/10.3390/su14020830
33. Gath, I., Geva, A.B.: Unsupervised optimal fuzzy clustering. IEEE Trans. Pattern Anal. Mach. Intell. **11**, 773–780 (1989). https://doi.org/10.1109/34.192473
34. Macqueen, J.: Some methods for classification and analysis of multivariate observations. In: 5th Berkeley Symposium on Mathematical Statistics and Probability, pp. 281–297 (1967)
35. Zadeh, L.A.: Fuzzy logic. Computer (Long Beach. Calif). **21**, 83–93 (1988). https://doi.org/10.1109/2.53
36. Bezdek, J.C.: Pattern Recognition with Fuzzy Objective Function Algorithms. Springer US, Boston (1981)
37. Halkidi, M., Vazirgiannis, M., Batistakis, Y.: On clustering validation techniques. J. Intell. Inf. Syst. **17**, 107–145 (2001). https://doi.org/10.1023/A:1012801612483
38. Charnes, A., Cooper, W.W., Rhodes, E.: Measuring the efficiency of decision making units. Eur. J. Oper. Res. **2**, 429–444 (1978). https://doi.org/10.1016/0377-2217(78)90138-8
39. Zhou, H., Yang, Y., Chen, Y., Zhu, J.: Data envelopment analysis application in sustainability: the origins, development and future directions. Eur. J. Oper. Res. **264**, 1–16 (2018). https://doi.org/10.1016/j.ejor.2017.06.023
40. Caiado, R.G.G., Heymann, M.C., Silveira, C.L.R., Meza, L.A., Quelhas, O.L.G.: Measuring the eco-efficiency of Brazilian energy companies using DEA and directional distance function. IEEE Lat. Am. Trans. **18**, 1844–1852 (2020). https://doi.org/10.1109/TLA.2020.9398625
41. Farrell, M.J.: The measurement of productive efficiency. J. R. Stat. Soc. Ser. A. **120**, 253–290 (1957)
42. Banker, R.D., Charnes, A., Cooper, W.W.: Some models for estimating technical and scale inefficiencies in data envelopment analysis. Manag. Sci. **30**, 1078–1092 (1984). https://doi.org/10.1287/mnsc.30.9.1078
43. Bernardo, M., Rodrigues, L.F.: Análise Envoltória De Dados: Aplicação Do Modelo Ccr E Do Modelo Bcc Para a Avaliação Do Desempenho De Bibliotecas Universitárias De Uma Ifes. Rev. Adm. Contab. e Econ. da Fund. **6** (2016) https://doi.org/10.13059/racef.v6i2.332
44. de Casa Nova, S.P.C., dos Santos, A.: Aplicação Da Análise Por Envoltória De Dados Utilizando Dados Contabeis - Rco. Rev. Contab. e Organ. **3**, 132–154 (2008)
45. Carvalho, A.N., Scavarda, L.F., Lustosa, L.J.: Implementing finite capacity production scheduling: lessons from a practical case. Int. J. Prod. Res. **52**, 1215–1230 (2014). https://doi.org/10.1080/00207543.2013.848484

MRO Inventory Demand Forecast Using Support Vector Machine – A Case Study

Guilherme Henrique de Paula Vidal, Rodrigo Goyannes Gusmão Caiado, Luiz Felipe Scavarda, and Renan Silva Santos

Abstract Today's world is living in the age of digital transformation, the so-called Industry 4.0, in which technological advances have revolutionized the decision-making process in supply chain management. In this domain, inventory management can represent 50% of all organizational costs, and still a challenging task to keep the trade-off between maintaining inventory levels as low as possible, meeting clients' demands, and maintaining satisfactory service levels. Forecasting the MRO inventory demand is even a more difficult task. To address this problem, machine learning (ML) applications, which deal well with nonlinear data, can predict irregular and intermittent demand with better accuracy than traditional approaches. This study employed the Support Vector Machine model to predict maintenance parts demand in a railroad logistic operator case study. This technique can deal with the nonlinear data encompassed by demand variations, avoid overfitting, and produce very accurate classifiers. Results indicated a considerable improvement in the demand forecast performance of the selected SKUs. This model can enhance the reliability of the purchasing and stock maintenance process and generate financial gains by reducing the need for large volumes of safety stock and greater assertiveness in meeting internal demands. It also contributes by showing a real case with an ML approach to predict inventory demands.

Keywords Inventory management · SVM · MRO inventory

1 Introduction

Supply Chain Management (SCM), is a field that has been assuming a strategic position in corporate environments and with increasing influence on the sustainable performance of any business [1–3]. One of the main pillars of SCM is the inventory

---

G. H. de Paula Vidal (✉) · R. G. G. Caiado · L. F. Scavarda · R. S. Santos
Pontifical Catholic University of Rio de Janeiro (PUC-Rio), Rio de Janeiro, Brazil
e-mail: guilherme.vidal@engenharia.ufjf.br

V. M. López Sánchez et al. (eds.), *Industrial Engineering and Operations Management*, Springer Proceedings in Mathematics & Statistics 400,
https://doi.org/10.1007/978-3-031-14763-0_18

management process, which can represent an amount as large as 50% of all SCM costs in an organization [4]. One of the main challenges in this field consists of managing the trade-off between cost efficiency, which is closely related to maintaining inventories as low as possible, and at the same time maintaining an adequate level of service, e.g., being able to meet clients' demands in order to avoid sales losses which could affect the profitability and even the image of a company with the public [5].

Modern inventory management demands from managers agile and reliable decisions. In this context, many tools have been developed, improved, and employed over the years. One of these approaches uses so-called Multi-Criteria Decision Making (MCDM) techniques, which consists of decision-making frameworks for situations involving multiple criteria [6]. Other techniques that have been applied to inventory management consist of Machine Learning (ML) techniques, which consist mainly of computer-implemented algorithms based on available data, which are capable of identifying patterns and making predictions for future demands [7].

The main applications of ML techniques are classification, clustering, association, selection, and regression-based on available data [8]. Among ML applications for inventory management, one that deserves attention is demand predictions. Compared to more traditional approaches to demand prediction, the main advantage of these methods is that they can predict irregular and intermittent with reasonable accuracy since the models can deal well with nonlinear data [9, 10]. A good demand prediction is of utmost importance for business performance since it is the basis for production planning [11]. Reliable and accurate predictions are vital to support SCM planning and decision-making, making these aspects of great interest to organizations [12].

This work presents a case study in which a machine learning technique is used to predict maintenance parts demand in a railroad logistic operator. The technique used is the Support Vector Machine (SVM), which can deal with the nonlinear data encompassed by the demand variations, and at the same time, avoids overfitting and can produce very accurate classifiers [13]. Thus, the present work contributes by showing a real use case for machine learning techniques in predicting inventory demands, which can be replicated for companies in other sectors or other theoretical studies.

This article is organized as follows: Sect. 2 is dedicated to the construction of the theoretical basis, in which the themes of Supply Chain Management and Machine Learning are presented. Section 3 is reserved for the methodology aspects; where the focus is on the SVM model. Section 4, in turn, delimits the case study by bringing the constitutive elements of the case and shows the results of the SVM model used to perform the inventory predictions, in addition to discussions regarding the study undertaken. Finally, in Sect. 5 the conclusions and indications of theoretical and practical implications are made.

2 Background

This section discusses the theoretical background of inventory management and machine learning.

2.1 *Inventory Management*

As stated in the Introduction section, inventory management is one of the most critical processes in a company and directly impacts an organization's performance. It is a challenging task to keep the trade-off between maintaining inventory levels as low as possible and at the same time being able to meet clients' demands. To this end, a series of management policies and methodologies can be employed to keep satisfactory service levels while controlling inventory costs [5].

The existing inventory management models can be divided into probabilistic and deterministic models [14]. Deterministic models are built on the assumption that the parameters that govern the inventory management process are known. On the other hand, probabilistic models consider that one or more parameters are uncertain, which is closer to real situations. Another important classification when dealing with inventory management tools is the inventory review period, which can be continuous, in which inventory status is continuously available, e.g., it is known at any desired time, or periodic review, in which its status is only known when it is reviewed [15]. Thus, four combinations can be achieved considering these two classifications, resulting in four different inventory management policies.

The choice of which inventory management policy to apply is directly related to the inventory item's importance for company processes. Companies have increasingly developed varied products [16], making the Stock Keeping Units (SKU) classification highly beneficial to inventory management and all production planning. Traditionally, SKUs are grouped into three classes, with A being the most important SKUs, B the moderately important SKUs and C the less important SKUs [17]. The so-called ABC SKU classification method stems from the Pareto principle, which states that a small part of the SKUs is responsible for most inventory demand in a given time, usually a year [18]. Thus, this method relies only on a criterion for classification, which is the item consumption in monetary units, obtained from the unitary cost multiplied by the consumption in item units.

There are different types of SKUs that organizations must manage, such as raw material inventories, work-in-process inventories, finished goods inventories, and maintenance repair and operation (MRO) item inventories. The latter focuses on keeping an organization's equipment running smoothly. MRO stocks include office supplies, items for asset maintenance, and spare parts [19]. The management of MRO-type inventories is a great challenge, as they usually present irregular demand patterns and are challenging to forecast. Despite this, the reliability of a good MRO inventory management is paramount because it ensures the smooth operation of a company's facilities [20].

For any chosen inventory management policy, demand forecasting plays a vital role. There are multiple costs associated with inventory management, such as storage costs, ordering costs, and even scarcity costs, related to the risk of keeping inventories lower than needed to meet demands [21]. Thus, inventory management heavily emphasizes accurate demand forecasting [21, 22]. Modern inventory management usually combines big data and machine learning techniques to tackle the challenge mentioned above [23–26].

2.2 *Machine Learning*

Machine learning is a field of artificial intelligence that employs algorithms capable of learning from data and solving problems [7]. Machine learning techniques are inspired by human nature, e.g., they try to emulate the way humans acquire knowledge and experience and evolve and share this knowledge [27].

The machine learning process can be supervised, unsupervised, or by reinforcement. Supervised learning uses input data and expected output data, called labeled data. The system iteratively makes predictions and learns as it adjusts automatically to expected outputs. Thus, in this method, the user defines a priori what the system should learn. In unsupervised learning, only input data is provided, called unlabeled data. The user does not specify a priori what the expected results are. It is up to the system which patterns and correlations can be extracted from data. In reinforcement learning, the system interacts with the environment, receiving negative or positive feedback, so that it learns by a succession of trial and error.

One of the main applications of ML techniques is making predictions. Given its learning properties, algorithms can recognize patterns in data and predict, with some precision, its future results. Compared to more traditional demand prediction approaches, the most significant advantage of ML models is their capability to deal with nonlinear data, which is mainly present when dealing with irregular and intermittent demands [9, 10]. Many ML techniques can be applied to this end, such as artificial neural network (ANN), support vector machine (SVM), k-nearest neighbors (KNN), and random forest (RF) [10, 11, 28].

Some authors address the application of these ML techniques in a specific type of inventory, the MRO type, due to some of its characteristics, such as the irregular and sometimes sporadic nature and the high impact of the lack of these items (lumpy demand) [29, 30]. The wrong decision making in this process can have several negative impacts, such as increased stocking costs, customer satisfaction, profitability, and even impact on their supply chain [31].

3 Research Methodology

This section presented and detailed the research methodology used throughout this study.

3.1 Case Study Definition

According to Yin [32], a case study is a suitable research methodology when (i) the research questions are of the form "how" and "why"; (ii) there is no need to, or it is impossible to control the events and (iii) studying contemporary events. It is also crucial that the research objectives are clear and can be evaluated in the study's conclusions.

A case study can be used to address several types of research, among which are (i) that which deal with the construction of a theory to explain a phenomenon; (ii) improvement of existing theories based on new evidence from the case; (iii) exploratory research, when the goal is to develop new ideas and research questions and (iv) theory testing, used to define the most suitable method for the majority of cases, relations or causal models [33]. Practical applications such as case studies allow you to observe the reality of companies and analyze the interaction between problem, actions, and results [34].

The main phases of a case study consist of [32]:

- planning and designing the case study;
- prepare the instruments for data collection and collect the data;
- analyze the data and share (report) the results.

In this work, the case study followed the phases of planning and designing, with support from the literature background shown in Sect. 2, which helped to identify the theoretical concepts to be applied, collecting the data directly from the company under analysis [35–37], both from databases and from key people involved in the process [38], analysis of data with the RapidMiner software and posterior reporting of the results.

The company under analysis consists of a Brazilian railroad logistic operator, with an inventory of approximately 16,000 SKUs distributed among 12 storage facilities. The choice of this company is due to the fact that inventory management plays a significant role since the costs of keeping and acquiring parts are high, and inventory levels play a major role in company operations.

The main categories of materials in this inventory are:

- spare parts for maintenance of the main assets (trains, railroads, pieces of equipment, and signalization);
- Individual Protection Equipment and uniforms provided to company personnel;
- other kinds of materials to support operation.

As the most critical parts of the inventory are the parts related to the MRO process, 6 SKUs highly used for periodic predictive maintenance of the trains were selected for this case study. Data for the study was obtained from the company, which provided historical records from the SKUs. The data consisted of the historical consumption of all SKUs from January 2010 to December 2020 and the predicted consumption for the year 2020. Also, parameters such as unit cost, lead time, inventory policy, and criticality of the item were also present in the database. Other parameters could be obtained from this primary data, such as the frequency of use of SKUs in 1 year.

Although, in this study, the model is applied to a sample of 6 SKUs, the research has the potential to be replicated for other SKUs in the company or even to predict other types of SKUs. This approach can be applied to MRO inventory items with the same characteristics and demand pattern. Time-series clustering, for example, can help group items with the same characteristics and historical demand patterns, ensuring replicability of the model, which can be treated as a suggestion for future research.

3.2 Data Analysis (Support Vector Machine – SVM)

Support Vector Machine (SVM) is a machine learning model originally developed for supervised classification. SVMs map the data in space, where it iteratively tries to find a separation hyperplane with maximum margin, e.g., with the highest distance between data points that are close, but belongs to different classes. The data samples in the border are called support vectors [39]. SVMs are one of many big data and machine learning models which, solely or in combination with other methods, have been used in the area of Operations Research for demand forecasting [40, 41].

A standard SVM can be defined as follows: given a labeled training dataset $\{x_i, y_i\}_{i=1}^{n}$, where $x_i \in \mathbb{R}^N$ and $y_i \epsilon \{-1, +1\}$ and a nonlinear mapping ϕ to a feature space, the method tries to solve the following problem [42]:

$$\min_{w,\xi_i,b}\left\{\frac{1}{2}\|w\|^2 + C\sum_{i=1}^{n}\xi_i\right\} \tag{1}$$

Restricted to:

$$y_i\left(\langle\phi\left(x_i\right)\rangle, w + b\right) \geq 1 - \xi_i,\ \ \forall i = 1, \ldots, n \tag{2}$$

$$\xi_i \geq 0, , \forall i = 1, \ldots, n \tag{3}$$

Where w and b define a linear classifier in $\mathbb{R}^N$ and ξ_i is a slack variable to enable dealing with allowed errors. C is a user selected parameter that controls the generalization capability of the classifier.

It can be noticed that, qualitatively, the minimization function in Eq. (1) corresponds to the maximization of the margin (which is equivalent to minimizing norm of model weights $\|w\|^2$) combined with minimization of the total error ($\sum_{i=1}^{n} \xi_i$).

4 Case Study and Results

This section presents the case study, applied to a railroad logistic operator, and the results found. The first subsection covers the detailing of the company's inventory management process and the SKUs chosen for the application of the SVM method. Finally, the following subsection presents the case study results of applying the SVM method for MRO inventory demand forecasting.

4.1 *Company Characterization*

Some gaps in the company's inventory management process motivated this work. The company keeps an excessive amount of low rotation materials in its inventories, contributing to increased storage costs and obsolescence. Approximately 20% of the total value stocked are considered non-rotating materials, which have not been used in the last 12 months. Besides this, the adherence between what is planned by the company, through the Bill of Material (BOM), and what is done on a monthly basis is low, harming the process of planning purchases of materials, which can generate excess inventory, causing costs for the company or even not meeting demands, which can directly affect its operation.

Figure 1 shows the consumption history, from January 2010 to December 2020, for each of the 6 SKUs selected for the application of the SVM. The graphs clarify the significant oscillation and variability of the data throughout the months, making it even more challenging to perform a demand forecast with high precision and low error rate. This data behavior justifies the application of the SVM model, given that they are models capable of capturing non-linear patterns, which is often not possible in traditional demand forecast models.

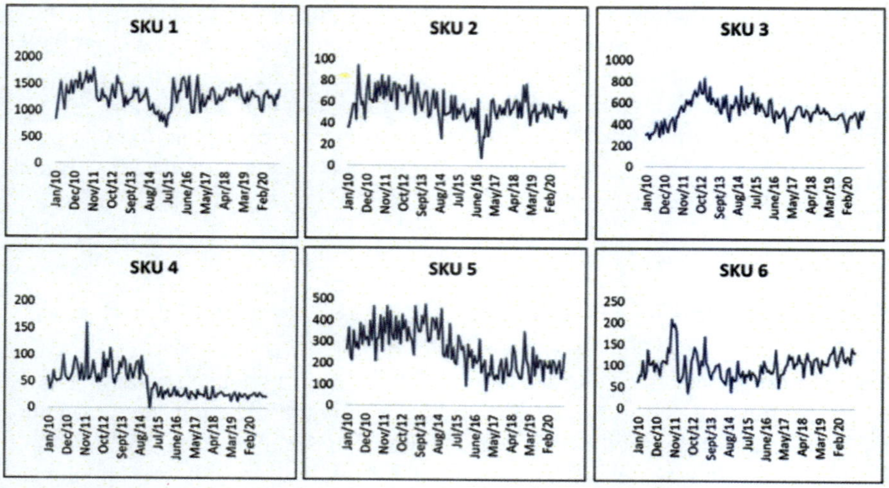

Fig. 1 Monthly consumption of the six selected SKUs, between January 2010 and December 2020

Table 1 SVM model performance measures – validation basis

SKU	RMSE	MAPE (%)
SKU 1	166.61	10.37
SKU 2	11.11	13.99
SKU 3	64.86	8.39
SKU 4	15.18	19.58
SKU 5	71.57	22.47
SKU 6	22.10	19.31

4.2 *Results and Analysis*

From the historical results from January 2010 to June 2019, the training and validation of the SVM model were performed, starting by splitting 80% of the data for training and 20% for validation. Table 1 presents the validation results of the SVM method found for each SKU concerning the two performance indicators used, which are root-mean-square error (RMSE) and mean absolute percentage error (MAPE) [43].

Analyzing the performance results of the SVM models' validation base, it is observed that the MAPE has large variability among SKUs, being 8.39% in the best model and 22.47% in the worst. These results show that the type of prediction and model performance is highly dependent on the type of data being predicted. Regarding RMSE, this comparative analysis between different SKUs for the same forecasting method is not applicable, as it is a method that calculates the root mean square between the errors of the forecasted values concerning the actual ones. Therefore, SKUs with higher data magnitude tend to have higher RMSE than SKUs of lower magnitude, which explains the significant difference in RMSE between SKUs.

Table 2 Performance measures – test basis

	SVM		Company model	
SKU	RMSE	MAPE (%)	RMSE	MAPE (%)
SKU 1	89.21	5.57	430.18	26.21
SKU 2	4.83	7.59	13.81	23.95
SKU 3	64.93	10.90	74.25	13.65
SKU 4	3.10	10.64	3.34	13.30
SKU 5	65.22	35.80	108.60	52.53
SKU 6	24.29	16.27	24.74	16.71

Fig. 2 Actual and forecast demand for the 6 SKUs – test basis

After training and validation, the trained model is applied to an unseen database, called a test, to test the learning generalization power. The outputs of the test correspond to the models' performance measure and the demand forecast for the period. Table 2 illustrates the performance measure of each SKU for the test basis in the proposed SVM model and the performance from the prediction made by the company.

Evaluating the results presented in Table 2, referring to the performance of the SVM test and comparing it with the company's forecast, it can be seen that the SVM has the best MAPE and RMSE measurements for all 6 SKUs. The average MAPE obtained from the 6 SKUs in the SVM model was 10.09%, while the average result from the company's forecast was 16.70%. This highlights the superiority of the model proposed in this paper and the considerable improvement it will bring to the company's results.

Figure 2 compares the actual and forecast demand by the company and from the SVM method for each of the 6 SKUs for the test basis.

Figure 2 makes evident the greater proximity of the forecast represented by the SVM to the actual result of the period concerning the forecast initially made by the company, corroborating the performance results that had already been shown in Table 2.

5 Conclusions

This paper presents an applied study of how SVM can aid decision-making in MRO inventory management by improving the demand forecasting process. Its objective was achieved by applying the SVM to a railroad logistic operator. The results found allow us to conclude that the application of this ML model can bring essential improvements to the inventory management process, with superior performance to the company's results in terms of forecast accuracy, which has, as a consequence, a higher probability of meeting maintenance demands and less need for high safety stocks that generate high costs to the company.

This work has important theoretical and practical implications. The application of the model has the potential to considerably improve the reliability of the purchasing and stock maintenance process, consequently generating financial gains by reducing the need for large volumes of safety stock and greater assertiveness in meeting internal demands.

Some future research opportunities may be directed. To further leverage its results, it is recommended to replicate the model for a broader range of SKUs, focusing on those considered critical and having the most significant financial and internal process impact. It is also recommended that other classical and ML forecast models be tested in addition to the applied one to confirm the SVM as the best demand forecast approach for the selected SKUs. Finally, it is recommended to integrate the SVM results with the company's systems to automatically obtain the results without manual intervention. The company is undergoing a digital transformation process, and this integration will represent a major step in its technological evolution.

Acknowledgments This work was supported by the Coordenação de Aperfeiçoamento de Pessoal de Nível Superior – Brasil – CAPES [Finance Code 001] & [Grant Number 88881.198822/2018-01]; Brazilian National Council for Scientific and Technological Development – CNPq [311757/2018-9]; Carlos Chagas Filho Foundation for Research Support of the State of Rio de Janeiro – FAPERJ [Grant number E-26/201.363/2021; E26/211.298/2021].

References

1. Machado, E., Scavarda, L.F., Caiado, R.G.G., Thomé, A.M.T.: Barriers and enablers for the integration of industry 4.0 and sustainability in supply chains of MSMEs. Sustainability. **13**, 11664 (2021)

2. Caiado, R.G.G., Scavarda, L.F., Azevedo, B.D., Nascimento, D.L.d.M., Quelhas, O.L.G.: Challenges and benefits of sustainable industry 4.0 for operations and supply chain management—a framework headed toward the 2030 agenda. Sustainability. **14**, 1–26 (2022). https://doi.org/10.3390/su14020830
3. Melnyk, S.A., Lummus, R.R., Vokurka, R.J., Burns, L.J., Sandor, J.: Mapping the future of supply chain management: a Delphi study. Int. J. Prod. Res. **47**, 4629–4653 (2009). https://doi.org/10.1080/00207540802014700
4. Lancioni, R.A.: New developments in supply chain management for the millennium. Ind. Mark. Manag. **29**, 1–6 (2000). https://doi.org/10.1016/S0019-8501(99)00106-6
5. Sustrova, T.: An artificial neural network model for a wholesale company's order-cycle management. Int. J. Eng. Bus. Manag. **8**, 1–6 (2016). https://doi.org/10.5772/63727
6. Sitorus, F., Cilliers, J.J., Brito-Parada, P.R.: Multi-criteria decision making for the choice problem in mining and mineral processing: applications and trends. Expert Syst. Appl. **121**, 393–417 (2019). https://doi.org/10.1016/J.ESWA.2018.12.001
7. Priore, P., Ponte, B., Rosillo, R., de la Fuente, D.: Applying machine learning to the dynamic selection of replenishment policies in fast-changing supply chain environments. Int. J. Prod. Res. **57**, 3663–3677 (2019). https://doi.org/10.1080/00207543.2018.1552369
8. Kartal, H., Oztekin, A., Gunasekaran, A., Cebi, F.: An integrated decision analytic framework of machine learning with multi-criteria decision making for multi-attribute inventory classification. Comput. Ind. Eng. **101**, 599–613 (2016). https://doi.org/10.1016/j.cie.2016.06.004
9. Gutierrez, R.S., Solis, A.O., Mukhopadhyay, S.: Lumpy demand forecasting using neural networks. Int. J. Prod. Econ. **111**, 409–420 (2008). https://doi.org/10.1016/j.ijpe.2007.01.007
10. Jaipuria, S., Mahapatra, S.S.: A study on behaviour of bullwhip effect in (R, S) inventory control system considering DWT-MGGP demand forecasting model. J. Model. Manag. **14**, 385–407 (2019). https://doi.org/10.1108/JM2-04-2018-0053
11. Kück, M., Freitag, M.: Forecasting of customer demands for production planning by local k-nearest neighbor models. Int. J. Prod. Econ. **231**, 107837 (2021). https://doi.org/10.1016/j.ijpe.2020.107837
12. Abolghasemi, M., Beh, E., Tarr, G., Gerlach, R.: Demand forecasting in supply chain: the impact of demand volatility in the presence of promotion. Comput. Ind. Eng. **142**, 106380 (2020). https://doi.org/10.1016/j.cie.2020.106380
13. Karamizadeh, S., Abdullah, S.M., Halimi, M., Shayan, J., Rajabi, M.J.: Advantage and drawback of support vector machine functionality. I4CT 2014: 1st Int. Conf. Comput. Commun. Control Technol. Proc. 63–65 (2014). https://doi.org/10.1109/I4CT.2014.6914146
14. Placido dos Santos, F.S., Oliveira, F.: An enhanced L-shaped method for optimizing periodic-review inventory control problems modeled via two-stage stochastic programming. Eur. J. Oper. Res. **275**, 677–693 (2019). https://doi.org/10.1016/j.ejor.2018.11.053
15. Babiloni, E., Guijarro, E.: Fill rate: from its definition to its calculation for the continuous (s, Q) inventory system with discrete demands and lost sales. Cent. Eur. J. Oper. Res. **28**, 35–43 (2020). https://doi.org/10.1007/s10100-018-0546-7
16. Scavarda, L.F., Schaffer, J., Scavarda, A.J., da Cunha Reis, A., Schleich, H.: Product variety: an auto industry analysis and a benchmarking study. Benchmarking. **16**, 387–400 (2009). https://doi.org/10.1108/14635770910961399
17. Soylu, B., Akyol, B.: Multi-criteria inventory classification with reference items. Comput. Ind. Eng. **69**, 12–20 (2014). https://doi.org/10.1016/j.cie.2013.12.011
18. Shamsaddini, R., Vesal, S.M., Nawaser, K.: A new model for inventory items classification through integration of ABC-FUZZY and fuzzy analytic hierarchy process. Int. J. Ind. Syst. Eng. **19**, 239–261 (2015)
19. Chen, J., Gusikhin, O., Finkenstaedt, W., Liu, Y.N.: Maintenance, repair, and operations parts inventory management in the era of industry 4.0. IFAC-PapersOnLine. **52**, 171–176 (2019). https://doi.org/10.1016/j.ifacol.2019.11.171
20. Bailey, G.J., Helms, M.M.: MRO inventory reduction – challenges and management: a case study of the Tennessee Valley Authority. Prod. Plan. Control. **18**, 261–270 (2007). https://doi.org/10.1080/09537280601127351

21. Silver, E.A.: Operations research in inventory management: a review and critique. Oper. Res. **29**, 628–645 (1981). https://doi.org/10.1287/opre.29.4.628
22. Kot, S., Grondys, K., Szopa, R.: Theory of inventory management based on demand forecasting. Pol. J. Manag. Stud. **3**, 148–156 (2011)
23. Aamer, A.M., Yani, L.P.E., Priyatna, I.M.A.: Data analytics in the supply chain management: review of machine learning applications in demand forecasting. Oper. Supply Chain Manag. **14**, 1–13 (2021). https://doi.org/10.31387/oscm0440281
24. Spiliotis, E., Makridakis, S., Semenoglou, A.-A., Assimakopoulos, V.: Comparison of statistical and machine learning methods for daily SKU demand forecasting. Oper. Res. **22**, 3037–3061 (2020). https://doi.org/10.1007/s12351-020-00605-2
25. Carbonneau, R., Laframboise, K., Vahidov, R.: Application of machine learning techniques for supply chain demand forecasting. Eur. J. Oper. Res. **184**, 1140–1154 (2008). https://doi.org/10.1016/j.ejor.2006.12.004
26. Tanizaki, T., Hoshino, T., Shimmura, T., Takenaka, T.: Demand forecasting in restaurants using machine learning and statistical analysis. Procedia CIRP. **79**, 679–683 (2019). https://doi.org/10.1016/j.procir.2019.02.042
27. Min, H.: Artificial intelligence in supply chain management: theory and applications. Int. J. Log. Res. Appl. **13**, 13–39 (2010). https://doi.org/10.1080/13675560902736537
28. Punia, S., Nikolopoulos, K., Singh, S.P., Madaan, J.K., Litsiou, K.: Deep learning with long short-term memory networks and random forests for demand forecasting in multi-channel retail. Int. J. Prod. Res. **58**(1), 1–16 (2020). https://doi.org/10.1080/00207543.2020.1735666
29. Dombi, J., Jónás, T., Tóth, Z.E.: Modeling and long-term forecasting demand in spare parts logistics businesses. Int. J. Prod. Econ. **201**, 1–17 (2018). https://doi.org/10.1016/j.ijpe.2018.04.015
30. Kourentzes, N.: Intermittent demand forecasts with neural networks. Int. J. Prod. Econ. **143**, 198–206 (2013). https://doi.org/10.1016/j.ijpe.2013.01.009
31. Aktepe, A., Yanık, E., Ersöz, S.: Demand forecasting application with regression and artificial intelligence methods in a construction machinery company. J. Intell. Manuf. **32**, 1587–1604 (2021). https://doi.org/10.1007/s10845-021-01737-8
32. Yin, R.k.: Case study research and applications. SAGE (2018)
33. Voss, C., Tsikriktsis, N., Frohlich, M.: Case research in operations management. Int. J. Oper. Prod. Manag. **22**, 195–219 (2002). https://doi.org/10.1108/01443570210414329
34. Carvalho, A.N., Scavarda, L.F., Lustosa, L.J.: Implementing finite capacity production scheduling: lessons from a practical case. Int. J. Prod. Res. **52**, 1215–1230 (2014). https://doi.org/10.1080/00207543.2013.848484
35. Caiado, R.G.G., Lima, G.B.A., Gavião, L., Quelhas, O.L.G., Paschoalino, F.F.: Sustainability analysis in electrical energy companies by similarity technique to ideal solution. IEEE Lat. Am. Trans. **15**, 675–681 (2017). https://doi.org/10.1109/TLA.2017.7896394
36. Nascimento, D., Caiado, R., Tortorella, G., Ivson, P., Meiriño, M.: Digital Obeya Room: exploring the synergies between BIM and lean for visual construction management. Innov. Infrastruct. Solut. **3**, 1 (2018)
37. Muniz, M.V.P., Lima, G.B.A., Caiado, R.G.G., Quelhas, O.L.G.: Bow tie to improve risk management of natural gas pipelines. Process Saf. Prog. **37**, 169–175 (2018). https://doi.org/10.1002/prs.11901
38. Caiado, R.G.G., Quelhas, O.L.G., Nascimento, D.L.M., Anholon, R., Leal Filho, W.: Measurement of sustainability performance in Brazilian organizations. Int. J. Sustain. Dev. World Ecol. **25**, 312–326 (2018). https://doi.org/10.1080/13504509.2017.1406875
39. Russell, S., Norvig, P.: Artificial intelligence a modern approach, 4th edn. Pearson (2021)
40. Wu, Q.: The hybrid forecasting model based on chaotic mapping, genetic algorithm and support vector machine. Expert Syst. Appl. **37**, 1776–1783 (2010). https://doi.org/10.1016/j.eswa.2009.07.054
41. Villegas, M.A., Pedregal, D.J., Trapero, J.R.: A support vector machine for model selection in demand forecasting applications. Comput. Ind. Eng. **121**, 1–7 (2018). https://doi.org/10.1016/j.cie.2018.04.042

42. Salcedo-Sanz, S., Rojo-Álvarez, J.L., Martínez-Ramón, M., Camps-Valls, G.: Support vector machines in engineering: an overview. Wiley Interdiscip. Rev. Data Min. Knowl. Discov. **4**, 234–267 (2014). https://doi.org/10.1002/widm.1125
43. Arunraj, N.S., Ahrens, D.: A hybrid seasonal autoregressive integrated moving average and quantile regression for daily food sales forecasting. Int. J. Prod. Econ. **170**, 321–335 (2015). https://doi.org/10.1016/j.ijpe.2015.09.039

A New Solution to an Old Problem: Inventory Control with Smart Glasses Riverstock

Claudio de Souza Rocha Junior, Adilson Vilarinho Terra, Miguel Ângelo Lellis Moreira, Marcos dos Santos, and Carlos Francisco Simões Gomes

Abstract Assuming the great importance of the stock for organizations, we idealize a solution applicable on a large scale, making possible its implementation without the need to change the organizational structure or implement training with the staff. Based on the Problem Based Learning methodology, the present study proposed an alternative solution for a general logistics problem concerning inventory accounting. On a theoretical basis, a brief review of the literature relating to logistic management is explored, presenting the main points of inventory problems and technological models practiced. In this context, is proposed an alternative technology based on smart glasses devices, integrating hardware and software based on cloud technology in search of making feasible work as support in the management of stocks. The proposal results were positive, demonstrating the viability within the real world, not being limited only to the metaphysical. Therefore, society can benefit by optimizing its resources and avoiding unnecessary waste and expense.

Keywords IoT · Hardware · Inventory management · Product patent

1 Introduction

The use of contemporary technologies in the industry allows the advance of process regarding production models to be considerably more remarkable when compared

C. de Souza Rocha Junior (✉) · A. V. Terra · C. F. S. Gomes
Fluminense Federal University, Niterói, Brazil

M. Â. L. Moreira
Fluminense Federal University, Niterói, Brazil

Naval Systems Analysis Center, Rio de Janeiro, Brazil

M. dos Santos
Naval Systems Analysis Center, Rio de Janeiro, Brazil

Military Institute of Engineering, Rio de Janeiro, Brazil

V. M. López Sánchez et al. (eds.), *Industrial Engineering and Operations Management*, Springer Proceedings in Mathematics & Statistics 400,
https://doi.org/10.1007/978-3-031-14763-0_19

to the old ones, demonstrating in practice the value of creating and implementing new viable alternatives within the production environment [1].

As presented in [2], the industry 4.0 is one of the revolutions involving production methods, as well as a direct line concerning the analysis and manipulation of a large volume of data, where the terminology of companies is born "data-driven": companies moved and managed based on decisions made on data [3].

In this context, organizations based on the production and management of physical artifacts have a technically explicit production and delivery flow in business logistics [4]. Nevertheless, despite all the structured complexity of logistics systems and their studies, stocks with counting errors are a reality for all companies, which is acceptable comprehensibility. After all, to vary from the segment of the organization, they are minimal parts to large machines that need to be correctly quantified [5]. This significant dispersion results in two negative consequences originating from fundamental parameters for the sustainability of an organization: billing and efficiency [6].

According to [7], periodic inventory counting errors take shape in financial reports as significant, and sometimes even invisible, but often highly summed up all variations in usage. Then, seeking to increase revenue, the manager is responsible for the organization's financial success and seeks alternatives, sometimes traditional, such as an action plan to capture new customers or a cost reduction policy [8].

The solutions mentioned above are most effective but inefficient since the problem is found in areas not mentioned in the action plan, such as subsequent losses of products in stock for various reasons, thus resulting in more significant losses, and solving inefficiently. Therefore, seeking the parallel between both, we sought a solution capable of being simultaneously favorable and compelling, where decision-making parameters are better measured [9] from well-defined reports of sectors with the high volatility of variables, such as logistics.

In this scenario, this study presents an alternative technological solution named Riverstock, whose patent registration application is in progress at the National Industrial Property Institute (INPI), the Brazilian Self-Regulatory Body responsible for registering trademarks in Brazil.

Starting from the principle of Design Science, where it is possible to involve the industry and the profession in intellectually important projects [10], we structured Riverstock. Your essence is on using smart glasses and intuitive UX (user experience) to help the work of those people responsible for inventory management on the shop floor. In that regard, Riverstock is the "eyes" that will always be automatically monitoring inventory without the need for any training or change in the company's organizational structure in question.

The literature in Design Science says that the Relevance Cycle inputs requirements from the contextual environment into the research and introduces the research artifacts into environmental field testing [11]. Considering Riverstock's context, it represents the impact caused on organizations through a well-managed and accounted stock, or not, where much of the planning involving organizational success has the store as one of the fundamental pillars.

It is a systemic approach within organizations that have the concept of stock applied in practice, using physical spaces reserved in their layout for storage of possession of an economic agent.

Whether supplies, materials, finished or unfinished products, this approach infers the function of inventories beyond regulating the flow of materials or damping the difference between input and output of objects in focus. Furthermore, this approach is entering their influence and power in routine decision making until reaching strategic pillars, with strategies adopted by members of senior management.

2 Theoretical Background

When organizations do not manage inventory correctly, it can bring a series of losses. It has caused several adverse effects in the budgetary context, including maintenance reduction, spare parts stocks reduction, and training and labor loss [12]. In this sense, companies and governments worldwide have invested in this field of research to solve logistical problems and mitigate the losses mentioned above, as we can see in [13, 14].

What maintains a company is a demand for any product or service. As exposed in [15], the inventories are relevant items, part of any business, from the retail area (wholesaler, services) to homes, and the amount is an independent value. Inventory management is referred to the management of "idle resources" holders of economic value and intended to supply future needs about material within organizations [16].

In a complementary way, [17] infer inventory management as the verification of saved products, wherein your organization must-know dates for making purchases, identifying, and correctly classifying products. In summary, inventory control indicates which actions are required to achieve organizational objectives from the stock context.

In the case of products (particularly physical products), the stock must be well controlled. And this control function is crucial because it is related to the level of accuracy of the expected demand, regardless of the time or organizational level used as a parameter [18].

Not only as an exhibition, but it is also necessary that there is control in the inventory since the inventory control can be of great value to companies. It is possible to monitor and manage the entry and exit of products produced and sold, recording and controlling everything entering and exiting the company [19]. This control helps the company make estimates of production, sales, order from suppliers, and even figure out which products are shipped faster [20].

These concepts are also related to the physical location of the products or materials. Companies must understand that the stock holds a considerable share of capital, so anyways to avoid waste and make its use efficient are valid for the growth of the organization [21]. Inventory management is a vital point for any organization. So, it is possible to infer critical success factors from the planning carried out with the data obtained in this same management [22] by the moment that the planning is essential to get favorable results for the management of a company, where the results tend to be positive and satisfactory [23].

A well-managed stock can bring several advantages to companies in the same way that a neglected stock will mean several risks, such as lack of product or customer service or very high costs due to idle inventory. Therefore, requiring extremely accurate, detailed, and up-to-date information is essential for inventory management.

These theoretical concepts exist to increase efficiency in parallel with maintenance cost reduction. Inventory maintenance costs are related to all costs required to maintain an X quantity of goods for a period Y, where X and Y are variable, so the opportunity cost is an existing parameter on account of capital immobilization, which companies could use otherwise. There are also the costs associated with insurance, taxes, storage costs, and quantity themselves, all related to the risk of losing. These are related to deterioration, obsolescence, damage, or theft [24].

Therefore, to improve processes, there are tools aimed at various purposes, such as 5S, a tool for the work environment, organization of materials, and disposal of unnecessary components [25].

Information technology in the business environment allows the change of ideas and concepts about administration [26]. Its use should be applied in favor of inventory, prioritizing quality and quantity, aiming to control knowledge aligned with the fulfillment of demand [27].

For example, it is possible to cite the ERP system as software embedded in the management mechanisms of the company. The ERP brings together the administrative solutions of various departments of the institution in a single system, eventually dispersing the need to use several similar tools used in the organization [28]. The combination of information technology applied in stock with its correctly realized management allows customization of time and performance in sales, which influences billing and reduces high monetary costs [29].

3 Technology Proposal

The proposal has as a basis the construction of smart glasses for constant investigation of the current condition of the stock.

Using a standardized label containing the product name, barcode, and color of the corresponding card, separated into three colors set in red, green, or blue, we equipped the glasses with a color detector and reader. All the user needs to do is look at the label, and the database is updated, as exposed in Table 1.

Table 1 Table captions should be placed above the tables

Product name	Bar code	Card color

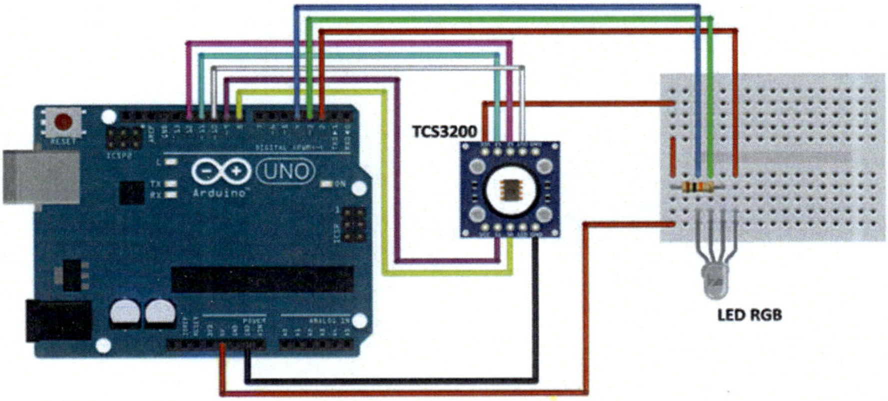

Fig. 1 Prototyping circuit

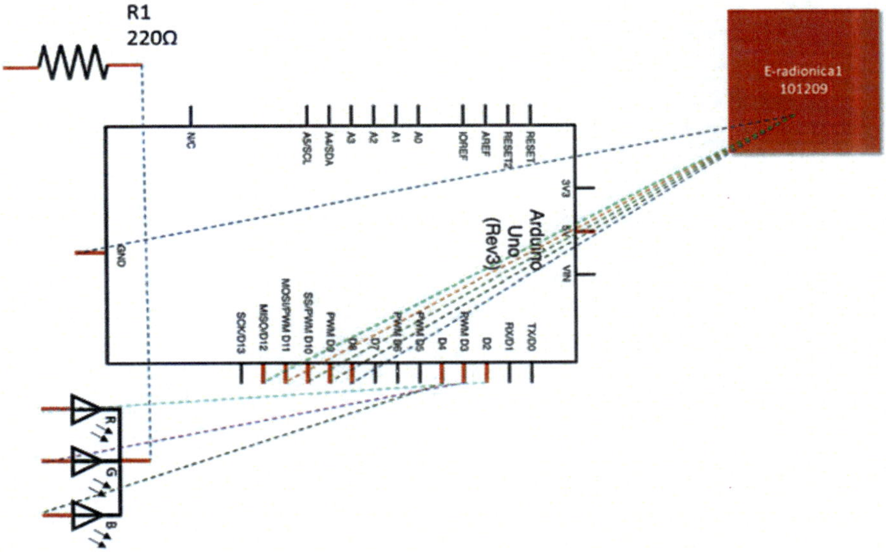

Fig. 2 Electrical circuit visualization

Structurally constructed from the synthetic polymer thermoplastic polyacid lactic, the material allows contact with the skin and is printed on 3D printing machines, reducing costs and enabling direct use.

We performed prototyping using an Arduino UNO R3, a protoboard, and an RGB TCS3200 sensor, as shown in Fig. 1.

As a technical organization, we also elaborated the electrical circuit to obtain the second level of visualization and the constructed prototype, as exposed in Fig. 2.

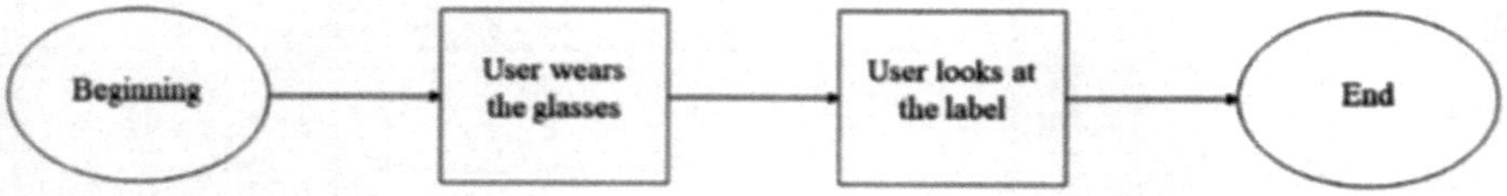

Fig. 3 UX flow

```
Captação de cores com base RGB
Entrada: Sensor TCS3200; Luz
Saída: String "nme da cor"

//Pinos de conexão do modulo
const int s0 = 8;
const int s1 = 9;
const int s2 = 12;
const int s3 = 11;
const int out = 10;

//Pinos do led RGB
int pinoledverm = 2;
int pinoledverd = 3;
int pinoledazul = 4;

//Variaveis cores
int red = 0;
int green =
0;int blue =
0;

void setup()
{
 pinMode(s0, OUTPUT);
 pinMode(s1, OUTPUT);
 pinMode(s2, OUTPUT);
 pinMode(s3, OUTPUT);
 pinMode(out, INPUT);
 pinMode(pinoledverm, OUTPUT);
 pinMode(pinoledverd, OUTPUT);
 pinMode(pinoledazul, OUTPUT);
 Serial.begin(9600);
 digitalWrite(s0,
 HIGH);digitalWrite(s1, LOW);
}

void loop()
{
 //Detecta a corcolor();
 //Mostra valores no serial monitor Serial.print("Vermelho
 :"); Serial.print(red, DEC); Serial.print(" Verde : ");
 Serial.print(green, DEC); Serial.print(" Azul : ");
 Serial.print(blue, DEC); Serial.println();

 //Verifica se a cor vermelha foi detectada
 if (red < blue && red < green && red < 100)
 {
  Serial.println("Vermelho");
  digitalWrite(pinoledverm, LOW); //Acende o led vermelhodigitalWrite(pinoledverd,
  HIGH); digitalWrite(pinoledazul, HIGH);
 }

 //Verifica se a cor azul foi detectada
 else if (blue < red && blue < green && blue < 1000)
 {
  Serial.println("Azul"); digitalWrite(pinoledverm, HIGH);
  digitalWrite(pinoledverd, HIGH);
  digitalWrite(pinoledazul, LOW); //Acende o led azul
 }

 //Verifica se a cor verde foi detectadaelse if (green < red &&
 green < blue)
 {
  Serial.println("Verde"); digitalWrite(pinoledverm, HIGH);
  digitalWrite(pinoledverd, LOW); //Acende o led verdedigitalWrite(pinoledazul,
  HIGH);
 }
 Serial.println();

 //Delay para apagar os leds e reiniciar o processodelay(50);
 digitalWrite(pinoledverm, HIGH);
 digitalWrite(pinoledverd, HIGH);
 digitalWrite(pinoledazul, HIGH);
}

void color()
{
 //Rotina que lê o valor das cores
 digitalWrite(s2, LOW);
 digitalWrite(s3, LOW);
 //count OUT, pRed, RED
 red = pulseIn(out, digitalRead(out) == HIGH ? LOW : HIGH);
 digitalWrite(s3, HIGH);
 //count OUT, pBLUE, BLUE
 blue = pulseIn(out, digitalRead(out) == HIGH ? LOW : HIGH);digitalWrite(s2, HIGH);
 //count OUT, pGreen, GREEN
 green = pulseIn(out, digitalRead(out) == HIGH ? LOW : HIGH);
}
```

Fig. 4 Color capture

The Simplified UX allows no need for any training or changes in the organization that you want to apply Riverstock to your corridors since the flow works, respectively, as shown in Fig. 3.

From the equation of color formulation with RGB uptake, we have the following Eq. (1).

$$C = \sqrt{2 * \Delta R^2 + 4 * \Delta G^2 + 3 * B^2} \tag{1}$$

3.1 Algorithm for Capturing Colors

We write the algorithm in C++ for reading colors in RGB using the TCS3200 sensor as input in the circuit, as shown in Fig. 4.

Therefore, the physical structure of Riverstock will couple the meeting of previous knowledge, which will perform the reading automatically, acting as a digital observer through an observer (person), where the flow of information runs as follows, represented in Fig. 5.

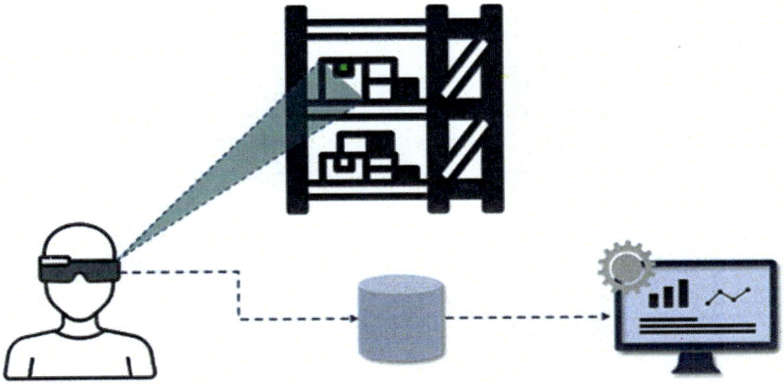

Fig. 5 Prototyping circuit

The observer, usually the stock handler, when using Riverstock, looks at the corresponding labels (these, following the pattern mentioned). The device captures the current situation of the product and lists it in a matrix present in the database.

This matrix is also the basis of a privileged access web page with up-to-date information about the stock situation of the mentioned product.

Therefore, Riverstock extends and facilitates the power of the shop floor. It makes intuitive the process of data recording since the act is reduced only to that of looking at the label on the shelf. Furthermore, it is integrated with those responsible for senior management since the data remains accessible on a web page, being possible to perform searches through any web language reader device.

4 Results

The objective of this work sustains the feasibility of applying an innovative product, low cost and value delivery also – relatively low in the market. The impacts are significant on revenue from the first month of use in the organization, without the need for changes in organizational culture, being an application of immediate use, training, and mutations in the organizational structure.

The first concept is in the feasibility of the solution, that is, if, in theory, it is possible to create a product capable of performing this type of activity. After studying the body of the problem, we verified the proposal's feasibility.

The next concept defined is technical feasibility. Therefore, if, in theory, there is a technically viable product that can be applied in the present times, using existing technologies, the application is sustainable and, later, scalable.

Next, we realized that this product could be made accessible to users without knowledge of information technology devices. The interface and user experience

(UI/UX) are so intuitive as not to require any user manual, simply using the product with human basics.

In meeting these concepts, we verify the replication capacity of this same product in a set of organizations, not being restricted only to one type of environment. Its application may be present in more conservative and inflexible settings, such as military organizations, and, at the same time, in more open media changeable, such as contemporary startups.

Riverstock was the product capable of passing all these criteria. Through this work, it is possible to prove its feasibility of application in all organizational environments holders of economic assets in stocks, from small stores to large industries with well-defined processes.

The intuition-oriented interface design allows companies to apply Riverstock in every type of environment without training or changes in the organization that implements it in its administrative routines.

5 Conclusion

The product in question meets all the requirements laid down in the basic premise of the research, being widely applicable, low cost, scalable and sustainable production, with process innovations and optimization of the flow of information between the central pillars.

Using the Arduino UNO plate and its color-identifying extension TCS3200, the product was able to meet the demands of computer-reproduced circuits, leaving its practical application for harvesting results in an uncontrolled environment.

Nevertheless, it is appropriate to remember the possibilities involving complete microcomputers rather than microcontroller boards, such as the Raspberry Pi, which already has artifacts such as an internet connection and a Linux-based operating system. Pricing discussions for the variability of its application and use.

Therefore, Riverstock proved to be the ideal product for implementing a flexible and scalable solution. The problematic situation within the symbolic universe demonstrates feasible applicability in the real world, between modeling and results, where the probable decisions are not derived from the metaphysical zone but grounded.

References

1. Dalenogare, L.S., Benitez, G.B., Ayala, N.F., Frank, A.G.: The expected contribution of Industry 4.0 technologies for industrial performance. Int. J. Prod. Econ. **204**, 383–394 (2018)
2. Lampropoulos, G., Siakas, K., Anastasiadis, T.: Internet of things in the context of industry 4.0: an overview. Int. J. Entrep. Knowl. **7**, 4–19 (2019). https://doi.org/10.2478/ijek-2019-0001
3. Costa, I.P.d.A., Basílio, M.P., Maêda, S.M.d.N., Rodrigues, M.V.G., Moreira, M.Â.L., Gomes, C.F.S., dos Santos, M.: Algorithm selection for machine learning classification: an application

of the MELCHIOR multicriteria method. Front. Artif. Intell. Appl. **341**, 154–161 (2021). https://doi.org/10.3233/FAIA210243
4. Novais, L., Maqueira Marín, J.M., Moyano-Fuentes, J.: Lean production implementation, cloud-supported logistics, and supply chain integration: interrelationships and effects on business performance. Int. J. Logist. Manag. **31**, 629–663 (2020). https://doi.org/10.1108/IJLM-02-2019-0052
5. van Heeswijk, W.J.A., Mes, M.R.K., Schutten, J.M.J., Zijm, W.H.M.: Evaluating urban logistics schemes using agent-based simulation. Transp. Sci. **54**, 651–675 (2020)
6. Verhoeven, P., Sinn, F., Herden, T.T.: Examples from blockchain implementations in logistics and supply chain management: exploring the mindful use of new technology. Logistics. **2**, 20 (2018)
7. Anantadjaya, S.P.D., Nawangwulan, I.M., Irhamsyah, M., Carmelita, P.W.: Supply chain management, inventory management & financial performance: evidence from manufacturing firms. Linguist. Cult. Rev. **5**, 781–794 (2021)
8. Happonen, A., Siljander, V.: Gainsharing in logistics outsourcing: trust leads to success in the digital era. Int. J. Collab. Entrep. **6**, 150–175 (2020)
9. Gomes, C.F.S., Santos, M.d., Teixeira, L.F.H.d.S.d.B., Sanseverino, A.M., Barcelos, M.: SAPEVO-M a group multicriteria ordinal ranking method. Pesqui. Oper. **40**, 1–20 (2020)
10. Arnott, D., Pervan, G.: Design science in decision support systems research: an assessment using the Hevner, March, Park, and Ram Guidelines. J. Assoc. Inf. Syst. **13**(11), 1 (2012)
11. Hevner, A.R.: A three cycle view of design science research. Scand. J. Inf. Syst. **19**(2), 4 (2007)
12. Santos, M.d., Costa, I.P.d.A., Gomes, C.F.S.: Multicriteria decision-making in the selection of warships: a new approach to the AHP method. Int. J. Anal. Hierarchy Process. **13**(1), 147–169 (2021). https://doi.org/10.13033/ijahp.v13i1.833
13. Jardim, R.R.J., Santos, M., Neto, E., da Silva, E., de Barros, F.: Integration of the waterfall model with ISO/IEC/IEEE 29148:2018 for the development of military defense system. IEEE Lat. Am. Trans. **18**(12), 2096–2103 (2020). https://doi.org/10.1109/TLA.2020.9400437
14. de Paula, N.O.B., de Araújo Costa, I.P., Drumond, P., Moreira, M.Â.L., Gomes, C.F.S., Dos Santos, M., do Nascimento Maêda, S. M.: Strategic support for the distribution of vaccines against Covid-19 to Brazilian remote areas: a multicriteria approach in the light of the ELECTRE-MOr method. Procedia Comput. Sci. **199**, 40–47 (2022)
15. Sodero, A.C., Namin, A., Gauri, D.K., Bhaskaran, S.R.: The strategic drivers of drop-shipping and retail store sales for seasonal products. Int. J. Retail. **97**, 561–581 (2021)
16. Muller, M.: Essentials of inventory management. HarperCollins Leadership (2019)
17. Makhija, A., Wickramasinghe, C., Tiwari, M.: Visual management. In: Lean tools in apparel manufacturing, pp. 131–208. Elsevier (2021)
18. Ganesha, H.R., Aitha, P.S., Kirubadevi, P.: Integrated inventory management control framework. Int. J. Manag. Technol. Soc. Sci. (IJMTS). **5**, 147–157 (2020)
19. Fernández-Caramés, T.M., Blanco-Novoa, O., Froiz-Míguez, I., Fraga-Lamas, P.: Towards an autonomous industry 4.0 warehouse: a UAV and blockchain-based system for inventory and traceability applications in big data-driven supply chain management. Sensors. **19**, 2394 (2019)
20. Muñoz-Villamizar, A., Velázquez-Martínez, J.C., Haro, P., Ferrer, A., Mariño, R.: The environmental impact of fast shipping ecommerce in inbound logistics operations: a case study in Mexico. J. Clean. Prod. **283**, 125400 (2021)
21. Klimek, D., Jędrych, E.: A model for the sustainable management of enterprise capital. Sustainability. **13**, 183 (2021)
22. Pacagnella Jr., A.C., da Silva, S.L., Pacífico, O., de Arruda Ignacio, P.S., da Silva, A.L.: Critical success factors for project manufacturing environments. Proj. Manag. J. **50**, 243–258 (2019)
23. Costa, I.P.d.A., Moreira, M.Â.L., Costa, A.P.d.A., Teixeira, L.F.H.d.S.d.B., Gomes, C.F.S., Santos, M.d.: Strategic study for managing the portfolio of IT Courses offered by a corporate training company: an approach in the light of the ELECTRE-MOr Multicriteria Hybrid Method. Int. J. Inf. Technol. Decis. Mak. **21**, 1–29 (2021). https://doi.org/10.1142/S0219622021500565

24. Abd Karim, N., Nawawi, A., Salin, A.S.A.P.: Inventory management effectiveness of a manufacturing company–Malaysian evidence. Int J Law Manag. **60**, 1163–1178 (2018)
25. Sremcev, N., Lazarevic, M., Krainovic, B., Mandic, J., Medojevic, M.: Improving teaching and learning process by applying Lean thinking. Procedia Manuf. **17**, 595–602 (2018)
26. Moreira, M.Â.L., Gomes, C.F.S., dos Santos, M., do Carmo Silva, M., Araujo, J.V.G.A.: PROMETHEE-SAPEVO-M1 a hybrid modeling proposal: multicriteria evaluation of drones for use in naval warfare. In: Springer proceedings in mathematics & statistics, pp. 381–393. Springer, Cham (2020)
27. Oztemel, E., Gursev, S.: Literature review of Industry 4.0 and related technologies. J. Intell. Manuf. **31**, 127–182 (2020)
28. Delen, D., Ram, S.: Research challenges and opportunities in business analytics. J. Bus. Anal. **1**, 2–12 (2018)
29. Mehmood, T.: Does information technology competencies and fleet management practices lead to effective service delivery? Empirical evidence from E-Commerce Industry. Int. J. Technol. Innov. Manag. (IJTIM). **1**, 14–41 (2021)

A Simplified Approach to the Process of Design and Development of Technological Products: Case Study of a Charpy Type Testing Machine

Mariannys Rodríguez-Gasca, Jorge Antonio Velasco-Parra, and Sanderson César Macêdo Barbalho

Abstract The development of new products leads companies to consider not only aspects related to the time and cost of producing the product, but also to implement innovative and technological features. Therefore, achieving all of these goals requires an efficient and systematic way of managing the product design and development process. In this context, this work presents the implementation of some basic concepts for the design and development of technological products, specifically a case study of the design of a charpy-type impact testing machine, with electronic components to automate it. To this end, a mechatronic product development methodological approach was implemented. This approach proved to be flexible for applications in different types of products with technological components and to facilitate the monitoring of the development process of a product as well as the fulfillment of user demands. The results also evidenced the improvements when applying a mechatronic product orientation in the optimization of a traditionally mechanical product.

Keywords New product development · Product design · Charpy tests · Automation

M. Rodríguez-Gasca (✉) · S. C. M. Barbalho
University of Brasilia, Research Group in Innovation, Projects and Process GPIPP, Brasilia, DF, Brazil
e-mail: mariannys.gasca@aluno.unb.br

J. A. Velasco-Parra
University of Pamplona, Research Group in Mechanical Engineering GIMUP, Pamplona, Colombia

V. M. López Sánchez et al. (eds.), *Industrial Engineering and Operations Management*, Springer Proceedings in Mathematics & Statistics 400,
https://doi.org/10.1007/978-3-031-14763-0_20

1 Introduction

With the advent of industry 4.0 technologies, in recent years the global economy has gone through a process of disruptive changes that have transformed everything from the way we live, work and manage, to the way we conceive systems and processes. Traditional creation of products and services. This new paradigm, in addition to representing a series of opportunities and benefits for companies, brings with it a series of challenges such as: reduction of the product life cycle, manufacturing of personalized products at the price of mass production, cost reduction and optimization of resources [1, 2].

Additionally, companies must overcome the growing competition in the market and cope with the heterogeneous and always volatile demands of customers to remain current in the market and increase their innovative potential. At the same time, this scenario requires companies to properly manage the product design and development process together with the ability to implement highly dynamic and flexible methodologies that allow identifying and satisfying market needs with integrated products and services [3].

On the other hand, this digitalization panorama has shown that there are serious limitations in products based solely on conventional mechanical systems, because they present restrictions in terms of process monitoring accuracy, data capture, signal processing accuracy of control, complexity of the design, speed, and efficiency of the system, among others [4]. Under this scenario, the application of concepts such as mechatronics favors the creation of more complex products from the approach of more elementary manufacturing systems, since it allows the multidisciplinary integration of principles such as electronics, mechanics, and software. Mechatronics promotes the convergence of disciplines and blurs the barriers between them, to develop products whose main essence is technology and therefore with innovative characteristics [5]. It has been shown that the implementation of mechatronic elements raises the characteristics of products and services, while contributing to the optimization of production processes within companies, since it promotes automation through the design of products and services that are applied to the industry making it faster, more productive, and safer.

Consequently, the development of new products leads stakeholders to consider not only aspects related to the time and cost of producing the product, but also to implement innovative and technological features. Achieving all these goals therefore requires an efficient and systematic way of managing the product design and development process. The product design and development methodology proposed by Ulrich, K. and Eppinger, S. (2012) is one of these tools, which provides the basic notions for the creation of the product concept, since it is considered as a theoretical reference of many projects in which it has been successfully implemented for the development of preferably physical, discrete and engineering products. The methodology divides the development process into six phases: planning, concept development, system-level design, detail design, testing and refinement, and start of production. It should be noted that each of this phase describes the activities to be developed and that they benefit from the joint participation of all the central functions of the company (marketing, design and manufacturing) [6].

Following the context exposed above, in this research it was proposed to use the basic concepts of the methodology proposed by Ulrich, K. and Eppinger, S. (2012) in their book design and product development [6], for the development of a case study of the design of a Charpy-type impact testing machine, for the characterization of composite materials with a polymeric matrix. The organization of this article was made in the following sections: introduction section, Sect. 2 methodology, which describes the step by step for the design of the Charpy machine; Sects. 3 and 4 where the results obtained, and the conclusions are shown.

2 Background

One of the driving forces for the development of new products is the implementation of methodologies or models that help guide this process from the conception of the idea to the launch of the product in the most efficient way. Consequently, various authors have made their contribution by developing a wide variety of frameworks and tools. Some of these reference models are specific to each case, however, there are also methodologies that are suitable for designing and developing a wide range of products. Below is a brief review of some antecedents of applications of the design and development methodology that will be used as a basis in this work, as well as the development and implementation of some other methodologies for product design.

Realpe, J. et al. (2018) presented a design and implementation of a photolithography system to fabricate thin film-based microdevices. The implementation was carried out based on the product design and development process proposed by Ulrich, K. and Eppinger, S. (2012), where they determined the technical specifications, generation and proof of concepts of the equipment modules of the system [8].

Mesa, D. et al. (2019) conducted a study with which they sought to promote the idea that the product development process can help bridge the gap between research and commercial applications. They proposed a framework to demonstrate how the product design and development methodology of Ulrich, K. and Eppinger, S. (2012) can be integrated with the STAM technology development model [9].

Mariawati, A. et al. (2019) carried out a study with the aim of designing a strength training tool for the feet and hands for patients with ischemic stroke. For this, they were based on previous observations and followed some of the phases of the product design and development cycle of Ulrich K. and Eppinger, S. (2012), the study yielded positive results in terms of meeting the requirements for customer satisfaction [10].

Grace, C. et al. (2021) proposed a methodology based on the fundamentals of collaborative, lean and agile design. For this, they carried out an analysis of the value flow of the product engineering process in a company in the retail sector, through lean tools, to implement the concept of product scalability. The results they obtained showed a reduction in product development time of approximately 10–20%, compared to the process they used at the time of implementation [11].

Likewise, Vari, M. et al. (2021) conducted research with the aim of developing a model for the renewal of product information systems and intelligent support tools for customization and integrating them into the design process. For this, the authors made a systematic analysis of the company. The results showed a considerable decrease in engineering changes, increased productivity and improved business competitiveness. It should also be noted that the framework proposed by these authors is applicable to similar business environments to promote competitiveness in unique product development processes [12].

Gogineni, S. et al. (2019) presented a study in order to present a systematic methodology for the development of personalized smart products based on the Internet of Things (IoT). They managed to identify, combine and improve existing methodologies, taking into account concepts such as variant management and intelligent service management to contribute to the efficient development of this type of product [13].

Carvalho, R. et al. (2021) describe a process for the design of innovative mechatronic products that integrates design thinking, concurrent engineering, and agility techniques into intellectual property management activities. In the early stages they implement design thinking to better explore creativity, while in the product development stage they apply concurrent engineering and agility to deal with emerging requirements and reduce development times [14].

Silva, E. et al. (2013) carried out an investigation where they used the hypothesis of lean startup in the health segment, this with the purpose of finding a way to improve the process of creation and development of new products and services in different areas of the industry [15].

Gremyr, I. et al. (2014) reviewed previous case studies of robust design methodology, which revealed their support for sustainability. However, the reviews also revealed that efforts so far have been more focused on the manufacturing and life cycle phases of a product. Therefore, the methodology requires adaptations in terms of conceptual and qualitative tools [16].

Riesener, M. et al. (2019) presented a methodology for the design of product development networks in order to increase organizational agility. For this they developed a Quality Function Deployment (QFD) model with two related Houses of Quality (HOQ). In this way, transparency was generated about the relationships between organizational agility attributes and network design characteristics. As a result, they showed that the methodology provides information on which network characteristics represent the most favorable levers for optimization measures that allow efficient prioritization in network management. As well as indications of the direction of optimization for each network characteristic depending on the actual state of the network [17].

Yilmaz, O. et al. (2020) developed a lean holistic fuzzy methodology for new product development projects, to deal with the uncertainties encountered in projects by emphasizing teams of cross-functional workers along with the impact of utility workers on the delivery time and total operating cost. To this end, they combined fuzzy design structure matrix (FDSM), fuzzy value stream mapping (FVSM), and a new fuzzy optimization model to visualize all processes, decrease lead time and operating cost, and determine improvement points. According to the results, the proposed methodology leads to a decrease in both the delivery time and the total operating cost [18].

3 Methodology

To carry out the implementation of the design of the Charpy testing machine, a literary search was carried out within different authors who deal with methodologies for product development, reaching the conclusion of bringing together innovation, flexibility and basic notions for the creation of product concept, provided by the development process proposed by Ulrich & Eppinger (2012) in their book. It should be noted that this methodology traditionally consists of 6 phases that were mentioned above, however, this research did not apply all these concepts, since it focused more on the product concept and design approach and did not reach a launch phase. Basically, the procedure that was proposed in this study for the Charpy machine design is shown below (see Fig. 1).

As can be seen, it begins with the identification of the client's needs, a stage in which those physical or functional characteristics that the client wishes to be reflected in the product are compiled. Subsequently, the objective specifications are established, which are measurable technical parameters focused on satisfying the client's demands. Consecutively, the concept generation process is carried out, which consists of exploring various solution alternatives, which results in a set of product concepts from which the team will make a final selection. Once this stage is finished, we proceed with the selection of the most optimal product concept to be developed. The subsequent stages are detail design and design validation, which includes geometry specifications, tolerances and description of all parts of the product, and geometry validation.

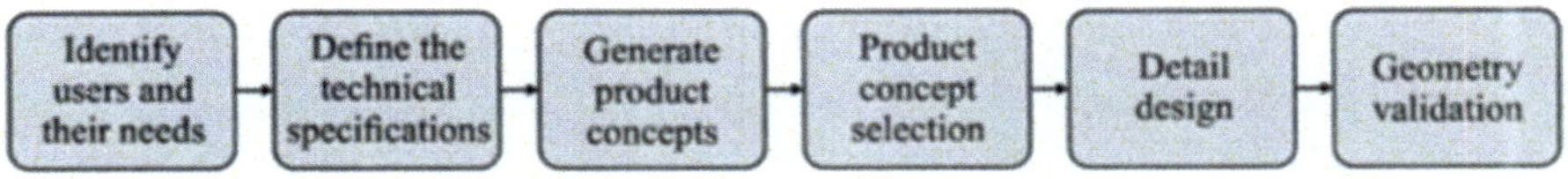

Fig. 1 Phases of the methodology. (**Fuente:** Autor)

4 Concept Development

4.1 Charpy Type Test

The charpy testing machine is due to its creator, the Frenchman Augustin Georges Albert Charpy. Its importance lies in the need to know the mechanical behavior of materials when they are exposed to different conditions. This is why many of the impact tests are carried out in states in which brittle fracture is favored. Through this test, the behavior of materials on impact can be known, and it consists of hitting a specimen that is placed on the support with a mass. The mass that is coupled to the end of the pendulum is dropped from a certain height, through which the speed of the load at the moment of impact is controlled [7].

The Charpy test is based on the theory of conservation of energy between two points. In this way it is possible to obtain the energy that is lost when the hammer collides with the sample of a certain material with standardized dimensions. By means of the Charpy test, it is possible to estimate the resistance to the deformation that causes the breakage in the material, the resilience and from the failure characteristics to deduce what type of material it has, if it is ductile or brittle and how much toughness it has [7].

Normally, the operation of the machine is completely mechanical. On the one hand, the positioning and release of the pendulum is manual and there is no temperature acquisition. On the other hand, the measurement of potential energy is performed by visually capturing the angle of the initial position and the angle of the final position after impacting the specimen. Therefore, the automation of the Charpy-type impact testing machine aims to increase the accuracy of data acquisition and improve the user-machine interface. For this purpose, the use of an electric motor controlled by means of a driver was proposed to position and release the pendulum. In addition, the use of an encoder was proposed for a more precise acquisition of the data, and its respective processing to obtain the final impact resistance of the polymeric matrix composite materials (PMCM). Finally, a temperature sensor was proposed to obtain this data and verify that the tests are carried out under normalized conditions.

4.2 Identify Users and Their Needs

The target market for the Charpy testing machine lies mainly in companies that manufacture new materials and university laboratories. That is why in this work a group of 12 professionals was brought together, including personnel who work in the materials laboratory, as well as university professors with experience and knowledge in this area.

The identification of the needs of the users was carried out based on the experience and knowledge of the users, through the focus group technique where the users were guided to communicate the real needs they find when carrying

Table 1 User requirements

Category	User requirements
1. Operation	1.1 Easy to operate control panel 1.2 Digital data collection 1.3 Easy maintenance
2. Design	2.1 Robust structure 2.2 Light weight 2.3 Easy to assemble 2.4 Visually modern
3. Costs	3.1 Low maintenance cost 3.2 Low manufacturing and assemble cost 3.3 Low-cost automation system
4. Reliability	4.1 Safe operation 4.2 Low probability of failure 4.3 Low probability of measurement errors

out an impact test in a traditional Charpy machine. In addition, the information collected through the literature review was considered, as well as the parameters established by the ASTM D6110 standard. In this way, a total of 13 requirements were identified, which were classified into 4 categories according to similarity, to facilitate their analysis. The needs are listed in the following Table 1.

4.3 Definition of Technical Specifications

The determination of the target specifications was made based on the needs collected in the previous step, which were translated into more technical terms, so that the product designer could better understand. Table 2 shows the specifications that must be considered in order to meet the customer's demands.

4.4 Generation of Product Concepts

To carry out the concept generation process, a decomposition of the problem into sub-problems or parts that are easier to understand was carried out, this process is called functional decomposition. For which it was necessary to develop a black box that operates in flows of material, energy and signals (see Fig. 2).

In this case, the thin solid lines denote the transfer and conversion of energy, the thick solid lines represent the movement of material within the system, and the dashed lines are the signals of control and feedback flows within the system. Starting from this result, a functional decomposition was carried out to divide the black box into sub-functions that would allow creating a more specific description of what the elements of the product could do to implement the general function of the product (see Fig. 3).

Table 2 Product technical parameters

User requirement	Technical parameters
1.1	The machine has a control panel
1.2	The machine has a microcontroller
1.3	The machine has a modular design
2.1	The machine is made of corrosion resistant materials
2.2	The machine is light
2.3	The machine has a strategic assembly
2.4	The machine has an attractive design
3.1	The machine has a low maintenance cost
3.2	The cost of manufacturing and assembling the machine is low
3.3	The machine has flexible hardware and software control system
4.1	The machine has safety standard
4.2	The machine has a low failure occurrence rate
4.3	The machine operates under the ASTM D6110 standard

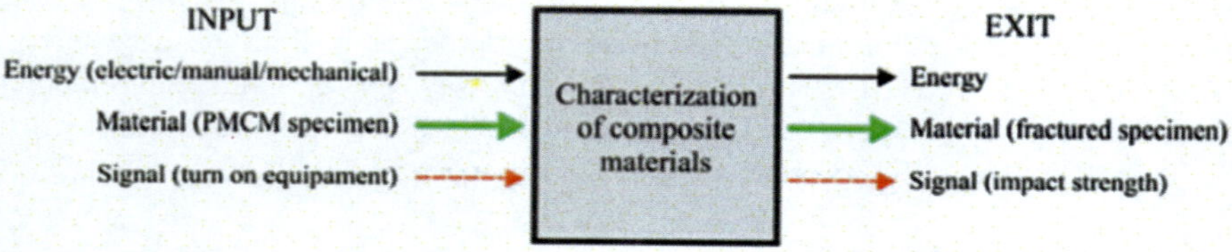

Fig. 2 General operating diagram of the Charpy type testing machine. (**Source:** Autor)

By carefully analyzing the functional diagram, subfunctions can be identified that can be grouped to give a more effective solution to each of these blocks. In this case they were classified into three groups (see Fig. 4). Subsequently, possible solutions were proposed for each of the groups of subfunctions, and it was determined which of them would be the ideal proposal to be used in the final design.

Solution for Group 1 The position control of the pendulum must be precise with a low percentage of error, according to the ASTM D 6110-02 standard. Due to this, the following solutions were proposed: the auxiliary axis to position the pendulum anchored to the hammer, by means of an extra axis anchored to the pendulum, a stepper motor is responsible for locating the pendulum at the required angle, to anchor the axis to the pendulum it is necessary to use manual energy. Clutch and transmission of movement through pulleys, the stepper motor transfers the rotation to a pulley-clutch system that positions the pendulum at a desired angle. When the electromagnetic stimulus is removed, the axis with the pendulum slides independently causing shock and breakage of the test tube.

Solution for Group 2 Support of two standardized supports: For Charpy-type impact tests, the placement and type of specimens are standardized, therefore, the solution for this function is unique and must comply with the requirements indicated in section 6, numeral 1.4 of the standard. ASTM D6110.

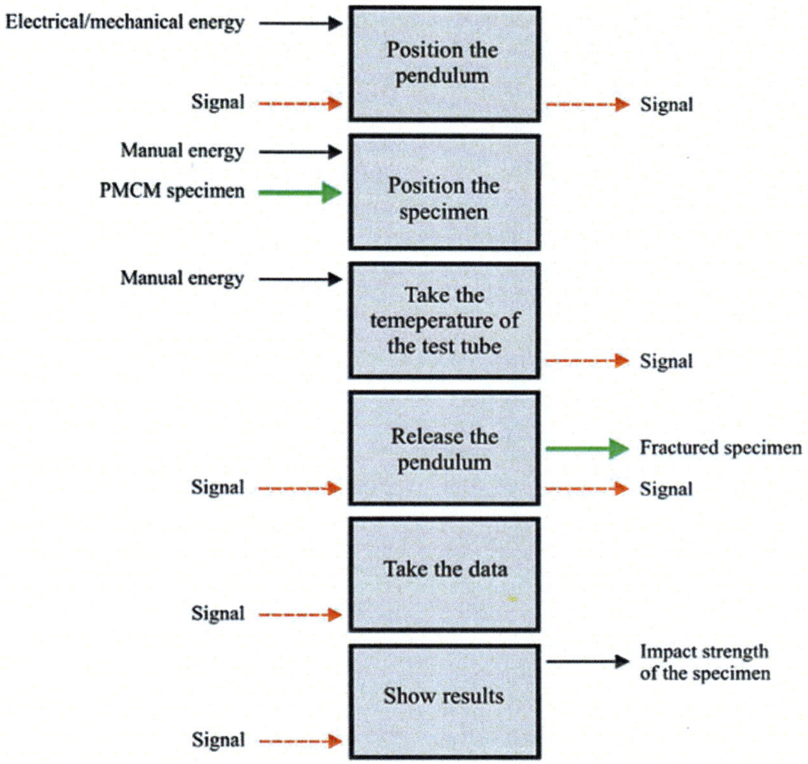

Fig. 3 Charpy-type testing machine sub-function diagram. (**Source:** Autor)

Fig. 4 Subfunction grouping (**Source:** Autor)

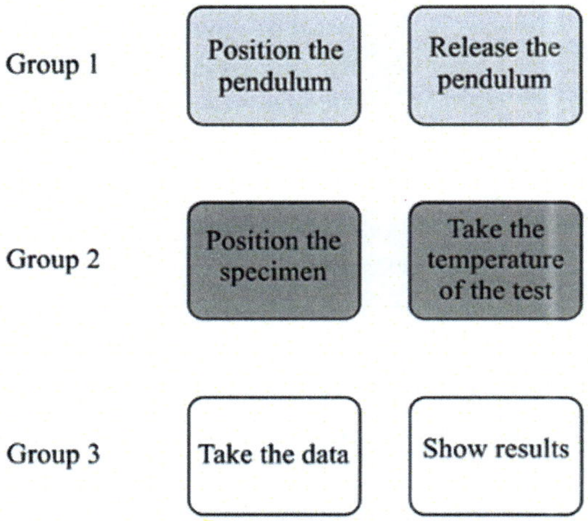

On the other hand, to measure the temperature of the specimen, an LM35 contact sensor is considered: The LM35 is an analog sensor, with a range of −55 °C to 150 °C, very popular due to the ease with which temperature can be measured. Another viable alternative would be to measure the temperature of the probe with the non-contact infrared temperature sensor MLX90614 GY-906: The MLX90614 GY906 Infrared Temperature Sensor is a silicon chip with a fine micro-machined membrane, designed to be sensitive to emitted infrared radiation. by a distant object. The sensor has a stage of amplification and digitization of the signal coming from the membrane. The sensor output is linear and is compensated according to variations in ambient temperature.

Solution for Group 3 Regarding the collection of test data, two solutions were proposed. The first would be to do it by means of an encoder, which consists of an electromechanical device that converts the angular position of an axis directly to a digital code and also delivers its speed and acceleration. The second alternative would be by means of a hand system that consists of a graduated board with a hand that follows the movement of the pendulum and indicates its final position.

While the interpretation and presentation of the results will be carried out by connecting the microcontroller to a computer. Both the microcontroller and the computer can be responsible for the interpretation and management of the data obtained in the test, for aesthetics and practicality the data can be presented on the computer using software.

4.5 Selection of the Product Concept

After generating each of the subfunctions and subsequently proposing possible solutions, a concept combination table was developed, which contains a way to systematically consider combinations of solution fragments. Potential solutions to the general problem are formed by combining a fragment from each column and then drawing a flow line (see Fig. 5). Each of the solutions was carefully studied

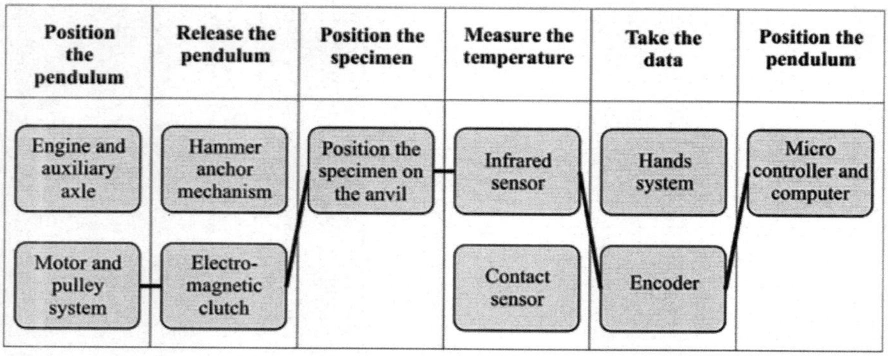

Fig. 5 Combination of concepts. (**Source:** Autor)

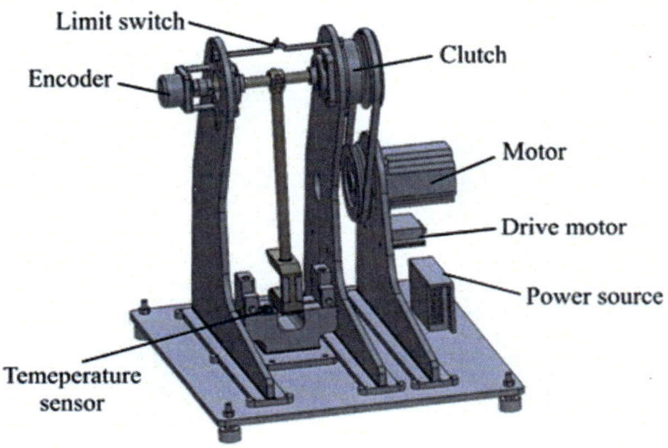

Fig. 6 Geometry of the Charpy testing machine. (**Source:** Autor)

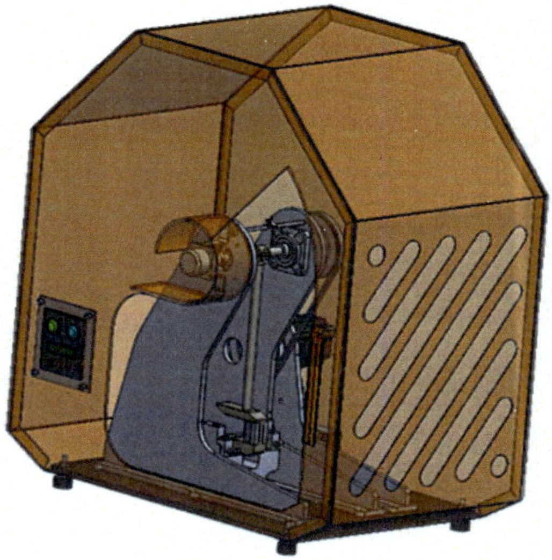

Fig. 7 Final design of the Charpy testing machine. (**Source:** Autor)

considering the advantages and disadvantages presented, in this way we have the general concept and specifications of what the equipment to be designed will be.

4.6 Detail Design

The Charpy test machine CAD was drawn using Solidworks software. The geometry was selected based on the target specifications (see Figs. 6 and 7).

4.7 Geometry Validation

The validation of the geometry was done through the finite element method, in the ANSYS simulation software. For this, each one of the pieces was analyzed separately and later the behavior of the complete machine, with the objective of verifying the reliability and identifying possible engineering changes before its manufacture.

For this, the critical operating conditions of each of the parts were simulated. The mechanical forces were obtained and compared with the resistance of the manufacturing materials, in order to obtain the values of the safety factor (mechanical force of the part / resistance of the material). As can be seen in Fig. 8, all the pieces have a minimum safety factor (MSF) above 1. This means that both the geometry and the manufacturing material are capable of withstanding the stresses generated

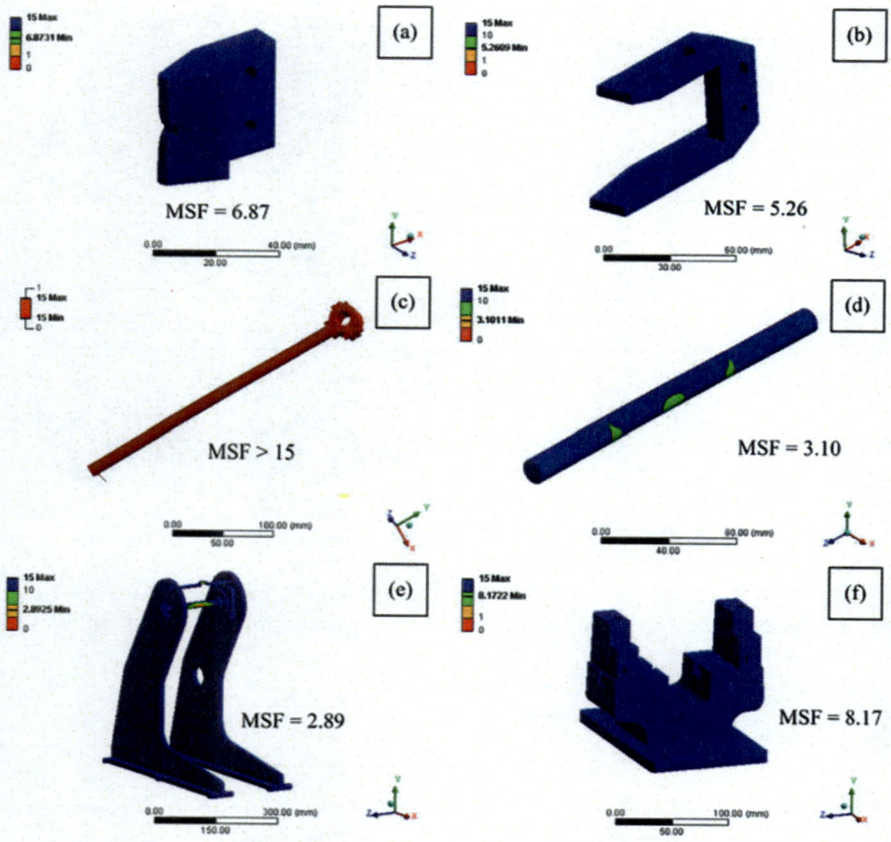

Fig. 8 Safety factor obtained in ANSYS: (**a**) pendulum striker, (**b**) pendulum hammer, (**c**) pendulum stem, (**d**) moving mass axis. (**e**) columns, and (**f**) Anvil. (**Source:** Autor)

in the execution of the tests. In addition, it can be seen that the columns are the critical component of the machine, with a minimum safety factor of 2.89. From a mechanical point of view, this value is completely reasonable and acceptable for this type of design.

5 Conclusions

This work presented a conceptual design and automation proposal for a Charpy-type impact testing machine. This machine had as a differential the incorporation of an automation system for digital data collection. This system shows a significant improvement in the accuracy of data collection since it reduces the chances of errors in data collection for impact tests.

The framework of ideas for the concept generation stage was expanded thanks to the initial interaction with the user, where he shared ideas that could solve one or more problems that the product was facing and from this fine-tune the project specifications.

The product development methodology was implemented for the initial stage of the development of the digital machine for Charpy-type tests. It was possible to define the client's needs and propose solutions to each of the different systems. It should be noted that, by implementing this methodology, the step-by-step to follow to develop a product that can be successfully introduced to the market becomes very clear. In this case, the development of the machine proposes an automated system with a lower cost than what can be presented on the market. It is recommended to continue studying the following stages of the development process to obtain the finished product.

Additionally, the implementation of CAD and CAE software in the detailed design and geometry validation stages represent a significant contribution to the development of the product concept. It was shown that its application contributes to the reduction of errors in the design proposal that may represent engineering changes and increased production costs.

References

1. Uemura Reche, A.Y., Canciglieri Junior, O., Estorilio, C.C.A., Rudek, M.: Integrated product development process and green supply chain management: contributions, limitations and applications. J. Clean. Prod. **249**, 119429 (2020). https://doi.org/10.1016/j.jclepro.2019.119429
2. Ozturk, M., Kocaoglan, S., Sonmez, F.O.: Concurrent design and process optimization of forging. Comput. Struct. **167**, 24–36 (2016). https://doi.org/10.1016/j.compstruc.2016.01.016
3. Hofmann, E., Rüsch, M.: Industry 4.0 and the current status as well as future prospects on logistics. Comput. Ind. **89**, 23–34 (2017). https://doi.org/10.1016/j.compind.2017.04.002
4. Gausemeier, J., Dumitrescu, R., Kahl, S., Nordsiek, D.: Integrative development of product and production system for mechatronic products. Robot. Comput. Integr. Manuf. **27**(4), 772–778 (2011). https://doi.org/10.1016/j.rcim.2011.02.005

5. Koch, T., Holtmann, J., Schubert, D., Lindemann, T.: Towards feature-based product line engineering of technical systems. Procedia Technol. **26**, 447–454 (2016). https://doi.org/10.1016/j.protcy.2016.08.057
6. Ulrich, K., Eppinger, S.: Product design and development. McGraw-Hill (2012)
7. Ortega, Y.: Prueba de impacto: ensayo Charpy. Rev. Mex. Fis. **52**(1), 51–57 (2006)
8. Realpe, J.J., Bolaños-Pantoja, G., Florez, J.F.: Design and implementation of a photolithography system to fabricate microdevices from thin films. DYNA (Colombia). **85**(205), 294–301 (2018). https://doi.org/10.15446/dyna.v85n205.68837
9. Mesa, D., Thong, C., Ranscombe, C., Kuys, C.: Integrating the product development process in scientific research. Bridging the research-market gap. In: Proceedings of the International Conference on Engineering Design, ICED, vol. 2019, pp. 2805–2814 (2019). https://doi.org/10.1017/dsi.2019.287.
10. Saffanah Didin, F., Sri Mariawati, A., Umyati, A., Fawaid, M., Nurtanto, M.: Identification customer need for redesign feet and hands strength training equipment for the post-ischemic stroke patients view project prevocational activity increasing quality of graduate of vocational high school view project identification customer need for redesign feet and hands strength training equipment for the post-ischemic stroke patients. Int. J. Sci. Technol. Res. **8**(10), 20–21 (2021) [Online]. Available: www.ijstr.org
11. Grijota, C.G., Acero, R., Yagüe-Fabra, J.A.: Product development methodology scalability. Procedia CIRP. **100**, 571–576 (2021). https://doi.org/10.1016/j.procir.2021.05.125
12. Varl, M., Duhovnik, J., Tavčar, J.: Customized product development supported by integrated information. J. Ind. Inf. Integr. **25**, 100248 (2021). https://doi.org/10.1016/j.jii.2021.100248
13. Gogineni, S.K., Riedelsheimer, T., Stark, R.: Systematic product development methodology for customizable IoT devices. Procedia CIRP. **84**, 393–399 (2019). https://doi.org/10.1016/j.procir.2019.04.287
14. de Carvalho, R.A., da Hora, H., Fernandes, R.: A process for designing innovative mechatronic products. Int. J. Prod. Econ. **231**, 107887 (2021). https://doi.org/10.1016/j.ijpe.2020.107887
15. Silva, S.E.P., Calado, R.D., Silva, M.B., Nascimento, M.A.: Lean startup applied in healthcare: a viable methodology for continuous improvement in the development of new products and services. In: IFAC Proceedings Volumes (IFAC-PapersOnline). vol. 6, no. PART 1, pp. 295–299 (2013). https://doi.org/10.3182/20130911-3-BR-3021.00054
16. Gremyr, I., Siva, V., Raharjo, H., Goh, T.N.: Adapting the robust design methodology to support sustainable product development. J. Clean. Prod. **79**, 231–238 (2014). https://doi.org/10.1016/j.jclepro.2014.05.018
17. Riesener, M., Rebentisch, E., Doelle, C., Kuhn, M., Brockmann, S.: Methodology for the design of agile product development networks. Procedia CIRP. **84**, 1029–1034 (2019). https://doi.org/10.1016/j.procir.2019.04.172
18. Yilmaz, O.F., Ozcelik, G., Yeni, F.B.: Lean holistic fuzzy methodology employing cross-functional worker teams for new product development projects: a real case study from high-tech industry. Eur. J. Oper. Res. **282**(3), 989–1010 (2020). https://doi.org/10.1016/j.ejor.2019.09.048

E-waste Reverse Logistics for Household Products and Its Regulation: Advances in Brazil

Danieli Braun Vargas, Lucila Maria de Souza Campos, and Mônica Maria Mendes Luna

Abstract The regulation for reverse logistics of household electrical and electronic products and their components published in 2020 raised the research gap as the absence of a detailed characterization which demonstrates the progress of the planned phases and their goals. Therefore, this work aims to analyze the advances in the agenda of structuring and implementing the reverse logistics for electronic products in the collective model. In the end, it presents the structured knowledge of the evolution and potential of this model to comply with legal requirements. The analysis was based on secondary data collected on the institutional pages of the companies which manages the collective model, not limiting the period, with data triangulation in the social networks of communication with society to obtain data intertextuality. Despite the challenge of waste management in a continental country like Brazil, this research showed a significant advance in operationalizing the reverse logistics system for electronic products, with industrial sector of electronic equipment joining at collective model and investing in infrastructure for reverse logistics. The study suggests to the entrepreneurs a market to act both in support of the logistics business, meet legal requirements, and set up recycling companies in regions that are lagging behind. Suggestions to fully meet the demands are simplification and tax incentives for the recycle industry and establish a relationship between government spheres to reduce tax differences between regions and adapt the system's operation to the reality of each location.

Keywords Reverse Logistics · Electrical and Electronic Equipment · Waste Management

D. B. Vargas (✉) · L. M. de S. Campos · M. M. M. Luna
Department of Production Engineering, Universidade Federal de Santa Catarina, Florianopolis, Brazil
e-mail: danieli.v@ufsc.br

V. M. López Sánchez et al. (eds.), *Industrial Engineering and Operations Management*, Springer Proceedings in Mathematics & Statistics 400,
https://doi.org/10.1007/978-3-031-14763-0_21

1 Introduction

As the population increased, the demand for products, especially electronics, grew [1]. Thus, electronic equipment waste has become the fastest-growing category of waste globally [1] due to planned obsolescence and, consequently, greater disposal of products, whether they are in operation or not [2].

In Brazil, the average per capita generated was 10.2 kg in 2019 [3]. In contrast, in 2016, the average per capita was 7.4 kg [4], an increase of practically 40% in the generation of waste electrical and electronic equipment (WEEE) per inhabitant. One way to capture the value of WEEE is through urban mining, it means to generate secondary raw material from urban waste, making the waste inputs to processes in a restorative and regenerative manner, as provided in the circular economy model [5].

The concept of urban mining includes reverse logistics as a tool for returning WEEE to the supply chain. According to the National Solid Waste Policy (NSWP), the reverse logistics enables the collecting and returning of waste to the business sector. The NSWP establishes the mandatory structuring and implementation of the reverse logistics system (RLS) by the business sector for some types of products, including electronic products and their components [6], whose regulation was carried out through Decree No. 10,240/2020.

Decree No. 10,240/2020 divides the evolution in two phases, the first for structuring the system, completed in December 2020, and the second for implementation and operationalization. The second phase, started in January 2021, stipulates collection and disposal goals. At the end of the second phase, the objective is to properly collect and dispose 17% by weight of household electrical and electronic products sold in the base year 2018, installing more than 5000 voluntary delivery points by 2025, considering all Brazilian municipalities with more than 80 thousand inhabitants. The collection percentages are based on the quantity by weight of electrical and electronic equipment (EEE) placed on the market in 2018, starting with 1% in 2021 and evolving to 17% by 2025 [7].

Although regulation and government defined the expected results, a study that provided a more detailed characterization and expressed the progress of the phases provided for Decree No. 10,240/2020 was not found in the literature. Thus, this work aims to analyze the advances in the agenda of structuring and implementing the WEEE reverse logistics in the collective model through manager entities. It means the paper does not deal with the individual model in which each company operates its reverse logistics system in isolation. Finally, it will present the structured knowledge of the evolution and the potential of the collective model to comply with legal requirements.

2 Literature Review

2.1 The Regulation of E-waste Reverse Logistics System

The NSWP provides, among several other aspects, on the mandatory structuring and implementation of the reverse logistics system by the business sector: distributors, retailers, manufacturers, and importers [6]. It is through the reverse logistics concept that the principle of shared responsibility for the product life cycle takes place, a concept also provided for the same policy, through the division and chain of responsibilities between the business sector, consumers, and government [8]. In this way, the NSWP enables the solid waste integrated management, encouraging the return of post-consumer waste to the production chain.

Decree No. 10,240/2020 applies to household WEEE, conceptualizing them as equipment of household origin that depends on electrical current, as well as their integral parts. To collect WEEE, although it is also possible to carry out door-to-door collection and periodic campaigns, Decree No. 10,240/2020 regulates the model of fixed receiving points, called voluntary delivery points. The goals and number of cities served started in 2021, with 24 cities served by the reverse logistics system and 1% of the collection and destination of household EEE in the base year 2018, to finally serve 400 cities and collect 17% of EEE. Thus, at the end of 5 years of expansion, the total population served by the reverse logistics system represents 61% of the country's population [7].

To operationalize the reverse logistics system, the responsibilities of each party are foreseen in the regulation, briefly shown in Table 1. Obligations are imposed on consumers to dispose of the waste generated properly. At the same time, the business sector must ensure an infrastructure that enables the return of products from the consumer to the business sector to dispose of the waste properly [6].

The decree offers two options for management models which the business sector can choose, the collective system and the individual system. In the collective system, the business sector hires management entities – namely, the Manager for Reverse Logistics of Electrical and Electronic Equipment (Green Eletron) and the Brazilian Association for Recycling of Electronic Waste and Appliances (ABREE) – to organize the reverse logistics system, with the costs divided among those associated with different values for each type of electronic product. In comparison, the individual system assumes that each company will operate its own reverse logistics system individually. The interactions between the parties, including the business sector, fiscal, and environmental control bodies, are summarized in Fig. 1, considering the collective model on which the manager mediates the interactions.

Table 1 Reverse logistics players' responsibility

Reverse channel member	Responsibilities under regulation
Consumer	Dispose e-waste at specific points.
Manufacturer and importer	Give proper destination; collaborate with the non-formal environmental education and communication plan; pay for RLS.
Distributor and retailer	Provide consolidation and collection point; inform consumers of responsibilities (only retailers); pay for the RLS; collaborate with the non-formal environmental education and communication plan.
Performance follow-up group	Follow-up, support, and control the RLS.
Managing entity	Perform RLS actions, being able to contract or subcontract third parties to provide services.
Cooperatives of waste pickers, Municipal Solid Waste Management Companies, Big generators and Non-governmental and related organizations.	They can join the RLS through a contract with companies or management entity.

Source by the authors based on [7]

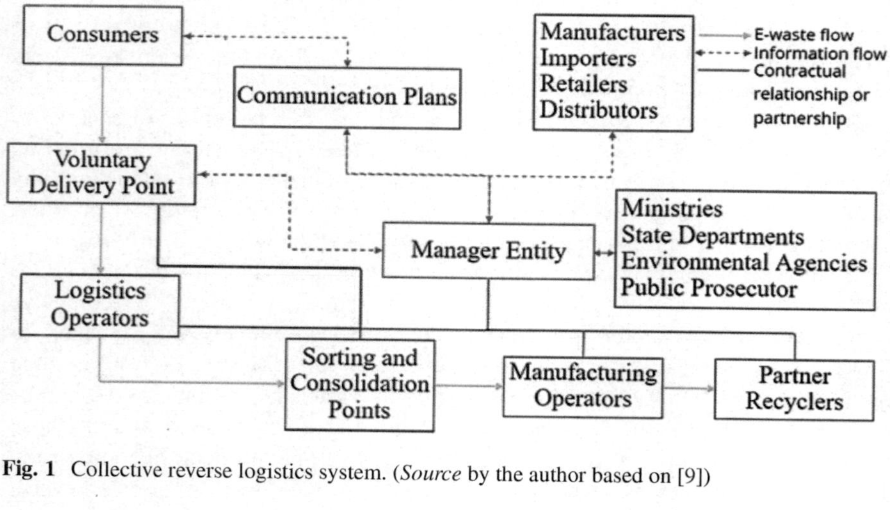

Fig. 1 Collective reverse logistics system. (*Source* by the author based on [9])

2.2 *Composition of Electronic Equipment Waste: Challenges and Opportunities*

WEEE has several peculiarities and a much-varied composition, from iron and plastic, even fractions of cadmium and lead, which have high toxicity despite being in smaller amounts by weight [10]. When their structure is unchanged, they are classified as non-hazardous waste. Nevertheless, if there is exposure due to dismantling or break, some of them are considered as hazardous waste and, therefore, need to be stored according to more standards safety restrictions.

The authors Bakhiyi et al. [11] bring examples of contaminants, sources, and applications. Although most studies reporting damage to the environment, human health, including WEEE recycling workers, are related to informal recycling, the same authors warn that there is little research focused on formal recycling impacts.

The sales value of materials for the recycling market varies according to their level of segregation [10], however the greater the decharacterization, the greater the exposure to hazardous components. In Brazil, manual disassembling of WEEE is quite common, and, despite requiring more processing time, it adds more value as more homogeneous fractions of the materials are obtained [12].

In spite of the challenges imposed on the WEEE segment due to its composition and heterogeneity in sizes and weights, as well as being an underexplored area, there are market opportunities for having high-value materials and an area in expansion nationally and internationally. Also, its regulation in Brazil brought legal certainty to entrepreneurs and fostered the market. Therefore, Giese et al. [13] suggest using the a business plan calculation tool to assist recyclers in making WEEE disassembly level decision making. The specific tool for the WEEE recycling area evaluates disassembly level scenarios between superficial, medium, or deep.

Another business opportunity is the repair of WEEE for reuse. This option may be more profitable than selling them for recycling. An example is the case of computers and their parts, when sold as a disassembled computer, it can earn around R$ 10.00 to R$ 30.00, while for the sale of a memory module, the price for each piece can be R$ 20.00 to R$ 100.00 [13]. Although there are challenges in WEEE management, there are also business opportunities to be explored, encouraging the return of resources to the supply chain through reverse logistics in collaboration for the transition to circular models.

2.3 *Urban Mining as a Driver for the Circular Economy*

The concept of circular economy emerges from the unsustainability of the traditional linear model of take-make-dispose and urban mining emerges as a business opportunity in which WEEE materials are extracted for coming back to supply chain as secondary raw material [5]. Through instruments already applied, such as reverse logistics, circular economy can lead us towards real change.

According to Forti et al. [3], the global scenario of environmental issues related to primary mining, market price fluctuations, scarcity of materials, and access to resources, made the environment favorable to urban mining to reduce the extraction of primary raw materials since the metal already comes benefited in WEEE. As such, urban mining can mitigate dependence on virgin raw materials across countries.

The Global E-waste Monitor 2018 estimated a generation of 52.2 Mt of WEEE for 2021 [4]. However, in the same report of the 2019, a generation of 53.6 Mt in 2019 was documented [3], that is, a generation of 1.4 Mt more than estimated for the next 2 years. It is estimated that the value of raw materials in the 2019 global generation is approximately $57 billion dollars (from 52.2 Mt), but only $10 billion dollars were recovered from this raw material in an environmental sound way, considering the 17.4% collected and recycled [3].

Consumers are responsible for disposal for large amounts of WEEE generated to be mined returning to the cycle through reverse logistics. However, studies show that consumers keep small EEE out of use, and even those that had problems, in the case of larger EEE, the tendency is donating or selling [14, 15], it means the consumers are stocking resources. Furthermore, it is shown that consumers do not know where to dispose of their EEE [14] and end up disposing of them in household waste, especially smaller EEE [15].

Therefore, a double challenge is posed for the government and the business sector in relation to WEEE. The first is to enforce environmental regulations, provide infrastructure and properly manage the reverse logistics system. The second is related to environmental education and consumer engagement in initiatives related to waste management.

3 Research Methodology

The methodological procedures adopted in this study is based on exploratory research, based on a qualitative approach, conducted in November of 2021. It is focused on the collective management model, in which companies hire a manager entity, which, in the specific case of WEEE, are Green Eletron and ABREE, for organization and fulfillment of the obligations set out in the decree. First of all, an in-depth study of Decree No. 10,240/2020 was carried out as a research strategy. It regulates the implementation of the reverse logistics system of EEE products and their components for household use.

Second, the information regarding the progress of the implementation of the WEEE reverse logistics was analyzed, through the collection of secondary data on the institutional pages of the manager entities, with data triangulation on the business social network LinkedIn and other social networks of the managers, seeking data intertextuality. Therefore, the data collection resulted from the communication sources of the manager entities with society. The main reason to choose this way is to match the current status of the WEEE reverse logistics system implementation with the regulation. There was no period limitation, it means that the companies' pages were analyzed in their entirety, all information since the creation date.

Figure 2 summarizes the applied methodological procedure. NSWP boosted the regulation of the WEEE reverse logistics system. Thus, this work fills the gap of the absence of a study characterizing the progress of the phases provided for in Decree No. 10,240/2020 to answer "How is the application of the principle of shared responsibility taking place?". The same figure also lists, in the dotted box, the topics targeted by this work.

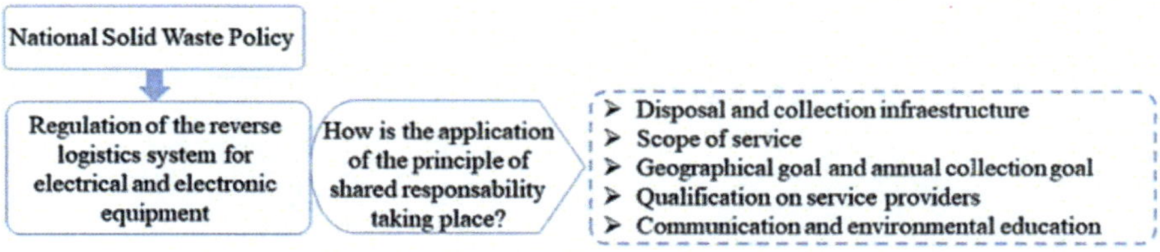

Fig. 2 Research scheme

One of the limitations pointed out in the work is the difficulty of collecting primary data directly with the managing entities of the reverse logistics. There was an attempt to contact in various ways, but unsuccessfully. Therefore, we opted for collecting secondary data.

4 Results and Discussion

4.1 Implementation of the Reverse Logistics System

The first phase of the WEEE reverse logistics system implementation started in 2020 with the structuring of the system, followed by the second phase started in 2021, which foresees the operationalization of equipment collection points and collection goals. In the first year of phase 2, it will cover at least 24 Brazilian municipalities [7], comprising 23%[1] of the population and 40%[2] of the total target population of the reverse logistics at the end of the 5 years of implementation.

Figure 3 shows the percentage of the population served by the reverse logistics system in each year of operation. Even after 5 years of reverse logistics system growth, 39% of the Brazilian population will not be covered. In contrast, Fig. 4 shows the implementation progress in relation to the target population served. Within the first year, 38%[3] of the target population will be served with the reverse logistics system. Initially, the advance occurs in terms of the geographic goal with the installation of voluntary delivery points. The annual collection goals increase as the planning progresses, with 1% collection in the first year and 17% at the end of 5 years.

[1] Considering the most populous cities in the state as priorities, defining themselves from the target states of phase 1 defined in Decree No. 10,240/2020. The implementation schedule with the municipalities served annually in each State was not disclosed on the Ministry of Environment website.

[2] Idem 1.

[3] Idem 1.

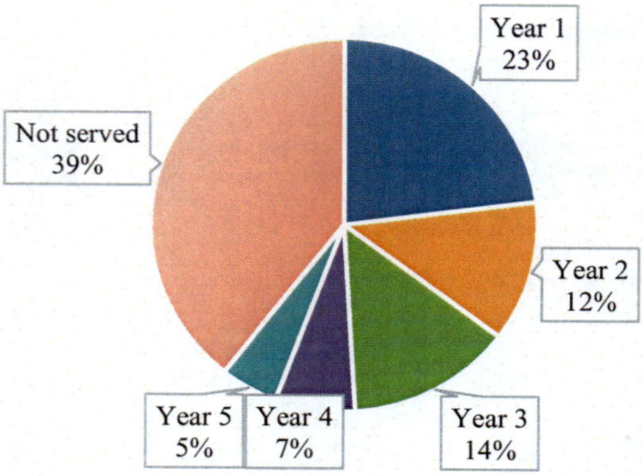

Fig. 3 Population served by reverse logistics

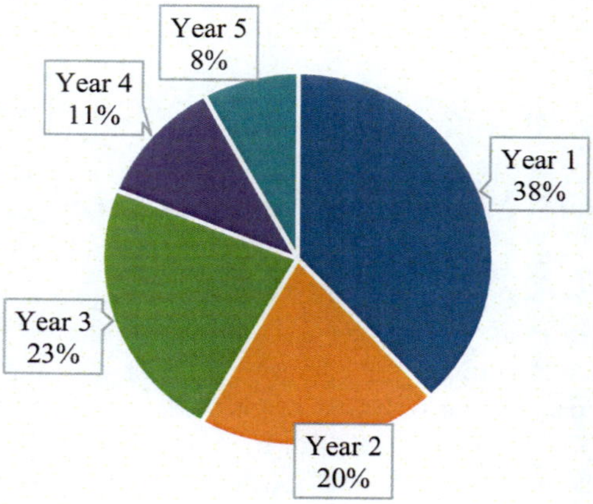

Fig. 4 Advance in implementing reverse logistics in relation to the target population

4.2 Geographic Distribution of the Reverse Logistics System

Figure 5 shows the target cities for reverse logistics system implementation by year of evolution. The first year of implementation mainly encompasses the South and Southeast regions of the country. They are emerging with a selective collection of municipal solid waste [16], and therefore they already have more structure to improve the reverse logistics system. Only in the fifth year, it is possible to see a significant expansion to the North and Northeast areas, which do not reach 13% of the municipalities with a selective collection – an essential instrument for achieving goals in solid waste management. The largest concentration of municipalities covered by this system will remain in the country's Southeast region.

All the cities covered in the first year provide at least one collection point from one of the two management entities. Nevertheless, in some cities, such as Salvador, Fortaleza, Brasília, there is no point for receiving household appliances (refrigerator, freezer, stove, etc.), only for other product lines. Such cities have over two million people, so their collection or even inadequate disposal results strongly impact the

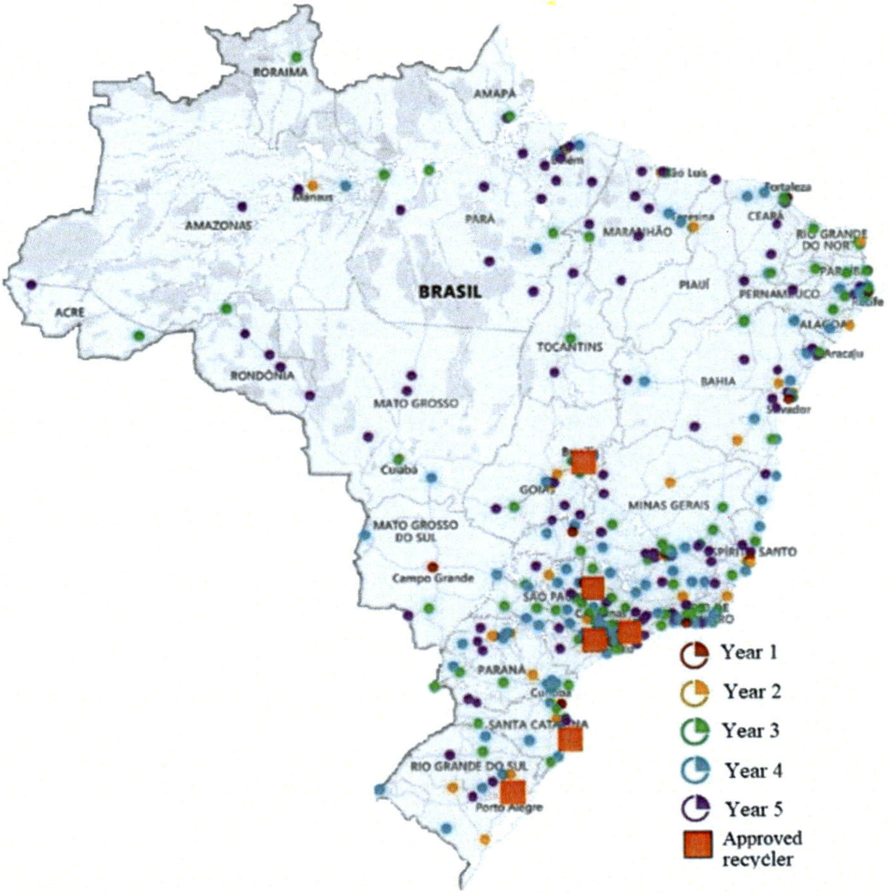

Fig. 5 Geographic distribution of the reverse logistics system

entire system. Besides, the lack of a collection structure favors informal recycling, as, according to Forti et al. [3], 31% of the world's WEEE generation is made up of large-scale equipment.

Although the country's top half has a lower population density, there may be greater dispersion of inadequate disposal due to having more areas for disposal. Likewise, the challenge is more significant when it comes to the government's role in inspecting these areas. It will require a strategic plan for partnerships with the government and reverse logistics facilitating agents to provide the return logistics for approved recyclers.

Green Eletron's disposal points can currently receive several categories of EEE, excluding only large household appliances such as refrigerators, washing machines, and others. Most companies associated with this manager operate in the market for

Table 2 Recyclers approved by Green Eletron

Approved recyclers	State	City
Reverse	Rio Grande do Sul	Novo Hamburgo
Weee.do	Santa Catarina	Florianópolis
Sinctronics	São Paulo	Sorocaba
GM&CLOG	São Paulo	São José dos Campos
Brasil Reverso	São Paulo	Nova Odessa
Programando o futuro	Distrito Federal	Brasília

small and medium-sized household appliances and electronics, so most voluntary delivery points are small and medium size to meet the goals of its own members. While in the case of ABREE, many of the members are active in the large household appliance market, so the company has signed partnership contracts with municipal governments for the installation of collection points, given that the challenge is greater due to the size and weight characteristics of the equipment [5].

4.3 Recycling Companies

Figure 5 also shows the location of Green Eletron approved recyclers, which are listed in Table 2. These companies meet specific technical qualification criteria and have become eligible to participate in the reverse logistics system. ABREE, on the other hand, did not disclose its approved recyclers, although some of the recyclers approved by Green Eletron report on their pages that ABREE also approves them.

The installation of recycling companies in the South and Southeast regions, mainly in São Paulo, is notorious. However, according to the mapping of Brazilian recycling, 272 WEEE recycling companies were identified in the country, noting the potential for expanding the industrial recycling park to other regions [17]. Therefore, Giese et al. [13] suggest a specific calculation tool for the WEEE recycling area which evaluates disassembly level scenarios between superficial, medium, or deep and which also helps the recyclers in the process of decision-making. As such, there are circular business opportunities to be explored.

4.4 Adhesion of the Business Sector

Data from Fig. 6 show an increase in the adhesion of the business sector to management entities over the years. Currently, 118 companies guarantee the environmentally friendly disposal of equipment placed by them on the market. Green Eletron and ABREE also have 58 retail partners that help implement the WEEE reverse logistics. According to ABREE, in 2020, its 29 associates represented 98

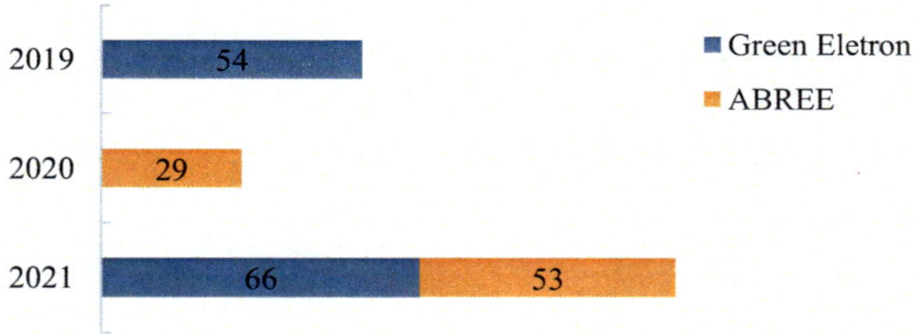

Fig. 6 Number of companies associated with managers entities

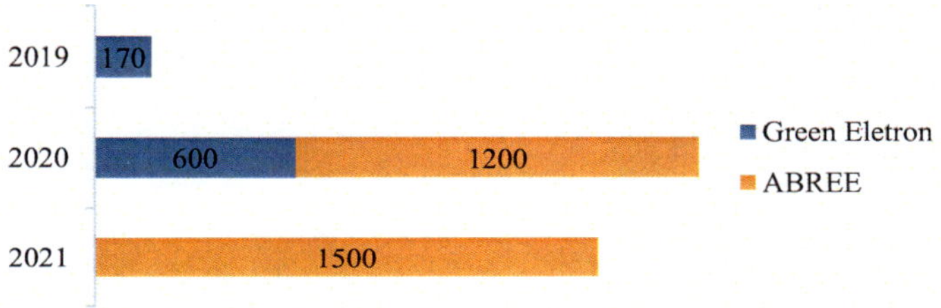

Fig. 7 Number of voluntary delivery points offered by each managing entity

brands and 85% of the total volume and weight of EEE placed on the Brazilian market. The relevance of the collection of large household appliances to achieving long-term goals is evident.

Despite of large household appliances have complex logistics due to their size and weight characteristics, they contribute to the goal in proportion to their weight. Cities such as Florianópolis and Brasília carry out the door-to-door collection of EEE weighing over 30 kg. This model could be adopted by more municipalities, together with collection by neighborhood and return of old products in the delivery of a new one.

4.5 Collection and Recycling Network

According to Decree No. 10,240/2020, at least one receiving point must be installed for every 25,000 inhabitants. To date, more than 2000 collection points (Fig. 7) have been installed in several cities across the country, not just those covered in the first year of collection. Therefore, it is impossible to state that the geographic goal was achieved. Green Eletron has appropriately disposed of more than 520 tons of EEE between periodic campaign actions and fixed collection points since the beginning of operations (Fig. 8). ABREE did not disclose the amount of EEE collected or respond to the researcher's contact attempts.

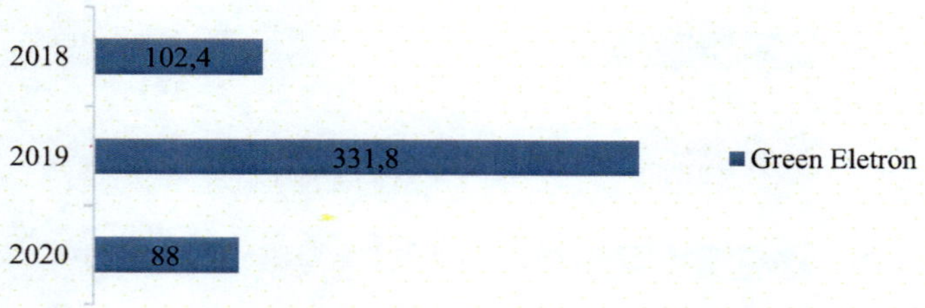

Fig. 8 Recycling of WEEE by Green Eletron (t)

Information was obtained through contact with the Ministry of Environment that reverse logistics' performance follow-up is carried out annually through a report prepared by the group composed of representative entities of the business sector and delivered by March 31, every year. However, in phase 1 (2020), there was no obligation to provide a report as it was a structuring phase. The first one will be delivered in March 2022, relating to the entire year of 2021. Thus, it was not possible to specify how much percent of the annual collection goal has been reached so far. Still, it is clear, through the data collected, that the operationalization of the reverse logistics is making progress.

There is no way to propose a mitigation plan if progress in goals is not happening, since the inspection is carried out through the report delivered after the end of the period defined in the goal.

4.6 *Communication and Environmental Education Plan*

Although the regulation imposes the minimum content of the communication and environmental education plan and also its availability on a website, only ABREE released the communication plan, which does not fully meet the legislation requirements. Green Eletron did not respond to inquiries about the researcher's contact attempts.

It is possible to see the efforts of managers to communicate information about environmentally friendly disposal, collection sites, life cycle aspects of EEE, among others. Collection campaigns are being promoted in partnership with state and municipal governments, participation of representatives of the managers entities in events in the area of reverse logistics, circular economy, and waste in general, as well as interviews with newspapers and media magazines. In this way, they act to mitigate the cultural barrier of the population's lack of environmental education in relation to reverse logistics [2].

Also, the social networks of the companies themselves and the managers, which is the main communication channel with the external environment, have been approaching consumers to promote collection initiatives and also informing members, opinion-makers, and political leaders in ways that managers entities can make solid waste management viable, both as a service and to engage partners.

5 Conclusion

The research reports the advances in the agenda of structuring and implementing the WEEE reverse logistics system in the collective model through manager entities. Thus, it seeks to fill the gap in the absence of a detailed study of the phases provided for in Decree No. 10,240/2020 and explain how the operationalization is taking place. The results showed that recent regulation in Brazil brought legal certainty to the business sector and fostered the formal recycling market. Thus, it is noted that waste management can mitigate the environmental impact caused by urbanization. The urban mining of WEEE has established itself as an enabler of the transition to the circular economy model.

Through the analyzed data, the first year with the collection goal is perceived as the most advanced in terms of serving the population. In addition, it represents the breaking of the initial barrier with environmental education work and the implementation of communication plans. Although the service to almost 40% of the Brazilian population is not included in the geographic goals, there is a significant advance in the WEEE reverse logistics system in Brazil. The challenges of the fifth largest country in the world are mainly logistical due to the operational costs of reverse logistics system, as well as the involvement of different players with different interests.

The installation of recycling companies in the South and Southeast regions, mainly São Paulo, highlights that the market is structured in places with greater economic potential. Therefore, the way to promote the system in less favored regions such as the North and Northeast may be simplification and tax incentives for the waste recovery industry, in addition to establishing a relationship between government spheres to reduce tax differences between regions and adapt the functioning of the system to the local reality.

Finally, the study evidenced the evolution in the involvement of the business sector with the adhesion of companies to managing entities and investment in infrastructure for the reverse logistics. It also highlights the challenge already shown in the literature regarding the return of large appliances due to size and weight characteristics since the voluntary delivery points currently available are mostly for medium and small-sized products. Therefore, it concludes that there is potential in adhering to the collective model to comply with legal requirements since the manager entity centralizes the business sector's efforts and even dilutes the costs among those involved.

The work generated a reflection on the lack of transparency in the disclosure of data both by managers and the Ministry of Environment, in the same way as reporting to society complying with legal norms, given the complexity of reverse logistics system operationalization. As a suggestion for future studies, it is proposed to involve managing entities that directly relate to the type of reverse logistics employed. In this way, there is a more consistent way of evaluating advances in the agenda for the operationalization of the WEEE reverse logistics system.

In terms of practical application, the research contributes as academic literature as it discusses reverse logistics based on empirical data on the progress of operationalization in the country. As for entrepreneurs in the area, such as business consultants, the study suggests a market to act both in support of the logistics business, meet legal requirements, and set up recycling companies in regions that are lagging behind. And in short, for the manager entities as an indication of performance improvements in the reporting of evolutionary data.

Acknowledgments The authors would like to thank CAPES (Coordination for the Improvement of Higher Education Personnel) and CNPq (National Council for Scientific and Technological Development) for the financial support of this research.

References

1. Awasthi, A.K., Cucchiella, F., D'Adamo, I., Li, J., Rosa, P., Terzi, S., Wei, G., Zeng, X.: Modelling the correlations of e-waste quantity with economic increase. Sci. Total Environ. **613–614**, 46–53 (2018)
2. Vieira, B., Guarnieri, P., e Silva, L.C., Alfinito, S.: Prioritizing barriers to be solved to the implementation of reverse logistics of e-waste in Brazil under a multicriteria decision aid approach. Sustainability. **12**, 4337 (2020)
3. Forti, V., Baldé, C.P., Kuehr, R., Bel, G.: The Global E-waste Monitor 2020: Quantities, Flows and the Circular Economy Potential. United Nations University, Bonn (2020)
4. Balde, C.P., Forti, V., Gray, V., Kuehr, R., Stegmann, P.: The Global E-waste Monitor 2017: Quantities, Flows and Resources. United Nations University, Bonn (2017)
5. Ottoni, M., Dias, P., Xavier, L.H.: A circular approach to the e-waste valorization through urban mining in Rio de Janeiro, Brazil. J. Clean. Prod. **261**, 120990 (2020)
6. BRASIL: Lei 12.305, de 02 de agosto de 2010. Presidência da República, Brasília (2010)
7. BRASIL: Decreto 10.240, de 12 de fevereiro de 2020. Presidência da República, Brasília (2020)
8. Cipriano, T.A.R.P.: O conceito de consumidor no direito brasileiro. Revista de Direito do Consumidor. **133**, 357–384 (2021)
9. Green Eletron: https://www.greeneletron.org.br/blog/esclarecemos-as-quatro-principais-duvidas-das-empresas-sobre-a-gestao-de-residuos-eletroeletronicos/, last accessed 21 Jan 2021
10. Cucchiella, F., D'Adamo, I., Lenny Koh, S.C., Rosa, P.: Recycling of WEEEs: an economic assessment of present and future e-waste streams. Renew. Sustain. Energy Rev. **51**, 263–272 (2015)
11. Bakhiyi, B., Gravel, S., Ceballos, D., Flynn, M.A., Zayed, J.: Has the question of e-waste opened a Pandora's box? An overview of unpredictable issues and challenges. Environ. Int. **110**, 173–192 (2018)
12. Souza, R.G.: E-waste situation and current practices in Brazil. In: Handbook of Electronic Waste Management, pp. 377–396. Elsevier, Butterworth-Heinemann (2020)
13. Giese, E.C., Xavier, L.H., Ottoni, M., Araujo, R.: Cooperativas e a gestão de resíduos eletroeletrônicos, 1st edn. CETEM/MCTI, Rio de Janeiro (2021)

14. Echegaray, F.: Consumers' reactions to product obsolescence in emerging markets: the case of Brazil. J. Clean. Prod. **134**, 191–203 (2016)
15. Rodrigues, A.C., Boscov, M.E.G., Günther, W.M.R.: Domestic flow of e-waste in São Paulo, Brazil: characterization to support public policies. Waste Manag. **102**, 474–485 (2020)
16. MDR, Ministério do Desenvolvimento Regional: Diagnóstico do Manejo de Resíduos Sólidos Urbanos – 2019, Brasília (2020)
17. CETEM: https://www.cetem.gov.br/antigo/reminare/recicladores, last accessed 21 May 2021

Application of Heijunka for Surgical Production Leveling

Thiago A. Souza, Guilherme L. R. Vaccaro, Rui M. Lima, and Erik T. Lopes

Abstract The increasing demand and the limitation of resources in health organizations raise the need for more efficient management. With demand higher than the capacity, patient queues increase, causing hospitals to need flexible management mechanisms to deal with the existing lists. In this context, adaptations of principles and methods of Industrial Engineering and Management, such as Lean Thinking and Heijunka, naturally arise. These adaptations contribute to the improvement of performance in health processes. This work aims to reduce the average resolution times of surgical cases by application of Heijunka to level the surgical production between different specialties of a large Brazilian hospital. With the application of Heijunka principles, the organization obtained an estimated mean reduction in the resolution time of elective surgeries from 17 to 10 months. In addition to the equity obtained between the queues of the specialties, reduction of costs with materials, consultations, and preoperative examinations, there are also estimated cost savings. In this context, Heijunka has advantages related to the leveling of surgical production, greater equity in resolution times, and serves as a tool for planning the distribution of operating rooms.

Keywords Lean Healthcare · Heijunka · Operating Room · Surgery Production Leveling

---

T. A. Souza (✉) · R. M. Lima · E. T. Lopes
Algoritmi Research Centre, Department of Production and Systems, School of Engineering of the University of Minho, Braga, Portugal

G. L. R. Vaccaro
Escola Politécnica – Programa de Pós-Graduação em Engenharia de Produção e Sistemas/Programa de Pós-Graduação em Gestão e Negócios – UNISINOS, Sao Leopoldo, Brazil

V. M. López Sánchez et al. (eds.), *Industrial Engineering and Operations Management*, Springer Proceedings in Mathematics & Statistics 400,
https://doi.org/10.1007/978-3-031-14763-0_22

1 Introduction

Healthcare systems coexist in a context of demand greater than capacity, increasing costs and dependence on public policies. High costs are caused by the introduction of new technologies that do not replace existing technologies [1], the isomorphic characteristic of health service delivery environments [2], and epidemiological and population change aspects [3]. These prevent the planning of services with large-scale immediate care capacity, increasing the need for efficiency in the use of resources available by health service providers [4]. This context directly affects the capacity of care: waiting lists for surgery are reported in several countries [5–7].

Although the revenue from hospital services in developed countries is increasingly based on measures of activity [8], the planning of its capacity remains defined by the number of beds existing in each organization [9]. Mendes et al. [10] reported the process of change that has been happening in healthcare and demonstrates an unplanned movement according to the requirements of the Brazilian moment of demographic, social, and epidemiological transition, in addition to a non-rational distribution of hospitalizations by specialty.

Strongly dependent on the criteria and context mentioned above, Surgical Centers (SC) are critical services responsible, under the essential aspect, for the delivery of a portion of relevant value to the patient and, from the administrative aspect, for a significant portion of revenue and costs in a hospital [11–13]. However, SC and beds are resources used in different hospital services and care cases, been demanded by many areas. In addition, they may have their level of activity reduced or limited by the availability of hospital beds [9].

Laganga [14] studied criteria related to the scheduling of surgical services and obtained an increase in the ability to admit new patients as a result of the transformation of service processes. Even though there are no systematic evaluations of waiting lines in SC in Brazil, this ignorance prevents the proper management of the public healthcare system, reducing the relationship between value and costs involved in the system [15].

Under the aspect of operations management, the administration of surgical services requires the application of concepts of efficiency, capacity management, and synchronization of production. In addition, the system needs to be managed to level its potential to the demand for surgeries. This aspect requires considering concepts such as the criticality of existing specialties and groups, demand patterns, availability of resources to perform care, and the logistic flow of surgical materials and supplies. For these reasons, the planning and programming of production in SC become essential in the search for continuous improvement and patient satisfaction [16]. This ability to apply operations management concepts and Lean principles has been increasingly highlighted in the literature [17].

Lean Thinking is an approach to operations management that considers spending resource that does not add value to the end customer as a waste of the process and emphasizes methods and tools to assist managers and workers in improving the delivery of value to the customer. This approach can support doctors, nurses, and other professionals to remove process barriers and allow them to focus on care delivery [1], especially in Brazilian public organizations, which have traditional management processes and low modernization [18]. Changes originated by Lean Thinking can improve the distribution and use of operating rooms, being recommended for its adoption by the healthcare institution with a broad approach to surgical process management [19]. A concept used in Lean is that of Heijunka, which in addition to the leveling of production, aims at the planning and control of its mix, and sequencing according to the total volume demanded at scheduling intervals [20].

Considering the above context, this study discusses how Heijunka production leveling can be applied to improve surgical management processes in hospitals, evaluating a case of application of Heijunka in SC of a large hospital, located in Brazil. It aims to discuss how to distribute SC work shifts among several specialties and how contextual differences can affect these decisions.

2 Theoretical Framework

2.1 Demand and Capacity Management in Surgical Services

Faced with the growth in demand and low capacity to offer services, hospitals should better manage their demand and thus make decisions to minimize risks. Some causes common to large centers related to the management of demand for healthcare services can be: the disordered flow of people, the unscheduled demand, the fragile agreement of procedures and services between managers, and the incipient regulation of access causing inefficiency and fragility to the system [21].

Healthcare demand depends on adequate management of population data in different time horizons. In the case of surgical centers, it is necessary to meet immediate (non-elective) and elective demands. In the context of the emergency department, controlled idleness of the SC is necessary to ensure an immediate response to critically ill patients. In addition, elective patients can be classified as inpatient and outpatient, and hospitalized patients need to stay one night or more in the hospital, while outpatients typically enter and leave on the same day. On the other hand, non-elective patients are divided between urgent and emergency patients, based on the response capacity of the patient's arrival: surgery in emergency patients (emergency) must be performed as soon as possible, while urgent patients (urgency) are sufficiently stable so that surgery can be postponed for a short period [16].

Efficiency improvements and surgery scheduling can translate into savings and significant benefits for the patient as well as the hospital [22], so managing the demand and productive capacity of surgical centers constantly demonstrates the importance of the theme for practical application of such improvements.

2.2 Lean Healthcare and Heijunka

Lean Healthcare represents the application of Lean principles and tools in healthcare organizations, achieving interesting results in surgical centers [23, 24]. However, there are still inconsistencies in the redesigns of health processes adopting Lean [25], which may result, in part, from the poor translation of models and methodologies developed in other contexts, which gives insufficient attention to existing practices and potential consequences.

Radnor et al. [26] highlight aspects of Lean's implementation in healthcare. First, there are specificities related to the way the value is defined; second, the disjointed approach to lean implementation across the organization; third, the understanding of Lean as an application of specific tools and techniques, not as a philosophy of change; fourth, project implementations "is likely to hit some low-lying glass ceiling." This finding presented by the authors suggests that health organizations are still embracing lean thinking more broadly.

The basis of Lean Thinking linked to Just-in-Time aims at leveling the workflow to optimize the process [27]. The leveling of production, or Heijunka, means meeting the demand of customers by creating a level schedule of orders and daily variations thereof, in addition to leveling the quantities and product types in the long term [28]. The objective is to achieve a constant flow, in the case of patient healthcare, in a mixed production model that supplies one or more processes with constant and stable flow [29]. For Sharma and Moody, to control the oscillation in production, organizations must analyze the monthly demand per product, and thus schedule daily their production level with the known demand [30]. In the case of surgeries, specifically talking about elective surgeries, the Surgical Center should monitor the issuance of surgical requests, and at the same time control its waiting list, ensuring this information is provided to the team to give agility and priority to appointments to be scheduled to level their production and reduce demand fluctuations.

The key to leveling is to find the smallest possible batch measure needed to stabilize production by increasing system efficiency and flexibility [31]. Smalley [32] exposes three critical activities in production leveling management: the need for continuous monitoring of customer demand, analysis and monitoring of process performance and stability indicators, and daily control of production and established standardized processes. For the case of service operations, Liker suggests that customer demand be leveled, such as having dentists with standard schedules, and thus, customers fit into these schedules [33].

3 Methodology

The research in question can be defined in relation to its nature as applied, because it generates knowledge for application and problem solving, and quantitative, because it translates into numbers demand information classifying and analyzing them. This research can also be classified as exploratory due to the objective of explaining a problem related to the process of distributing work shifts in surgical centers related to the demand for existing specialties.

From the point of view of technical procedures, the present study is guided by action research being carried out in association with an action or resolution of a real problem, where the researcher and the team are involved in the improvement in a participatory way. According to Gil, there are nine phases in Action Research, starting with an exploratory phase and the formulation of the problem, until the elaboration of the action plan and the dissemination of the results obtained [34].

The present study started with bibliographical research involving the themes: Lean Healthcare, Heijunka, and demand and capacity management in surgical services. This stage sought to analyze a year of existing reports in the information systems used, and thus raise the average number of surgical requests issued per month by the specialties. At this stage, the existing waiting list was also collected.

After the demand study, we sought to analyze the productive capacity of each specialty as well as the average of surgeries produced per work shift, in addition to the comparison with the collected demand. With demand and capacity, the ideal scenario was designed with the number of work shifts needed to meet the average monthly demand for surgeries and the waiting list accumulated up to the time of the research.

As Optimal Leveling proved to be unfeasible for application, the study calculated leveling based on capacity constraints considering the maximum number of work-shifts of the hospital studied. And so, several alternatives of application of Heijunka were modeled according to the Monthly Demand, the Waiting List, and the Monthly Demand + Waiting List.

After proposing these alternatives, the researchers together with the managers involved in the study estimated results impacting the process mainly on the variables waiting list resolution, time, and costs. Finally, interviews were conducted between the researchers and hospital managers to discuss the models and their actual application or develop new approaches.

4 Heijunka Application Modeling

In the model used, according to Fig. 1, a systematic demand for new patients requiring surgery (Monthly Demand – MD) is added to the system. These patients compete with pre-existing patients from the system's Waiting List (WL) to the utilization of the production capacity.

Considering the criticality characteristics of the demands for specialty, capacity and demand, WL will behave dynamically: if the capacity is intrinsically lower than the demand, the WL will grow indefinitely; if priorities are adapted so that certain specialties are privileged, the list may grow in some specialties and disappear into others; if the capacity is effectively higher than the demand, or adequately

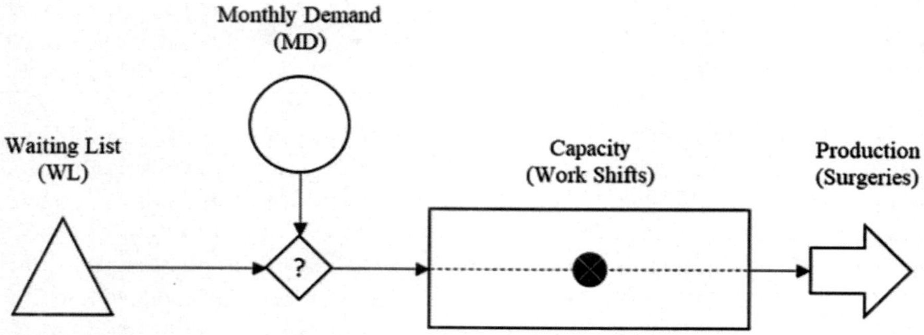

Fig. 1 Used model

synchronized, the WL can be reduced or stabilized adequately to acceptable socioeconomic criteria.

Continuing the analysis of Fig. 1, it is considered that the installed capacity of the SC is divided into work shifts and that each specialty has a recognized level of production. According to distributed capacity and production/shift, the total production of the SC is established. The net capacity of the SC, however, depends on the efficiency of the services, which can be differentiated and subject to different technical and technological aspects, however, this model considers a constant efficiency according to the history of the study hospital.

Considering the previous model, four alternative Heijunka solutions were elaborated: considering infinite capacity (ideal), considering finite capacity and demand priority (Heijunka 1), considering finite capacity and priority to the waiting list (Heijunka 2), and considering finite capacity and the balance between monthly demand and waiting list (Heijunka 3).

The ideal leveling (considering infinite capacity) can be calculated by dividing the monthly demand or the waiting list for production per work shift, which results in the number of ideal work shifts required according to the variable used, as shown in Eq. 1.

$$Ideal\ Work\ Shifts\ in\ (MD\ or\ WL) = \frac{Monthly\ Demand\ (MD\ or\ WL)}{Production/Shift} \tag{1}$$

For the other leveling performed, the number of proposed work shifts is calculated by multiplying the necessary capacity by the ratio between the total installed capacity and the ideal total capacity, according to Eq. 2.

$$Proposed\ Work\ Shifts\ (y) = Necessary\ Capacity\ (Ideal\ y) * \frac{Total\ Capacity\ (Installed)}{Total\ Capacity\ (Ideal\ y)} \tag{2}$$

5 Case Study

This study was carried out in a public hospital, based in southern Brazil. The study hospital is defined as a university under public-private administration, containing approximately 400 hospitalization beds, 11,000 consultations and 600 surgeries per month. Several specialties work in its operating room, such as Cardiovascular Surgery, Head and Neck Surgery, and Urology. A peculiarity of the application refers to the analysis of the leveling of the production of surgeries from the SUS (Brazil's Unified Health System), given that the demand for surgery of agreement and private individuals was smaller and with few restrictive variables. Another peculiarity is that, as the hospital has the essence of a back-up hospital, the surgeries demanded are elective in nature originated by the outpatient clinic or hospitalization, and emergency cases are rare. The overall production capacity of the operating room was based on four rooms available daily to perform procedures from the SUS, working 11.5 hours per day in each room. With the existence of 2 daily work shifts per room, the total number of work-shifts per week would be 40 (and monthly around 160, with small monthly variations).

The historical context of the hospital's surgical center is based on a significant increase in surgeries, operating with seven rooms in parallel, however, due to budgetary constraints, it was decided to retract it to for four rooms, turning efficiency a priority among the teams. In addition to the financial context, the demand for surgeries remained and as the real demand for specialties was not known until this study, a frequent routine was to perform surgical task forces when the specialties felt, qualitatively, peaks in demand generating instability in the process and increasing the monthly operational cost. Another common process was to distribute work shifts according to the requests of the teams and especially among surgeons, also professors linked to the training of residents, without analyzing the actual demand of patients. This fact led to uneven waiting lists, floating bottlenecks and specialties, and a complex flow of patients difficult to manage.

5.1 *Demand and Capacity Analysis*

First, we sought to know the average demand of monthly surgeries and the waiting list per specialty. According to the survey based on the hospital's information system regarding reports issued per specialty per month, an average of 617 surgery reports per month was obtained. In addition, the information system of the Municipal Health Department collected the waiting list at the time of the study, resulting in a total of 3636 patients. This information is presented in Table 1.

In addition to the demand analysis, the study of production capacity by specialty was carried out by analyzing the monthly production history generating the average production per nominal work shift for future calculations. With the average monthly production and the current waiting list quantity, an average list resolution time (or

Table 1 Monthly demand and waiting list

Specialty	Monthly demand (MD)	Waiting list (WL)	Monthly production (average)	Work shift production (average)	Makespan (months)
Oral and Maxillofacial Surgery	9	27	7	2	3.9
Cardiovascular Surgery	19	58	2	1	29.0
General Surgery	122	1.305	38	3	34.3
Pediatric Surgery	21	106	33	3	3.2
Plastic Surgery	34	177	4	1	44.3
Thoracic Surgery	16	10	11	3	0.9
Vascular Surgery	12	88	4	2	22.0
Coloproctology	7	57	4	4	14.3
Gynecology	66	587	15	4	39.1
Neurological Surgery	4	17	3	1	5.7
Otolaryngology	47	292	28	3	10.4
Traumatology	227	720	110	2	6.5
Urology	33	192	18	3	10.7
Total	**617**	**3.636**	**277**	**2**	**17.3**

makespan) can be estimated in months per specialty. Note that the overall average of makespan by a patient, in months, in this system is 17.3 months.

It is observed that the monthly production, based on the history of the Surgical Center and with the current mix, is 277 monthly surgeries. Although this study was limited in the analysis of efficiency and productivity of the specialties, using the historical basis of production confronted with the standard times of the procedures performed, the average production per work shift of each specialty, used to calculate the leveling, was estimated. In addition to the average production per work shift, Table 1 presents the estimate waiting time of patients in the specialty.

5.2 *Heijunka (Leveling)*

With the information of demand and production capacity of the specialties, the number of ideal work shifts needed to perform the surgeries were calculated using the regular monthly demand and the waiting list. This first Heijunka demonstrates the need for a higher production capacity than the current one, according to Table 2. Using Eq. 1 for the specialty Oral and Maxillofacial Surgery, for example:

$$Ideal\ Work\ Shifts\ (MD) = \frac{9}{2} = 4.5 \tag{3}$$

Table 2 Ideal leveling

Specialty	Ideal work shifts (MD)	Ideal work shifts (WL)
Oral and Maxillofacial Surgery	4.5	13.5
Cardiovascular Surgery	19.0	58.0
General Surgery	40.7	435.0
Pediatric Surgery	7.0	35.3
Plastic Surgery	34.0	177.0
Thoracic Surgery	5.3	3.3
Vascular Surgery	6.0	44.0
Coloproctology	1.8	14.3
Gynecology	16.5	146.8
Neurological Surgery	4.0	17.0
Otolaryngology	15.7	97.3
Traumatology	113.5	360.0
Urology	11.0	64.0
Total	**278.9**	**1.465.5**

$$Ideal\ Work\ Shifts\ (WL) = \frac{27}{2} = 13.5 \quad (4)$$

It is observed that to meet the average monthly demand, about 279 monthly work shifts are required, and the available capacity is 160. This would represent an increase of 3 operating rooms operating with the same workload as the other. Another analysis view is the need for 1465 work shifts to perform the existing general waiting list, being about 9 times larger than the current production capacity (160 work shifts). In addition to the need for capacity increase, the cost involved with nursing staff, surgeons, materials, and others should be analyzed with hospital discharge.

Another possible scenario is to level production according to the average monthly demand for specialties, making the production rate according to such demand and thus estimate the makespan. Furthermore, it is possible to level the work shifts according to the existing waiting list up to the time of the study, noting a different leveling than the previous ones and the fixing of the makespan in 9.2 months, on average. Finally, the last Heijunka analyzed in this article aims at leveling according to the Monthly Demand and the Waiting List, taking as a premise that the production of specialties reaches the same production rate in the two variables. The three scenarios are presented in Table 3 below.

Using the Eq. 2 and the presented Heijunka and makespan concepts for the specialty Oral and Maxillofacial Surgery, for example:

$$Heijunka\ 1 = 9 * \frac{2}{7} = 2.6 \quad (5)$$

Table 3 Leveling performed

Specialty	Average demand leveling		Waiting list leveling		Average demand and waiting list leveling	
	Proposed work shifts (MD)	Makespan (months)	Proposed work shifts (WL)	Makespan (months)	Proposed work shifts (MD + WL)	Makespan (months)
Oral and Maxillofacial Surgery	2.6	5.2	1.5	9.2	2.0	6.7
Cardiovascular Surgery	10.9	5.3	6.3	9.2	8.6	6.7
General Surgery	23.3	18.6	47.5	9.2	35.4	12.3
Pediatric Surgery	4.0	8.8	3.9	9.2	3.9	9.0
Plastic Surgery	19.5	9.1	19.3	9.2	19.4	9.1
Thoracic Surgery	3.1	1.1	0.4	9.2	1.7	1.9
Vascular Surgery	3.4	12.8	4.8	9.2	4.1	10.7
Coloproctology	1.0	14.2	1.6	9.2	1.3	11.1
Gynecology	9.5	15.5	16.0	9.2	12.7	11.5
Neurological Surgery	2.3	7.4	1.9	9.2	2.1	8.2
Otolaryngology	9.0	10.8	10.6	9.2	9.8	9.9
Traumatology	65.1	5.5	39.3	9.2	52.2	6.9
Urology	6.3	10.1	7.0	9.2	6.6	9.6
Total	**160.0**	**9.6**	**160.0**	**9.2**	**160.0**	**8.7**

$$Heijunka\ 2 = \frac{WL}{9.2} = 1.5 \tag{6}$$

$$Heijunka\ 3 = \frac{H1 + H2}{2} = 2.0 \tag{7}$$

With the leveling based on average demand, there is a proportional reallocation between specialties and, mainly, a variation between makespan. Previously, the estimated average was 17.3 months and with the proposed Heijunka, the estimate reduces to 9.6 months of waiting.

However, using the second Heijunka, with the specialties maintaining the estimated level of production per shift, in 9.2 months the current queue should reach zero, disregarding the queue created after the implementation of the scenario.

Finally, there is a significant reduction in makespan, with and alternative Heijunka considering simultaneously monthly demand and waiting list, from 17.3 to 8.7 months. It is noteworthy that with this model, the specialties have the productive capacity to partially meet their monthly demand and part of their waiting list, with the exit rate being proportionally leveled between all of them. So, the general conclusion would be that with low capacity and high demand, we have to design a system that is not ideal in the number of work shifts needed, however, that can meet optimally with the capacity we have in the organization.

However, it is worth mentioning that the calculation does not consider some aspects such as urgency of procedures, needs of the municipality through the Health Department and costs with orthotics and prostheses, which can interfere in the dimensioning. Thus, with the analysis of specialists and the technical direction of the hospital, the final Heijunka leveling was defined as described in Table 4.

Table 4 Leveling implemented in the University Hospital

Specialty	Proposed work shifts (MD + WL)	Makespan (meses)
Oral and Maxillofacial Surgery	4.0	3.4
Cardiovascular Surgery	4.0	14.5
General Surgery	36.0	12.1
Pediatric Surgery	8.0	4.4
Plastic Surgery	4.0	44.3
Thoracic Surgery	8.0	0.4
Vascular Surgery	4.0	11.0
Coloproctology	4.0	3.6
Gynecology	16.0	9.2
Neurological Surgery	8.0	2.1
Otolaryngology	20.0	4.9
Traumatology	44.0	8.2
Urology	4.0	16.0
Total	**164.0**	**10.3**

With the change in work shift distribution, the administration had an increase of 4 monthly work shifts, resulting in a total of 164 shifts available. The major modifications made between the initial model and the last one proposed refer to qualitative questions analyzed by experts in relation to the variables mentioned above, leading to a reduction from 17.3 to 10.3 months of makespan.

6 Final Considerations

Given the need for hospitals to increase their efficiency and reduce their costs [35] and the focus of the development of studies related to the management of service operations [36], this article proposed to analyze the demand for surgeries of a Brazilian hospital in a context of high monthly demand, existing waiting list, and low production capacity for surgeries. Some forms of leveling were presented with different calculation assumptions, and the ideal leveling was discarded due to budgetary issues and qualitative care analyses.

Possible scenarios were based on monthly demand, waiting list and the two variables together. The latter presented itself as an optimal point estimating an average resolution for surgeries of 8.7 months, however, an adaptation based on care and cost criteria was proposed by hospital management (10.3 months). The estimated financial impact with this implementation is about US$750,000.00 per year, of which approximately US$600,000.00 with orthotics, prostheses, and materials, in addition to the increase in surgeries with lower cost, and US$150,000.00 with reduction of consultations and repetitive preoperative examinations.

It is suggested that the leveling study, accompanied by efficiency analyses in the use of work shifts by specialties and thus, can adapt the capacity and demand setting goals to also increase the efficiency in the installed capacity. It is also suggested that future research using computer simulation for Studies of Heijunka in Surgical Centers, as cited by Robinson et al. [37]. Such use is suggested by the speed in the comparison of scenarios and dynamics of animations to understand those involved who do not have experience based on Operations Management.

Acknowledgements This work was partially supported by FCT – Fundação para a Ciência e Tecnologia within the R&D Units Project Scope: UIDB/00319/2020.

References

1. Graban, M.: Lean Hospitals: Improving Quality, Patient Safety, and Employee Satisfaction. CRC Press, Boca Raton (2011)
2. Dimaggio, P.J., Powell, W.W.: The iron cage revisited institutional isomorphism and collective rationality in organizational fields. In: Economics Meets Sociology in Strategic Management, pp. 143–166. Emerald Group Publishing Limited, Bingley (2000)

3. Lima-Costa, M.F., Veras, R.: Aging and public health. Rep. Public Health. **19**(3), 700–701 (2003)
4. Lamiri, M., et al.: A stochastic model for operating room planning with elective and emergency demand for surgery. Eur. J. Oper. Res. **185**(3), 1026–1037 (2008)
5. Conill, E.M., Giovanella, L., de Almeida, P.F.: Waiting lists in public systems: from expanding supply to timely access? Reflections on Spain's National Health System. Ciên. Saúde Colet. **16**(6), 2783–2794 (2011)
6. Lemos, M.S.A., et al.: The impact of the project elective surgeries in Belo Horizonte on the waiting list. Enferm. Rev. **16**(3), 159–174 (2013)
7. Martins, R.A.S., Nolasco, B.C.A., Severino, R.R.: Estratégias de efetivação do acesso à saúde: a busca ativa de pacientes mediante demanda reprimida na saúde no HC/UFTM. III Simpósio Mineiro de Assistentes Sociais (2012)
8. Rechel, B., et al.: Hospital capacity planning: from measuring stocks to modelling flows. Bull. World Health Organ. **88**(8), 632–636 (2010)
9. ANAHP, Livro Branco: Brasil Saúde 2015 | A sustentabilidade do sistema de saúde brasileiro (2014)
10. Mendes, A.D.C.G., et al.: The public healthcare system in the context of Brazil's demographic transition: current and future demands. Rep. Public Health. **28**(5), 955–964 (2012)
10. Cima, R.R., et al.: Use of Lean and Seis Sigma Methodology to Improve Operating Room Efficiency in a High-Volume Tertiary-Care Academic Medical Center. J. Am. Coll. Surg. **213**(1), 83–92 (2011). https://doi.org/10.1016/j.jamcollsurg.2011.02.009
12. Dexter, F., et al.: Use of operating room information system data to predict the impact of reducing turnover times on staffing costs. Anesth. Analg. **97**(4), 1119–1126 (2003)
13. Marjamaa, R., Vakkuri, A., Kirvelä, O.: Operating room management: why, how and by whom? Acta Anaesthesiol. Scand. **52**(5), 596–600 (2008)
14. Laganga, L.R.: Lean service operations: reflections and new directions for capacity expansion in outpatient clinics. J. Oper. Manag. **29**(5), 422–433 (2011)
15. Marinho, A.: A study on organ transplantation waiting lines in Brazil's Unified National Health System. Rep. Public Health. **22**(10), 2229–2239 (2006)
16. Cardoen, B., Demeulemeester, E., Belien, J.: Operating room planning and scheduling: a literature review. Eur. J. Oper. Res. **201**(3), 921–932 (2010)
17. Lima, R.M., Dinis-Carvalho, J., Souza, T.A., Vieira, E., Gonçalves, B.: Implementation of lean in healthcare environments: an update of systematic reviews. Int. J. Lean Six Sigma. **12**(2), 399–431 (2021). https://doi.org/10.1108/IJLSS-07-2019-0074
18. Souza, T.A., das Chagas Marques, M., Tarrago, F.C., Harzheim, E., Lima, R.M.: Lean demand management: application in a National Health Department. In: Thomé, A.M.T., Barbastefano, R.G., Scavarda, L.F., dos Reis, J.C.G., Amorim, M.P.C. (eds.) Industrial Engineering and Operations Management. IJCIEOM 2020. Springer Proceedings in Mathematics & Statistics, vol. 337. Springer, Cham (2020). https://doi.org/10.1007/978-3-030-56920-4_24
19. Collar, R.M., et al.: Lean management in academic surgery. J. Am. Coll. Surg. **214**(6), 928–936 (2012)
20. Matzka, J., Di Mascolo, M., Furmans, K.: Buffer sizing of a Heijunka Kanban system. J. Intell. Manuf. **23**(1), 49–60 (2012)
21. Pires, M.R.G.M., et al.: Supply and demand of the medium-complexity/SUS: relation with primary health care. Cien. Saude Colet. **15**(Suppl 1), 1009–1019 (2010)
22. Min, D., Yih, Y.: Scheduling elective surgery under uncertainly and downstream capacity constraints. Eur. J. Oper. Res. **206**(3), 642–652 (2010)
23. Souza, T.A., Queiroz, G.A., Lima, R.M., Pimenta, L.V.: Application of the Operating Room Effectiveness Indicator (ORE) in a Brazilian Hospital. International Joint Conference on Industrial Engineering and Operations Management (IJCIEOM) (2020)
24. Souza, T.A., Vaccaro, G.L.R., Lima, R.M.: Operating room effectiveness: a lean health-care performance indicator. Int. J. Lean Six Sigma. **11**(5), 973–988 (2020). https://doi.org/10.1108/ijlss-12-2017-0141

25. Waring, J.J., Bishop, S.: Lean healthcare: rhetoric, ritual and resistance. Soc. Sci. Med. **71**(7), 1332–1340 (2010)
26. Radnor, Z.J., Holweg, M., Waring, J.: Lean in healthcare: the unfilled promise? Soc. Sci. Med. **74**(3), 364–371 (2012)
27. Liker, J.K.: The Toyota Way: 14 Management Principles from the World's Greatest Manufacturer, p. 350. McGraw-Hill, New York (2004)
28. Ghinato, P.: Elementos fundamentais do Sistema Toyota de Produção. In: Almeida, A.T., Souza, F.M.C. (eds.) Produção e Competitividade: Aplicações e Inovações, pp. 31–59. UFPE, Recife (2000)
29. Furman, K.: Models of Heijunka-levelled Kanban-Systems. In: Papadopoulos, C.T. (ed.) 5th International Conference on Analysis of Manufacturing Systems – Production and Management 2005 Zakynthos Island, Greece, pp. 243–248, Publishing Ziti, Thessaloniki, Greece (2005)
30. Sharma, A., Moody, P.E.: A máquina perfeita: como vencer na nova economia produzindo com menos recursos. Prentice Hall, São Paulo (2003)
31. Araujo, L.E.D.: Nivelamento de Capacidade de Produção utilizando Quadros de Heijunka em Sistemas Híbridos de Coordenação de Ordens de Produção. 2009. Dissertação (Mestrado) – Escola de Engenharia de São Carlos, Universidade de São Paulo, São Carlos (2009)
32. Smalley, A.: Criando o sistema puxado nivelado. Tradução de Diogo Kosaka. Lean Institute Brasil, São Paulo (2005)
33. Liker, J.K.: The Toyota Way: 14 Management Principles from the World's Greates Manufacturer. Tradução de Lene Belon Ribeiro. Bookman, Porto Alegre (2005)
34. Gil, A.C.: Como elaborar projetos de pesquisa, 4th edn. Atlas, São Paulo (2002)
35. Litvak, E., Bisognano, M.: More patients, less payment: increasing hospital efficiency in the aftermath of health reform. Health Aff. **30**(1), 76–80 (2011)
36. Johnston, R.: Operations: from factory to service management. Int. J. Serv. Ind. Manag. **5**(1), 49–63 (1994)
37. Robinson, S., et al.: SimLean: utilising simulation in the implementation of lean in healthcare. Eur. J. Oper. Res. **219**(1), 188–197 (2012)

Effects of the Sars CoV-2 Pandemic on the Quality of Work: Relations of Effects on Worker Health

Carlos Jefferson de Melo Santos, Ângelo Márcio Oliveira Sant'Anna, and Joilson Nascimento Paim

Abstract The new coronavirus pandemic has brought changes in the lives of the world population. The perception of the need for social isolation became a determining factor for creating new work models with known characteristics, such as the home office and mixed work (face-to-face and telework). There was a significant increase in illnesses caused by abrupt introduction of these models in the context to pandemic. There was a strong relationship of the working model with 4 variables: area of work ($p = 0.000$), work equipment used during quarantine ($p = 0.000$), illnesses found during the pandemic ($p = 0.001$), problems reported during the pandemic like sleep, stress, etc. ($p = 0.010$) and future expectations regarding the work performed and the pandemic ($p = 0.022$). There were also relationships between illnesses before the pandemic and those perceived reported during the pandemic period studied ($p = 0.000$) and the sleep profile before and during ($p = 0.010$). The study results added physical and psychological aspects, which have as premises the worker's perception of the dynamics of companies, as well as their social structures and conditions of work execution. These events interfered with the quality of work of the interviewees and may lead to further studies in the clinical, ergometric or productivity areas. The results presented can help companies make more informed decisions when changing work processes and guide workers in labor operations.

Keywords Quality work · Ergonomics · COVID-19

1 Introduction

Severe acute respiratory syndrome coronavirus 2 (SARS-CoV-2) spread rapidly across multiple countries in early 2020 [1]. With a purpose to reduce the trans-

C. J. de Melo Santos (✉) · Â. M. O. Sant'Anna · J. N. Paim
Polytechnic School Federal University of Bahia, UFBA, Salvador, Bahia, Brazil
e-mail: cj.ufba@gmail.com; angelo.santanna@ufba.br; jnppaim@gmail.com

V. M. López Sánchez et al. (eds.), *Industrial Engineering and Operations Management*, Springer Proceedings in Mathematics & Statistics 400,
https://doi.org/10.1007/978-3-031-14763-0_23

mission of SARS-CoV-2, an estimated 4·5 billion people globally have been placed under some form of lockdown restriction. Modeling and early empirical investigations indicate that physical distancing measures are crucial to reducing transmission of the virus, consequent pressure on health systems, and the number of deaths [2].

In addition to the millions of people infected by SARS-CoV-2, hundreds of thousands have died—and will die—from COVID-19 [3]. An anticipated wave of mental and behavioral problems is beginning to be documented, compounded by social and economic stressors and uncertainties, which will likely present a population health burden for months and years [4]. Isolation can bring unpredictable consequences, with potential illnesses and domestic, mental, and work problems [5, 6]. Working in quarantine can be a challenge for several reasons. The new home office work models, even the face-to-face model with all the sanitary protocols and the hybrid models, diversify how work can be done [7, 8]. However, bodily, and mental injuries and various problems can occur in these adaptations. A government agency recognized the term injury in 1992 to be an inaccurate and misleading term, and the term work-related upper limb disorder was suggested as an alternative [9]. In the field of ergonomics, the work of occupational medicine can play a fundamental role in identifying these problems. Clinicians must understand that work's physical and psychosocial characteristics (and nonworkplace settings) can profoundly affect their patients' wellbeing and produce various musculoskeletal disorders. Physicians have a crucial role in the early recognition of these disorders and in stimulating appropriate ergonomic interventions to reduce their frequency and severity [10]. Also understanding the quality of work developed in the adaptation of work to the pandemic can be an important factor in implementing public policies and legislation for companies to help in this process [11].

These studies were scarce at the beginning and throughout the pandemic. However, when carried out in the form of research, they can bring data that influence the decision-making of public bodies and the private sector itself. This paper aims to assess the quality of work during pandemic time, considering medical and ergonomic aspects, and introduces qualitative variables to understand the adaptations of the population's work during the new coronavirus pandemic. No company was involved in the research, with an exclusively academic purpose.

2 Materials and Methods

2.1 Participants and Main Objectives

Eight hundred and one participants from various regions of Brazil (North, Northeast, Midwest, Southeast and South) with different age groups. All participants were volunteered to response the research, and the answers were not for convenience. The questionnaire link was sent to more than 200 groups of people who have virtual work environments, such as social media and emails. The research was applied between April to June 2020, and the analyzes were carried out between july to december of the same year.

The research's main objective is to evaluate people's quality of work during the new coronavirus pandemic. In addition, to unfold as secondary objectives: (i) understand which aspects most influence the loss of productivity; (ii) evaluate the impacts of the transformation of jobs in the 3 existing models; (iii) understand the main problems that arose during the pandemic in work routines and; (iv) direct better decision-making by the public and private authorities in the actions taken to face the pandemic in ergonomic, behavioral aspects and general assistance to the worker.

2.2 *Instruments*

The questionnaire consisted of 24 questions about quality of work in times of pandemic. There is a reservation that these 24 questions have different types of response as Likert scale (five points) or options scale to allow more information of the issues and correlate certain types of data that we consider useful for the analysis development of the research. The 24 variables defined for modelling from the survey were planned with focus on the work environment and productivity. The statistical analysis was performed using the SPSS 25® software.

2.3 *Statistical Analysis*

We conducted statistical analysis to assessment the quality of work, considering some medical and ergonomic variables aspects. The data were tested for assumptions that should be met when using a correlation test between paired questions, which means that the medical and ergonomic variable should be statistically significance, and there should be correlation between the investigated aspects.

The Cronbach's alpha coefficient is an internal consistency measure, which adds a set of items as a group by the average correlation among them – items. Therefore, it is considered a statistic that measures the reliability of a questionnaire on a numerical scale from zero to one, where 0 would represent no consistency and 1 high reliability [12]. Although it was obtained through software in this study, the calculation of Cronbach's alpha coefficient is given by,

$$\alpha = \left(\frac{k}{k-1}\right) * \left(1 - \frac{\sum_{i=1}^{k} S^2}{S_t^2}\right) \tag{1}$$

where: k is the number of questionnaire items; S_i^2 corresponds to the variance of each item; S_t^2 is the total variance of the questionnaire, determined as the sum of all the variances.

3 Results and Discussions

3.1 Respondents Profile

The number of interviews who completed the questionnaire was 801, considering the sample size with a confidence level of 95% and a maximum error of 5%. The sample size was 801 responses, and the respondents characterized a response profile through some basic questions, such as age, region of residence and work, height, area of profession and model of work. Some questions were multiple-choice, which allows an N greater than the total number of people in some cases. Regarding age, the profile of the respondents m is mostly in the age group of 18 to 39 years, with 82.49% of the total, as well as more than 90%, are in the northeast and southeast regions of Brazil and the height profile corresponds, in meters, from 1.51 to 1.90 m (corresponding to 96.38% of the answers). The profile of the area of a regular occupation, be diversified, with a high number of respondents in industrial area, as chemistry, administration, and engineering.

As for the work model during the pandemic, most people were in the non-face-to-face model, characterized by the home office and workshifting, corresponding to 64.49% of respondents. This gives us a parameter of association with the ability of companies to reduce the impacts of the pandemic by directing their employees to care for social distance, avoiding crowds in offices and other workspaces. As for equipment and work tools, more than 80% of the responses showed a remote work profile, with the extensive use of notebooks, headsets and microphones. This equipment provides a tool profile that enables a routine work structure with team meetings and tele-assistance.

Regarding the consumption profile of respondents, there was an increase of 18.01% for fast food orders, compared to a 12.87% increase for healthy food consumption. The questionnaire also showed that 17.73% of respondents had an increase in some type of addictive substance (alcohol or some type of psychoactive substance) and 17.32% demonstrated the need for greater coffee consumption against 16.83% of increased consumption of water. All the details and information about the demographic data and profile of the studied population are in Table 1.

In the analysis of the health problems reported by the respondents, some comparison evaluations were carried out to verify the trend and relationship with the work profile. This profile is divided into 3: face-to-face, home office and workshiting. Even though the main analysis was based on these profiles, other strong relationships implied significant behavioral and physiological health changes. The highest amount of illnesses reported before the pandemic were (n = 1206): back

Table 1 Demographic description, work profile and consumption of the studied population

Characteristics	n	%	Characteristics	n	%
Age group (n = 801)			*Height in meters (n = 801)*		
14–15 (apprentice)	3	0.37	1,40–1,50	13	1.62
16–17 (internships, fellows, etc.)	4	0.5	1,51–1,60	160	19.98
18–29	422	52.68	1,61–1,70	308	38.45
30–39	238	29.71	1,71–1,80	203	25.34
40–49	87	10.86	1,81–1,90	101	12.61
50 or more	44	5.49	1,91–2,00	10	1.25
No reply	3	0.37	No reply	6	0.62
Country region (n = 801)			*Working model (n = 842)*		
Northeast	587	1.75	Non-presential	543	64.49
Southeast	136	73.28	Presential	288	34.2
South	44	2.0	Others	11	1.31
Midwest	16	5.49	*Equipment used at work (n = 1423)*		
North	14	16.98	Fixed Computer	255	17.92
No reply	4	0.5	Laptop/Notebook	432	30.36
General practice area (n = 850)			Pen, notebook and other writing materials	257	18.06
Food / Restaurants / Diners	27	3.18	Headphones	164	11.52
Industry / Engineering / Chemistry / Administration	229	26.94	Microphone	67	4.71
Teaching and research	122	14.35	Headseat	52	3.65
Health Sector	104	12.24	Landline	45	3.16
Self-employed	63	7.41	Smartphones and Cell Phones	39	2.74
Marketing and Communication	56	6.59	Sound and sound materials	29	2.04
Law / Advocacy	45	5.29	TV and Radio	23	1.62
Banks / Financial	29	3.41	White board	19	1.34
Trade	24	2.82	Overhead projector / Image projector	15	1.05
Tourism / Hospitality	23	2.71	Others	26	1.83
Social media	17	2.0	*Consumption items during the pandemic (n = 842)*		
Telemarketing / Telephone Service	16	1.88	Fast food	259	18.01
MEI at home	14	1.65	Coffee	249	17.32
Security Sector	13	1.53	Water	242	16.83
Arts / Cultural Artistic Production	12	1.41	Alcohol	208	14.46
Investment / Economy	12	1.41	Healthy food	185	12.87
Public service	12	1.41	Drinks with high sugar content as soft drinks	175	12.17
Others	32	3.76	Psychoactive substances	47	3.27
			Candy	46	3.2
			Nothing	12	0.83
			Chocolate	9	0.63
			Others	6	0.42

pain (26%, n = 316), pain in the cervical region (17%, n = 205), pain (16%, n = 192) and tendonitis in the wrist (12%, n = 146 and shoulder (12%, n = 144). The effect of the pandemic brought illnesses in people who previously had no work-related problems. For this new analysis, there was the same categorized profile of prior illnesses, with incidences of (n = 801): 26% for back pain, 21% for neck pain, 13% for pain, 10% for shoulder tendonitis, and 7% for wrist tendonitis. For workers, it reflected in psychological illnesses contracted in the same period, with incidences of (n = 801): 50.7% for anxiety attacks, 45.1% for high stress, 44.9% for insomnia and 43.3% for constant changes of mood. In the list of these psychosocial illnesses and whether there was a search for treatment for them (n = 801), 43.5% reported that they know your problems, but they did not seek specialized help (n = 366). The only psychosocial illness relevant to these findings was anxiety crisis, where 13.1% (n = 110) of the people sought specialized help. Regarding the sleep study, there was an incidence of non-regularity and non-continuity of sleep during the pandemic (39.8%, n = 801), followed by the incidence of some type of irregularity at sleep time without periodicity (32.7%, n = 801).

For the profile studied, the research showed significant relationships with some characteristics of respondents' profile and health-related problems. There was a strong relationship of the working model with 4 variables: area of work ($p = 0.000$), work equipment used during quarantine ($p = 0.000$), illnesses found during the pandemic ($p = 0.001$), problems reported during the pandemic like sleep, stress, etc. ($p = 0.010$) and future expectations regarding the work performed and the pandemic ($p = 0.022$). There were also relationships with illnesses before the pandemic and those perceived and reported during the pandemic period studied ($p = 0.000$) and the sleep profile before and during ($p = 0.010$) (Table 2).

There are other significant relationships in the profile of the interviewees that overlap occupational problems that add to the pandemic. The equipment that was used in the remote, face-to-face and mixed work contract, whether donated by the company or purchased by the workers themselves, stimulates an influence on the perception of safety in the workplace ($p = 0.0000$). In addition to the increased consumption of items considered addictive or harmful to health, such as drugs, sugar, etc. ($p = 0.0004$). The area of activity also influences the appearance of new illnesses ($p = 0.0000$) and the sleep profile ($p = 0.007$). The latter seems to have a strong relationship with the appearance of other psychosocial problems ($p = 0.000$) and the search for professional help for these same problems ($p = 0.000$), in addition to increasing consumption ($p = 0.074$).

3.2 Discussion

Social distancing [13] was very important to reduce the spread of the pandemic in the world and, especially in the period studied (initial time of the blockade in several countries), it had a great impact on the reduction of new cases and management of hospital beds [14]. However, there is an aggravation in other human spectra that

Table 2 The frequency of patterns of occurrence of physical and psychosocial health effects, a prevalence rate with a 95% confidence interval (CI) and comparisons of trends in work intensity

Health effects	n	%	Incidence	Prevalence
Problems related to the extension of the elbow to the wrist/hand, lower back or muscle pain before the pandemic				
Low back pain	316	0.26	–	0.39
Pain in the cervical region	205	0.17	–	0.26
Malgias (muscle pain)	192	0.16	–	0.24
Tendonitis (on the wrist)	146	0.12	–	0.18
Tendonitis (on the shoulder)	144	0.12	–	0.18
Nothing diagnosed yet / I had no problems	55	0.05	–	0.07
DeQuervain's tenosynovitis (pain at the base of the thumb)	41	0.03	–	0.05
Medial epicondylitis (pain in the elbow)	35	0.03	–	0.04
READ.	31	0.03	–	0.04
Tendonitis (in the Elbow)	22	0.02	–	0.03
Knee pain	5	0	–	0.01
Others	16	0.01	–	0.02
Problems related to the extension of the elbow to the wrist/hand, lower back or muscle pain after the pandemic				
Low back pain	207	0.17	026	0.653
Pain in the cervical region	168	0.14	0.21	0.466
Malgias (muscle pain)	103	0.09	0.13	0.368
Tendonitis (on the shoulder)	79	0.07	0.1	0.281
I had no problems	72	0.06	0.09	0.27
Tendonitis (on the wrist)	60	0.05	0.07	0.144
DeQuervain's tenosynovitis (pain at the base of the thumb)	34	0.03	0.04	0.094
Medial epicondylitis (pain in the elbow)	27	0.02	0.03	0.077
Repetitive Strain Injury (RSI)	22	0.02	0.03	0.066
Tendonitis (in the Elbow)	17	0.01	0.02	0.049
Others	24	0.02	0.03	0.036
Symptoms and other illnesses during the pandemic				
Anxiety crisis	406	0232	0.507	–
High stress	361	0.206	0.451	–
Insomnia	360	0.205	0.449	–
Constant mood swings	347	0.198	0.433	–
I didn't develop any kind of problem	158	0.09	0.197	–
Elevated bad mood	113	0.064	0.141	–
Discouragement / Tiredness	3	0.002	0.004	–
Melancholy	3	0.002	0.004	–
Headache	2	0.001	0.002	–
Search for professional help for the detected illnesses				
I had problems but did not seek expert help	366	0.435	–	–
I don't need any expert help	183	0.217	–	–
Anxiety crisis	110	0.131	–	–

(continued)

Table 2 (continued)

Health effects	n	%	Incidence	Prevalence
High stress	70	0.083	–	–
Constant mood swings	54	0.064	–	–
Insomnia	50	0.059	–	–
Elevated bad mood	7	0.008	–	–
Others	2	0.002	–	–
Regular and continuous behavior of sleep before the pandemic				
Yes	445	0.556	–	0.56
Sometimes	223	0.278	–	0.28
Not	127	0.159	–	0.16
No reply	6	0.007	–	–
Regular and continuous behavior of sleep after the pandemic				
Not	319	0.398	0.398	0.557
Sometimes	262	0.327	0.327	0.286
Yes	216	0.27	0.27	0.825
No reply	4	0.005	–	–

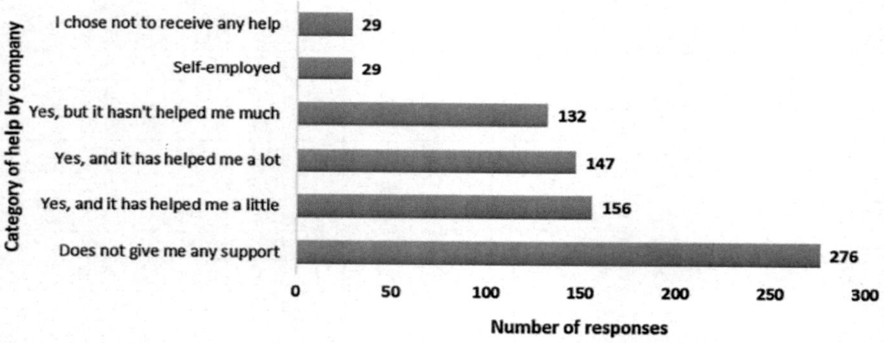

Fig. 1 Frequency distribution to Help from companies during the pandemic

must be observed. In the model of work during a pandemic in the period studied, most were in kind of remote work. The need to adapt remote work at home or in another location that was not conventional in-person, in most cases, enabled the worker to create their work routines, as well as to understand the pace of delivery of their results in light of the activities demanded by the companies [15]. Many people did not have help from the companies to adapt the equipment and infrastructure of the work, either due to the company's restrictions, the worker or imposed on both sides (Fig. 1).

The prevailing height indicates compliance with the average height of the 5 regions of Brazil ($p = 0.000$), consistent with the manufacturing profile of most furniture and electronic equipment in production in Brazil [16], which does not interfere with great influence on the onset of illnesses and adequate ergonomics for daily work ($p = 0.761$). However, these choices of equipment and infrastructure studied be done in fact evaluating the productive capacity of professionals and their physical structure and limitations [17]. The consequence of this lack of ergonomic planning is evidenced in the appearance of several occupational problems, including those that already had the physical infrastructure to enter remote work due to the great demand existing in the pandemic period. All the prevalences and incidences shown in the research reinforce the trends of appearance of illnesses and psychosocial problems relevant to workers' health, configuring occupational problems. Increased consumption, disturbed sleep or irregular sleep, as well as increased consumption of substances harmful to health demonstrate the care that companies must take when categorizing remote work routines.

These same with the preservation of personal life with social distancing to delay a pandemic caused by a virus with community transmission [18], new work has entered people's lives and implemented irreversible consequences as evidenced in diseases and health and psychosocial problems [19]. Generally with no way out of home and the need for work, you can have new health problems and a definitive and no longer transitory work profile. However, these adjustments must be reviewed by both parties, as the issues for most companies that have entered remote work will no longer return models to traditional face-to-face work [20] The study of social and ergonomic behavior at home must be carried out to enter in which people accept, from the execution of the best work routines, the solution that best fits the delivery of results, considering the work or the employee.

3.3 General Expectations About the Deliverables of the Activities

Regarding the expectations of the future of work during the new corona virus pandemic, 34.08% of the respondents reported not having much difference in deliveries and 30.46% had very high expectations of delivering a good result (Fig. 2).

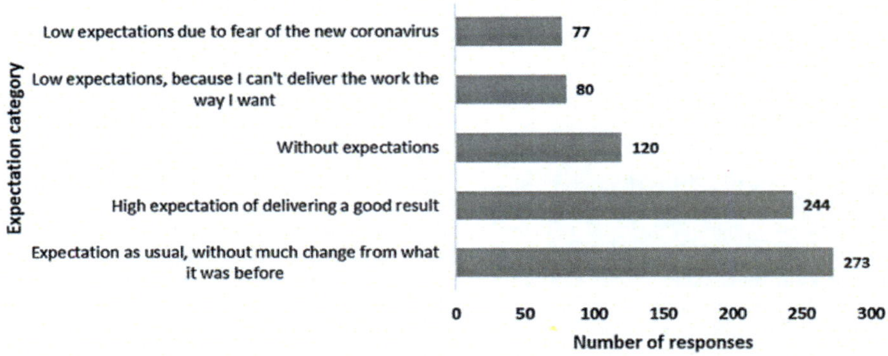

Fig. 2 Frequency distribution to Respondents' future expectations

4 Conclusions

Work routines were severely affected by the pandemic. In various aspects, professionals suffered from the drastic changes that a pandemic can bring, both in the pre-established molds and the need to adapt to the labor market. The results show how important is for public authorities to act on the health to working class in general. The companies that hire them, offering the necessary technical and medical support for them to adapt to the routines of social isolation (specifically for those working remotely) and follow protocols when it comes to hybrid and face-to-face work.

In this first moment of the pandemic of the studied period, the unpreparedness of the entire production and service chain for adaptations to the continuity of work during the pandemic is notorious through the study presented.

It is important that the scientific community can better understand the pandemic labor and the cooperation of public authorities, together with private organizations, carry out continuous improvement plans to better adapt their employees, the routines of their activities, and as the long-term consequences can be much larger than those currently presented.

References

1. Gudbjartsson, D.F., Helgason, A., Jonsson, H., et al.: Spread of SARS-CoV-2 in the Icelandic population. N. Engl. J. Med. (2020). https://doi.org/110.1056/NEJMoa2006100. Published online April 14
2. Drake, T.M., Docherty, A.B., Weiser, T.G., et al.: The effects of physical distancing on population mobility during the COVID-19 pandemic in the UK. The Lancet, Lancet Digital Health (2020). https://doi.org/10.1016/S2589-7500(20)30134-5. Published Online June 12, 2020
3. World Health Organization (New York): WHO Director-General's opening remarks at the media briefing on COVID-19 – 11 March 2020. [S. l.], (2020, March 11). Available in: https://www.who.int/director-general/speeches/detail/whodirector-general-s-opening-remarks-at-the-media-briefing-on-covid-19%2D%2D-11-march2020 (2021, November)
4. Galea, S.: The art of medicine: compassion in a time of COVID-1. The Lancet **395**, (2020, June 20). https://doi.org/10.1016/S0140-6736(20)31202-2. Published Online May 22, 2020. School of Public Health, Boston University, Boston, MA 02118
5. Mann, S., Holdsworth, L.: The psychological impact of teleworking: stress, emotions and health. New Technol. Work Employ. **18**(3), 196–211 (2003)
6. Morganson, V.J., Major, D.A., Oborn, K.L., Verive, J.M., Heelan, M.P.: Comparing telework locations and traditional work arrangements: differences in work-life balance support, job satisfaction, and inclusion. J. Manag. Psychol. **25**(6), 578–595 (2010). https://doi.org/10.1108/02683941011056941
7. European Commission: Telework in the EU Before and After the COVID-19: Where We Were, Where We Head To – April 2020. Joint Research Centre, European Union. https://ec.europa.eu/jrc/sites/jrcsh/files/jrc120945_policy_brief_-_covid_and_telework_final.pdf (2020)
8. OECD: Tackling Coronavirus (Covid-19): Contributing to a Global Effort. Productivity Gains from Teleworking in the Post COVID-19 Era: How Can Public Policies Make It Happen? OECD Policy Responses to Coronavirus (COVID-19). https://read.oecd-ilibrary.org/

view/?ref¼135_135250-u15liwp4jd&title¼Productivity-gains-from-teleworking-in-the-post-COVID-19-era (2020)
9. Cucinotta, D., Vanelli, M.: WHO declares COVID-19 a pandemic. Acta Biomed. **91**, 157–160 (2020)
10. Yassi, A., Hagberg, M., Silverstein, B., Wells, R.: Work Related Musculoskeletal Disorders (WRMSDs): A Reference Book for Prevention, vol. 349, pp. 5–184. Taylor and Francis, London (1995). 7 June 1997
11. Windeler, J.B., Chudoba, K.M., Sundrup, R.Z.: Getting away from them all: managing exhaustion from social interaction with telework. J. Organ. Behav. ***38***(7), 977–995 (2017). https://doi.org/10.1002/job.2176
12. Bohrnstedt, G.: A quick method for determining the reliability and validity of multiple-item scales. Am. Sociol. Rev. **34**(4), 542–548 (1969)
13. Francisco, M.G., Canciglieri Junior, O., Sant'Anna, A.M.O.: Design for six sigma integrated product development reference model through a systematic review. Int. J. Lean Six Sigma. **11**(4), 767–795 (2020). https://doi.org/10.1108/IJLSS-05-2019-0052
14. Cirrincione, L., Plescia, F., Ledda, C., Rapisarda, V., Martorana, D., Moldovan, R.E., Theodoridou, K., Cannizzaro, E.: COVID-19 pandemic: prevention and protection measures to be adopted at the workplace. Sustainability. **12**(9), 3603 (2020). https://doi.org/10.3390/su12093603
15. Collins, M.: The (not so simple) case for teleworking: a study at Lloyd's of London. New Technol. Work Employ. **20**(2), 115–132 (2005)
16. Iida, I.: Ergonomia: projeto e produção. Edgar Blücher, São Paulo (1990). 465p
17. Fadel, M., Salomon, J., Descatha, A.: Coronavirus outbreak: the role of companies in preparedness and responses. Lancet Public Health. **5**, e193 (2020). https://doi.org/10.1016/S2468-2667(20)30051-7. http://www.ncbi.nlm.nih.gov/pubmed/32119831
18. Bouziri, H., David, R.M.S., Descatha, A., Dab, W., Jean, K.: Working from home in the time of COVID-19: how to best preserve occupational health? BMJ J., Occup. Environ. Med. (OEM). https://doi.org/10.1136/oemed-2020-106599. https://oem.bmj.com/content/77/7/509
19. Nunes, F.: Most relevant enablers and constraints influencing the spread of telework in Portugal. New Technol. Work. Employ. **20**(2), 133–149 (2005)
20. Standen, P., Daniels, K., Lamond, D.: The home as a workplace: work–family interaction and psychological well-being in telework. J. Occup. Health Psychol. ***4***(4), 368–381 (1999). https://doi.org/10.1037/1076-8998.4.4.368

Consequences on Supply Chain Performance in Times of Scarcity During a COVID-19 Pandemic: A Case Study in an Automotive Industry

Alexandre Milkiewicz Sanches, Lourival José de Souza, Siliana Lemes da Silva, Edson Pinheiro de Lima, and Sergio E. Gouvea da Costa

Abstract This work aims to present the elements related to production and logistics that companies have used in the last four years and that can lead managers to a correct decision in times of uncertainty based on the literature review. During the pandemic caused by COVID-19, industrial companies in the automotive sector had to adapt to the sharp drop in demand, varying negatively by up to 40% from March to November 2020. As of December 2020, demand increased again, recovering the fall from the pandemic's beginning. Thus, a supply chain did not have enough time to recover and meet that speed. Even with international air freight, there is still a halt in automaker's production lines in the automotive sector. Illustrating that context, we studied a case in a company that operates in Brazil, which produces automotive components for a manufacturer of heavy cargo and passenger vehicles. The company was impacted by the COVID-19 pandemic due to its low level, heightened by its suppliers' delays. The results indicated the need for the company to be prepared for the increase in costs and the search for strategies that minimize the negative impacts still caused by the pandemic context.

Keywords Inventory Management · Performance · Quality · Costs · Industrial Processes.

A. M. Sanches (✉) · S. L. da Silva
Instituto Federal do Paraná, Palmas, Brazil
e-mail: alexandre.sanches@ifpr.edu.br

L. J. de Souza
Industrial and Systems Engineering Graduate Program (PPGEPS), Pontifical Catholic University of Parana (PUCPR), Curitiba, Brazil

E. P. de Lima · S. E. Gouvea da Costa
Federal University of Technology – Parana (UTFPR), Pato Branco, Brazil

V. M. López Sánchez et al. (eds.), *Industrial Engineering and Operations Management*, Springer Proceedings in Mathematics & Statistics 400,
https://doi.org/10.1007/978-3-031-14763-0_24

1 Introduction

One of the main pillars of supply chain management is logistics; essential activities are order processing, stock maintenance, and transporting raw materials or finished products. When one of these activities presents a problem, the entire management is compromised, especially in delivering to the customer [1, 2].

The corporate supply chain has been facing the challenges of a pandemic for more than a year. Due to this situation, industrial companies in the automotive sector had to adapt to the fall in demand [3]. This context of the pandemic in the industry caused a negative variation of up to 40% from March to November 2020 in order.

This paper aims to present the impacts on the automotive supply chain sector during a COVID19 pandemic. For this, the literature review was sought as the primary reference on the theme of the supply chain and what has been discussed in the literature on the challenges in the area, the main contributions that the authors brought to the topic and how these contributions they can help, directly or indirectly as associations.

A case study was carried out in a multinational company in the automotive sector with a branch in Brazil, analyzing the decisions taken by its managers to ensure the continued delivery of its products to its customers within the stipulated deadlines.

Therefore, the rationale for choosing this sector was to understand how a dynamic industry such as automobiles ended up experiencing severe material restrictions for a considerable period, seriously affecting the balance between material supply and demand fulfilment.

2 Research Method

Initially, the methodology proposed for this paper was a literature review addressing the terms that interface with the industrial environment, reflecting the objective of this research. The use of the SCOPUS database guided the previously found and refined results.

The choice of articles published from 2017 onwards was due to the need to understand whether the most up-to-date publications would bring newer elements about supply chain management in organizations and the literature on this topic itself.Table 1 shows the two groups of search terms used.

Based on the parameters indicated in Table 1, after performing the search, 2457 documents were found, refined in stages, until the selection of articles to be analyzed for this study. Words related to the subject: considering the 2457 documents found initially based on the search terms, there was a refinement, delimiting within Scopus some terms closer to the matter in question, as indicated in Table 1.

Journals related to the subject: looking only at journals related to the topic, there was a new refinement, leaving 659 selected articles. Title selection: even with the search sentence pre-selected in the database, some titles, after analysis, were not related to the search. Some focused on areas that did not address the area studied. Thus, 98 articles remained. Abstract selection: in the same way as the titles, the

Table 1 Criteria used to conduct the literature review

Keywords	("raw materials") AND ("performance" OR "quality" OR "cost*" OR "industrial process" OR "covid")
Document	AND (LIMIT-TO (SUBJAREA, "ENGI") OR LIMIT-TO (SUBJAREA, "BUSI"))
Year	AND (LIMIT-TO (PUBYEAR, 2021) OR LIMIT-TO (PUBYEAR, 2020) OR LIMIT-TO (PUBYEAR, 2019) OR LIMIT-TO (PUBYEAR, 2018) OR LIMIT-TO (PUBYEAR, 2017)
Language	AND (LIMIT-TO (LANGUAGE, "English"))

abstracts showed that the approach of most of the remaining articles had another focus. Finally, 16 articles were selected as relevant for the literature review. The papers [1], [4] and [5] were included due to their importance in the theme addressed in this article.

Interestingly, it shows that the terms "COVID-19" and "pandemic" do not appear in the highlighted words. However, understanding how this phenomenon occurs in these supply chain and industrial production variables will be sought in the case study section.

Therefore, this article includes a case study in the automobile industry. The case study aims to answer "how" or "why" research questions through a contemporary phenomenon [6]. The data were obtained from the logistics and production managers of the company studied. Interviews with the three managers linked to logistics, finance and production were carried out to understand the impact of the COVID-19 pandemic in each of their respective sectors, as well as the managerial lessons learned during this period.

In addition, company data was analyzed with the financial manager on costs incurred with international freight and its impact after the beginning of the pandemic. Company data on revenues obtained from January 2019 to October 2021 and inventory turnover for the same period were also collected to observe the behaviour of values, and a possible relationship impacted by the pandemic can be seen in Sect. 4 [4].

3 Literature Review

To understand what has been published between 2017 and 2021 with the supply chain and pandemic themes, Table 2 expresses the main contributions of the authors of the articles selected for this research.

Based on Table 2, the supply chain care issues have had several topics addressed in recent years. It was possible to observe that there was no mention of COVID-19 on the subject. Therefore, there is no definite way to deal with this reality of the pandemic. Thus, Sect. 4 will address how a multinational company has dealt with this reality.

Table 2 Contributions found in supply chain research

Author(s)	Theme	Method	Contributions
[1]	Uncertainties in supply chain management	Literature review	To present the effects of parallel interaction, decision complexity, and IT complexity in environments of uncertainty in supply chain management.
[4]	Risks in supply chain management	Literature review	The authors proposed five common risks in various supply chains, including macro risk, demand risk, manufacturing risk, supply risk, and infrastructure risk.
[7]	Inventory management	Model proposition	Shows a model in which uninterrupted production is desired for some periods to fulfil an urgent order. This can be due to a tight shipping schedule, urgent demand, special events, etc.
[2]	Planning and production control	Model proposition	The proposed method was able to plan production so that the continuity of production throughout the entire process chain was guaranteed through the supply of material, and the level of service expected from the customer was maintained.
[8]	Material supply management	Case study	The article proposed a precise delivery schedule, including information about the material and its exact quantity that must be delivered in each cycle and for each vehicle produced, ensuring no shortage of material, based on a massification logic to customize the demand's service.
[9]	Multi-level closed-loop supply chain (CLSC)	Model proposition	The proposal of this work concluded that when the revenue from the sale of newly produced items (remanufactured items) increases, the quality level of acceptance of the returns decreases (increases).
[10]	Environmental impact of supply chain operations	Case study	The study results reveal that the location decision has a relevant impact on the environmental sustainability of the machining process and should be considered in future strategic location planning.
[11]	Disruptions on the supply chain	Case study	The article shows that the demand for each product depends on its price, the price of substitutable products in the market, and the availability of the products. As a result, in the event of a shortage of a specific product, it is possible to shift demand from the stocked product to its substitutable products through appropriate pricing strategies.
[12]	Additive manufacturing	Case study	Bringing manufacturing closer to the customer solves the physical supply chain risk to a large extent. On the other hand, manufacturing taking place upstream alleviates many cyber threats associated with the virtual supply chain.

(continued)

Table 2 (continued)

Author(s)	Theme	Method	Contributions
[13]	Additive manufacturing	Literature review	This study identified that there are still contradictions regarding the effects of AM on competitiveness, cost, supplier dependence, environmental impact, health and safety, product quality, production efficiency, responsiveness, complexity and flexibility of the supply chain, qualification of workers, number of workers and workload.
[14]	Supply chain	Case study	The best locations for biorefineries are located mainly close to existing refineries, indicate consistency between total costs and greenhouse gas emissions due to less material transport. Most importantly, due to reducing transport-related costs and greenhouse gas emissions, the proposed model prefers to establish fewer collection centres with higher capacity.
[15]	Decision-making methodology to warehouses	Case study	Scheduling jobs using flow-shop priority rules provides solutions for distribution centres facing timeout pressures. By prioritizing each position, warehouses can efficiently maintain responsiveness without increasing the workforce to meet deadlines for filling compressed orders.
[16]	Warehouse management system (WMS)	Case study	This work proposed integrating data for decision-making activities in purchasing, production, management, and logistics.
[17]	Sales and operations planning	Literature review	In the end-to-end coordination between purchasing, production, distribution, and sales, it may become relevant to create the role of the S&OP coordinator, responsible for generating plans and aligning functions and supply chain metrics.
[18]	Sales and operations planning	Literature review/ Framework proposition	When all retailers share their information confidentially, the chain's ideal balance wholesale price is lower with a higher degree of confidentiality.
[5]	Supply chain management	Literature review	Contributions of leading journals, publications, authors, institutions, and keywords regarding supply chain disruption were identified.
[19]	Supply chain	Case study	With an environmental approach, the authors consider that robustness resides in the adequate logistical application of the strategic performances of the supplier's marketing executive so that the integrated supply chain cost and the total amount of pollution significantly decrease.
[20]	Supply chain management	Model proposition	Although factors such as market conditions and contract terms also affect supplier and manufacturer preference for inventory strategies, affecting order quantity, the risk attitudes of these two participants are the most important influencing factors.
[3]	Sustainable product development	Literature review	To achieve sustainable fashion, the sustainable fashion product development process is one of the critical areas that all fashion apparel supply chains must be aware of.

4 Case Study

4.1 Company Studied

The case study was carried out in a company that manufactures components for commercial vehicles, mining, construction, agriculture, general industry, and telecommunications. The unit studies is located in Brazil. Its headquarters are located in Europe. In addition to the branch in Brazil, the company has eight branches in Europe, a branch in the United States, and one in China.

In Europe, the company is significant. However, in Brazil, a unit is compact and uses great technology [10]. Another important fact to highlight is that 85% of its raw material is imported from Europe, with only 15% being domestic. About 80% of its production is for the domestic market, where the company has a robust supply contract with truck assemblers. Its service policy works with 100% service within the desired period [14, 16].

4.2 Context of the Pandemic in the Studied Company

The demand for the studied company was on the rise until March 2020. After this period, there was a fall of 45%, which lasted until November 2020. During this period, there was a lack of raw materials, mainly steel and electronic components, on a world scale [3]. From December 2020, demand started to grow again until October 2021. Furthermore, the worsening of the shortage of electronic components and the global crisis of lack of containers forced the Brazilian industry to reduce its demand by 50% in November, December 2021, and January 2022. However, as of February 2022, this reduced demand will resume, as these bottlenecks will tend to decrease.

However, the global supply chain has not had time to recover to meet this pent-up demand. In the case of Brazil, companies have had difficulties in regularizing their production due to this instability in the supply chain and the crisis in maritime transport [15].

Based on the data provided by the company, the round trip cycle of a container from Brazil to Sweden before the pandemic was up to about 4.9 times a year. Today, the process is 1.9 times a year. This is due to the lack of availability of containers in the world. The production of the month of August waited about 45 days for the availability of containers.

Before, these cargoes were shipped in dedicated containers. However, due to scarcity, these cargos were being divided into up to three parts, being sent in shared containers. Thus, the costs of transport, storage, and customs expenses are considered higher. Therefore, in the coming months, the company understands the need to work in this way so that its customers do not run out of products [14, 16].

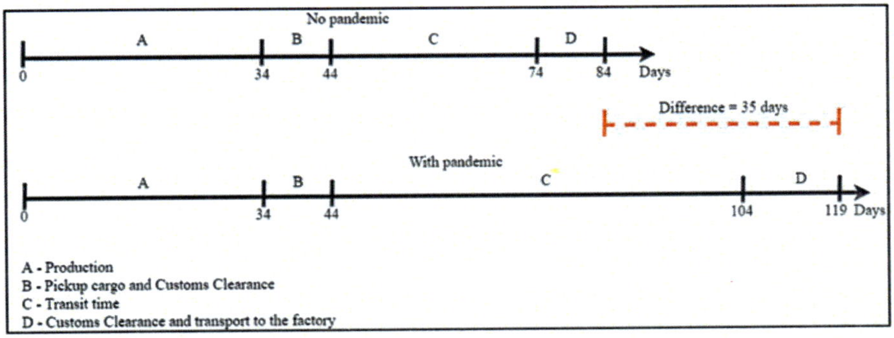

Fig. 1 Stock policy context without the pandemic and with the pandemic

COVID-19 has directly influenced all decisions and actions taken by the company. The absence of employees in the industries caused the production capacity of the supply chain to decrease [13, 18]. In addition to the lack of raw material, which the company waited for six months for resupply, absenteeism in the global supply chain at the main suppliers of steel products reached 35% during the year 2021.

4.3 Inventory Management in the Studied Company

The stock policy of the studied company is to have a maximum of 30 days of stock turnover. But, currently, this value is around 56 days due to the sudden drop in demand for the final months of 2021 and the beginning of 2022 [5, 13, 15]. In addition, it is worth noting the value of stock in transit has increased due to the flow that can be seen in Fig. 1:

With firm orders and demands delivered to the supplier, the possibility of postponing deliveries beyond the desired deadline is analyzed due to the shortage of raw materials, labor, and the crisis in maritime transport (affecting transit time, as seen in Fig. 1). This makes the company act with the stock policy in a committed way because the service term of the suppliers is beyond the desired date [1, 4, 12].

The internal processes, from the lead time to the arrival of the raw material, and the planning and control of production, have more compromised deadlines than before the pandemic (this period from the supplier to the company was 84 days). This supplier lead-time time impacts the company's production process precisely because these materials arrive practically at the time of production or even late (after the pandemic, this period was increased to 119 days). This delay compromises the internal production lead time, making the company work overtime on weekends to ensure the delivery of the product to the assembler that considering that the studied company has an agreement with the assembler (customer) for 100% of the deliveries

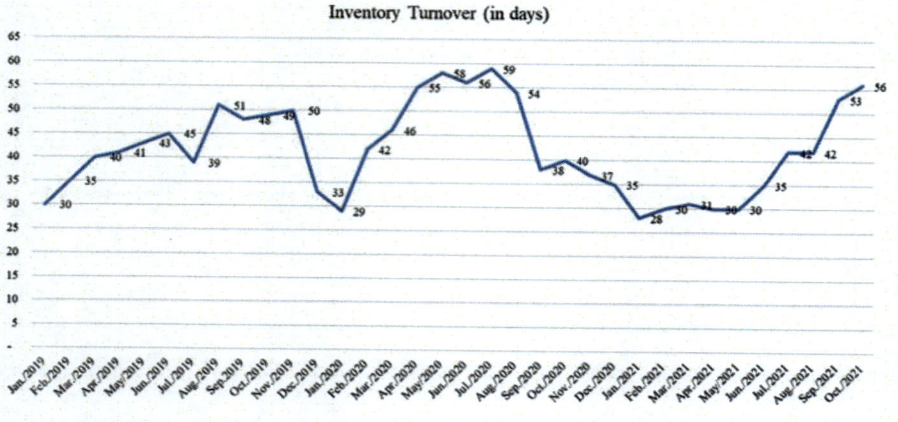

Fig. 2 Inventory Turnover

on the desired date (the company's relationship with its customer remained within 30 days) [12, 17].

Then it is possible to notice that there is an imbalance between demand and supply chain capacity, which, in turn, affects the performance of all operations involved. The lead time of 85% of imported raw materials is 119 days, on average [20]. This has deficient raw material stock levels, as shown in Fig. 2.

In addition, there is a problem with the logistics of imported items, such as the lack of containers for transport and the lack of space on the ships. This situation delayed the receipt of materials to be used by the production line [7, 10, 20]. Figure 3 indicates the ships in their transit time. The green color refers to cargo ships. The red color for oil tankers and the blue color is for passenger ships:

Another reason was the liquidity crisis of the companies generated by the COVID-19 pandemic, affecting the receipt of raw materials due to the increase in the time of stock in transit. The time taken to collect material from the supplier added to the maritime transit has increased the value of this stock in transit. This situation causes a delay in the processing of raw material and, consequently, in the sale of its products [4, 10, 17].

4.4 *Management of Logistical Costs and Production Planning in the Company*

Based on the current maritime reality, according to the company's direction, there is a fear that this scenario could worsen throughout 2022, forcing the use of air freight again, which is much more expensive. Part of this scenario is represented in Fig. 4.

In the domestic market, there were two significant problems during 2021. In Brazil, there was a great shortage of special steel. When in 2020 the steel companies

Fig. 3 Flow of ships in the world

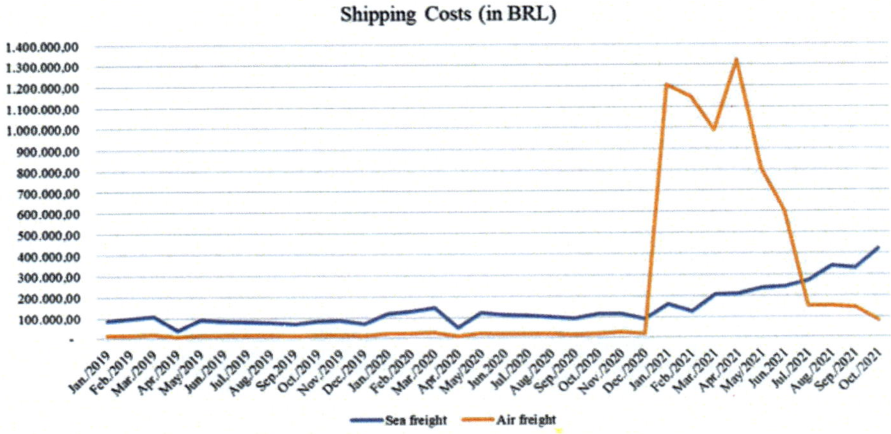

Fig. 4 Freight costs in different modes

had a drop in demand, they stopped part of their production process. And, in the resumption of operations in the industries, there was a lack of material. In the first months of 2021, as a result, steelmakers struggled to resume production [4, 7, 17].

Within this context, the company studied, consequently, had difficulties in consolidating cargo. In Brazil, the company works through consolidated routes through the milk run as the company's parts quantity was low due to a lack of

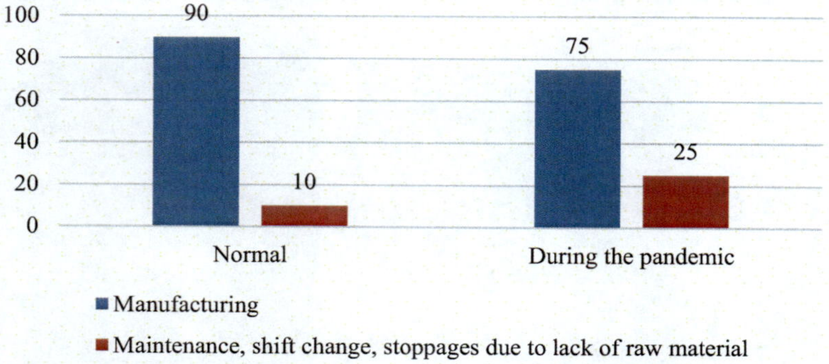

Fig. 5 Use of productive capacity

parts at the suppliers, the need to turn to them more frequently made the operating cost for these operations higher. This forced the company to reduce its contribution margin to honor commitments to its customers [8, 13, 20].

As for production planning and control and maintenance planning and control, the company receives its actual and anticipated demands from customers. This data is processed by the company's MRP. This data is then sent by EDI to the supply chain [13, 18]. However, currently, suppliers are returning with a supply proposal within 30 days of the desired date. Firm orders are planned up to two months in advance of the selected date and with a forward forecast of up to 52 weeks.

As for the planning and control of maintenance, to guarantee 90% availability of equipment, some preventive maintenance is postponed in order to use this time for production flow. Today, 10 to 15% has been lost in the performance of the production line due to the lack of raw materials [4, 20]. The on-time delivery rate of the company's suppliers before the pandemic was 98% on-time deliveries, and today it has dropped to 65%, delaying the supply of raw materials (Fig. 5).

As for the use of alternative sources of raw material sources, the company studied does not have it. The approval of a new basis of an application takes a long time to develop and approve, making its use unfeasible in times of emergency [9, 11]. To address this situation, some parts are transferred from one unit to another between the different countries where the company has its branches to connect more smoothly the backlogs of suppliers, seeking to maintain the quality levels of the manufactured products. Thus, there is a need to use air freight, and cost costs are due in Fig. 4.

The company's senior management does not believe that the situation will normalize before the second half of 2022. This is due to the shortage of raw materials in the European and world market and the crisis in world maritime transport caused by the lack of ships and containers. The supply chain of automobile companies was put to the test by an unexpected scenario with immediate impacts on the chain and with a slow recovery in the face of a fast recovery of demand [14].

5 Conclusions

The results obtained in the research corroborate the impacts that the COVID-19 pandemic has still caused around the world and which the literature addresses the management of the supply chain. This is because the company studied has a supply contract where the goal of on-time service is 100%. In the industrial environment, the supply chain also suffers from these consequences, causing area managers to make more proactive decisions even if the production cost is higher than planned [9, 13].

The COVID-19 pandemic interfered directly in all stages of the operation. Therefore, the challenge of balancing a supply chain involves planning and risk management, seeking to minimize the negative impacts of an adverse moment [4].

Inventory turnover was also impacted by the fact that the difficulty of keeping material in stock was due to the collapse of the company's supply chain, added by the time in transit of the cargo. As shown in Fig. 2, inventory turnover went through several fluctuations over almost two years, analyzed in a pandemic situation. The search for solutions is based on load splitting, mainly due to the uncertainty of material supply, leading to stoppages in the production line [4, 7].

Due to the imbalance between demand and supply of ships (see Fig. 3) and containers, both the researched company and numerous others worldwide chose to work with fractional cargo. Uncertainty in the logistical environment, especially in terms of transport, increased operating costs. The lack of containers and ships forced the researched company to use air freight, known to be more expensive. However, it was the mechanism that prevented the production line from stopping during the most critical periods of the pandemic [7, 8, 16].

One of the points that were observed during the research was the fact that the company is prepared through its production planning and control systems, integrated with customers and suppliers. This approach and a good flow of communication have alleviated the pandemic's negative impacts, as well as the maintenance of the quality of the products produced by the company [17, 18, 19].

The contributions of this research are aimed at understanding the effects of a pandemic on industrial activity and the impacts it causes throughout the supply chain and cargo handling in all modes. This situation provokes a reflection to the academic world on how to prepare for possible scenarios that could happen in the coming years.

References

1. Simangunsong, E., et al.: Supply-chain uncertainty: a review and theoretical foundation for future research. Int. J. Prod. Res. **50**(16), 4493–4523 (2012)
2. Gyulai, D., et al.: Robust production planning and control for multi-stage systems with flexible final assembly lines. Int. J. Prod. Res. **55**(13), 3657–3673 (2017)

3. Fung, Y.N., et al.: Sustainable product development processes in fashion: supply chains structures and classifications. Int. J. Prod. Econ. **231**, 1–15 (2021)
4. Ho, W., et al.: Supply chain risk management: a literature review. Int. J. Prod. Res. **53**(16), 5031–5069 (2015)
5. Xu, S., et al.: Disruption risks in supply chain management: a literature review based on bibliometric analysis. Int. J. Prod. Res. **58**(11), 3508–3526 (2020)
6. Yin, R.K.: Estudo de caso: planejamento e métodos, 5th edn. Bookman, Porto Alegre (2015)
7. Chan, C.K., et al.: An integrated production-inventory model for deteriorating items with consideration of optimal production rate and deterioration during delivery. Int. J. Prod. Econ. **189**, 1–13 (2017)
8. Fathi, M., et al.: An optimization model for material supply scheduling at mixed-model assembly lines. Procedia CIRP. **72**, 1258–1263 (2018)
9. Taleizadeh, A.A., Sadegh, M.M.: A consignment stock scheme for closed loop supply chain with imperfect manufacturing processes, lost sales, and quality dependent return: multi levels structure. Int. J. Prod. Econ. **217**, 298–316 (2019)
10. Sihag, N., et al.: The influence of manufacturing plant site selection on environmental impact of machining processes. Procedia CIRP. **80**, 186–191 (2019)
11. Khalilabadi, G., et al.: A multi-stage stochastic programming approach for supply chain risk mitigation via product substitution. Comput. Ind. Eng. **149**, 106786 (2020)
12. Gupta, N., et al.: Additive manufacturing cyber-physical system: supply chain cybersecurity and risk. IEEE Access. **8**, 47322–47333 (2020)
13. Franco, D., et al.: Consolidated and inconclusive effects of additive manufacturing adoption: a systematic literature review. Comput. Ind. Eng. **148**, 1–29 (2020)
14. Akbarian-Saravi, N., et al.: Development of a comprehensive decision support tool for strategic and tactical planning of a sustainable bioethanol supply chain: real case study, discussions and policy implications. J. Clean. Prod. **244**, 1–19 (2020)
15. Kim, T.Y.: Improving warehouse responsiveness by job priority management: a European Distribution Centre field study. Comput. Ind. Eng. **139**, 1–12 (2020)
16. Frontoni, E., et al.: Optimal stock control and procurement by reusing of obsolescences in manufacturing. Comput. Ind. Eng. **148**, 1–12 (2020)
17. Liu, Z., et al.: Supply chain coordination with risk-averse retailer and option contract: supplier-led vs. retailer-led. Int. J. Prod. Econ. **223**, 107518 (2020)
18. Pereira, D., et al.: Tactical sales and operations planning: a holistic framework and a literature review of decision-making models. Int. J. Prod. Econ. **228**, 1–28 (2020)
19. Bhattacharya, K., De Sujit, K.: A robust two layer green supply chain modelling under performance based fuzzy game theoretic approach. Comput. Ind. Eng. **152**, 107005 (2021)
20. Song, Z., et al.: Inventory strategy of the risk averse supplier and overconfident manufacturer with uncertain demand. Int. J. Prod. Econ. **234**, 1–16 (2021)

Technological Disruption in Grocery Retail: An Overview of the Last Decade (2012–2021)

Roberto Rivera, Marlene Amorim, and João Reis

Abstract After grocery retail experienced a radical change with the emergence of the Internet and the use of dot-com in electronic commerce, this sector is once again in a gradual and accelerated process of adaptation and implementation of disruptive technologies. These technologies are the result of automation processes and connectivity implemented by Industry 4.0 tools and recently by the new regulations that brought the impact generated by the COVID-19 pandemic. This research paper uses a systematic literature review, this work was carried out from 50 scientific papers collected in the last 10 years (from 2012 to 2021) and that portray the planning of prototypes and implementation of disruptive technologies in grocery retail. The results gather information from technologies such as the Internet of Things, Artificial Intelligence, Machine Learning, Blockchain, Virtual Reality, Augmented Reality, use of Drones, Big Data, Robotics and Cloud Computing, and their use in the grocery retail industry in specific areas such as Logistics, Warehousing, Customer Service and Assistance, Freight Transport, Supply Chain and Supplier Management.

Keywords Disruptive technology · Grocery retail · Digital transformation · Brick-and-mortar · Proximity retail

R. Rivera (✉)
University of Aveiro, Aveiro, Portugal
e-mail: r.rivera@ua.pt

M. Amorim
Department of Economics, Management, Industrial Engineering and Tourism, GOVCOPP. University of Aveiro, Aveiro, Portugal

J. Reis
Department of Economics, Management, Industrial Engineering and Tourism, GOVCOPP. University of Aveiro, Aveiro, Portugal

Industrial Engineering and Management, Faculty of Engineering, Campo Grande, Lusófona University and EIGeS, Lisbon, Portugal

V. M. López Sánchez et al. (eds.), *Industrial Engineering and Operations Management*, Springer Proceedings in Mathematics & Statistics 400,
https://doi.org/10.1007/978-3-031-14763-0_25

1 Introduction

In recent years, one of the most influential contributions to the literature in terms of innovation and digitalization is the concept of disruptive technology [1]. A theory developed by Clayton Christensen and presented in his book "The Innovator's Dilemma" in which such technologies or innovations creates new markets providing different values able to overtakes incumbents at long last [2].

In global production and industrial practices, during the current Industrial Revolution (I4.0) based on high interconnectivity and smart automation, some technologies have evolved and improved, and are currently considered as disruptive innovations such as Internet of Things (IoT), Blockchain, Artificial Intelligence (AI) and Machine Learning (ML), Unmanned Aerial Vehicles (UAV) or Drones, Virtual Reality (VR), Augmented Reality (AR), Big Data, Robotics Services and Cloud Computing to mention some examples [3–5].

Based on a systematic literature review, this research aims to analyses the current state-of-the-art of the applicability and development of those technologies in the retail business, specifically in the processes arising from the grocery market over the last 10 years. Preliminary results point to the transition of traditional product-centered approach to a customer/service perspective [6].

From 2020, when the global economy took an unprecedented impact due to the spread of the COVID-19 pandemic, the restrictions implemented by governments in different countries redirected people's daily activities, including their shopping habits. In this regard, grocery retail was forced to adapt to new regulations, converting e-commerce into a social necessity by replacing physical purchases with virtual ones. A great opportunity for retailers to break with existing consumption patterns and revolutionize the market again as happened in 2002 as the repercussion that took the dot-com [7].

Despite the fact that no previous research was found that encompasses the use of disruptive technologies in grocery retail processes, this work intends to highlight what is being planned, what is being done and what is already being used in this sector. The results offer a wide range of possibilities that involve specific areas such as logistics, warehousing, customer service and assistance, goods transport and supply chain and supplier management. The paper is structured into six sections. In Sect. 1 is presented the introduction of the research; Sect. 2 mentions a brief definition of disruptive technologies and their interaction on the retail industry; Sect. 3 depicts the used research methodology; Sect. 4 presents the findings regarding to the reciprocal action between the grocery retail and the use of disruptive technologies; Sect. 5 examines the discussion, challenges and limitation; and finally, in Sect. 6 the conclusions are presented.

2 Disruptive Technology and Retail

Disruptive Innovation is described as the radical technology-driven change in productions models, generating alteration in previous approach models, which transforms capital, labor, information and materials in products or services of greater value [8, 9]. These changes allow the satisfaction of less-demanding customers or stablishing new markets where none existed previously [10]. The concept of disruptive technology was first presented by Bower and Christensen [11], under the idea of successful businesses that envision customers future needs and not just their present ones.

According with McKinsey & Company [12], the retail industry bears the greater potential to create value by disruptive technologies mainly with technologies such as Artificial Intelligence (AI) and Machine Learning (ML), Robotics, AI-based self-checkout, drones delivering last-mile goods. However, those technologies, together with Blockchain, Internet of Things (IoT), Virtual Reality (VR), 3D printing, Big Data, Cloud Computing among others are widely used in the sector, mostly observed in the automotive industry [13], oil industry [14], distribution/delivery [15, 16], fashion [7], finances [17], food [18] and Communication services [19].

3 Methods

This research uses a systematic literature review, by adopting a technical process, rational and standardized in order to demonstrate objectivity and transparent process to the reader, defining search approaches, stating inclusion and exclusion criteria [20].

Being the largest abstract and citation database of peer-reviewed literature [21], Scopus was selected as source database for this work, for seeming the most appropriate option as it increases the quality of the research and the coverage of publication compared to similar databases such as Web of Science, Science Direct and Google Scholar, the latter not restricted to peer-reviewed articles [22].

In order to guarantee coverage of the current state of the art, in January 2022, the process of searching for information regarding the application of "disruptive technologies" in "proximity retail" in the last 10 years (2012–2020) started. However, the specific keywords used in the search resulted in only 1 paper entitled "Surviving disruption: the grocer's tale" [5], which describes the transformation of grocery retail in the United States in the last 2 decades, mainly with the arrival of the Internet and its path to digital transformation. According to Kalyani [23], this sector belongs to the largest retail category in the United States, and even so, it only has a 5% share in online sales, remaining predominantly in the in-store format [24].

Therefore, in the fields of Title, Abstract and Keywords, new alternatives were evaluated, divided in two independent searches, which lead the study considerably closer to its objectives. For the first research group, the keywords "disruptive

Table 1 Key terms and results

Criteria	Filters	Docs
Keywords	"disruptive technology" AND retail	16
	"technological disruption" AND retail	4
	grocery AND retail AND IoT	13
	grocery AND retail AND Blockchain	3
	grocery AND retail AND Artificial Intelligence	10
	grocery AND retail AND Machine Learning	18
	grocery AND retail AND UAV	1
	grocery AND retail AND Virtual Reality	4
	grocery AND retail AND 3D	2
	grocery AND retail AND Big Data	10
	grocery AND retail AND Robotics	2
	grocery AND retail AND Cloud Computing	2
Fields	Article title, Abstract, Keywords	
Language	English	
Source Type	Journals and Conferences	
Document Type	Articles and Conference Paper	
Years	Since 2012 until 2021 included	

technology" OR "technological disruption" AND "retail" were used. On the other hand, disruptive technologies were listed and used as keywords for the second research group. Those technologies were identified specifically by their link with grocery retail, based on information collected through reports issued by large business consulting and professional services companies such as Deloitte [25] and McKinsey [26, 27], as well as intergovernmental organizations such as the World Customs Organization [3]. For a better understanding and credibility of the results, only documents written in English were selected, being only articles published in journals and conferences. Therefore, the selection resulted in 85 peer-reviewed papers as shown in Table 1.

After the records completed, the process of observation, analysis and description of data began. This process was focused on descriptive research, which is characterized by being restricted to the factual record, being objective or neutral, showing reality as it is, defining the state of nature of the problem at a point in time [28, 29].

The goal identifies use cases that could integrate disruptive technologies (e.g. Blockchain, IoT, AI and ML, UAV (Drones), VR, 3D printing, Robotics, Big Data, Autonomous Vehicles, Advanced Genomics, Cloud Computing, Renewable Energy, Biometrics to name a few [3–5], and its adaptation to support the grocery retail business. For the purposes of this research, only 10 of the main disruptive technologies were considered as search terms: IoT, Blockchain, AI, ML, UAV, VR, 3D, Big Data, Robotics, Cloud Computing.

From the 85 selected articles, the filtering process started based on PRISMA (Preferred Reporting Items for Systematic reviews and Meta-Analyses), an evidence-based minimum set of checklists and phase flow diagram to help ensure the clarity

Table 2 PRISMA protocol

Records identified through database searching	85
Records after duplicates removed	73
Records without access	7
Articles excluded with reasons	16
Full-text articles assessed for eligibility	50

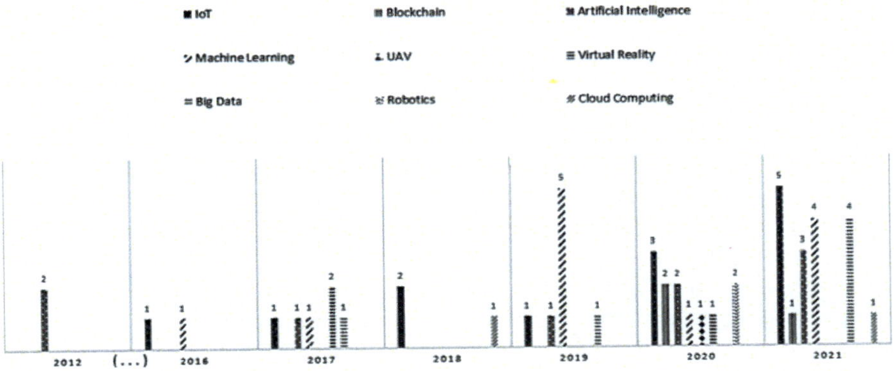

Fig. 1 Disruptive technologies in grocery retail identified based on the literature review between 2012 and 2021

and transparency of reporting of systematic reviews [30, 31]. Table 2 summarized this process, in which 12 articles were identified as duplicated, representing 14.11% of the total of selected articles. Likewise, during the data collection process, 8.23% of the records could not be accessed and 18.82% of the total of the selected papers did not meet the minimum requirements to be considered in this research.

By analyzing the content of the 50 resulting articles, patterns were identified allowing the categorization of information based on disruptive technologies and their relationship with the grocery retail market. Through this analysis, it was possible to identify the variation in the production and article publications in journals and conferences in the last 10 years. Figure 1 shows this trend, evidencing the lack of publications referring to any type of disruptive technologies between the years 2012 and 2016, finding in 2012, only two publications referring to the link between AI and grocery retail, and being from 2016 that there was a significant increase in published articles. The list of the 50 manuscripts retrieved from Scopus can be provided upon request to the first author.

4 Disruption Technology in Grocery Retail

Given the accelerated need for technological adaptation that humanity has been experiencing in recent years, a new reality in the buying and selling processes has been gradually developing. A phenomenon that continues to revolutionize grocery retailing through the application of tools to collect consumer data through Big Data, application of Robotic Services, VR/AR, blockchain, AI and IoT which is taking over the management of product inventories through the cloud. The disruptive technologies identified for this research are presented below, which will certainly have a major impact on society as a whole in near future [32–34].

4.1 Internet of Things (IoT) in Grocery Retail

According with the literature, IoT is one of the most common disruptive technologies that has been used in retail for the last years. Defined as an ecosystem of interconnected physical entities embedded with technology able to communicate with other devices and external environment [14, 35], IoT systems are rapidly growing and being applied in grocery retail. This technology presents opportunities for the development of infrastructures that allow interaction with consumers in real time both in physical and virtual stores through tools such as applications for mobile devices, store shelves, shopping basket, digital displays and even smart products [36].

From 2020, when the COVID-19 pandemic reached alarming levels of transmission worldwide, several projects were developed for the retail market, in order to contribute to the fulfilment of criteria recommended by the World Health Organization (WHO), in particular with regard to social distancing. Some examples found in the literature are based wearable devices alerts for customer to maintain distancing while shopping, detection of face masks, smart shopping cart and pre-booking for shopping to try to reduce the consumption of time at checkout [37]. For the purposes of this research, the elements found were categorized into three areas: In-store shopping, queue/billing and out-store.

In-store Shopping Based on the advantages that IoT brings to retailers when interacting with their customers, location-based beacon technology has gained a great acceptance in grocery retail in recent years. This is due to the sharing real-time personalized information, which provides valuable insights through analytics that offer the possibility to improve consumer shopping experiences [38]. Another project identified in the literature review refers to the development of motorized racks. The prototype presented by Pal et al. [39] has three fundamental parts: the motorized vertical shelves, the invoicing system and the stock update system. The whole system allows the identification of products for sale with a unique entry code, updating in real time the amount of similar products remaining on the shelf and consequently in stock [39].

Focused on consumer trends, models of customer tracking systems in grocery retail stores have been developed, making the shopping process safe and simple. The SysMART for instance, offered several features for a fast-shopping experience, including indoor navigation though cart monitoring, fast check out, branches alternatives identification, and food analysis to monitor the status of perishable goods [40]. On the other hand, with an accuracy of 94% and supported by IoT, the interaction of customer and products was identified in the study of Radhakrishnan et al. [41], in which an environment of grocery store were simulated using data sensed from customer's personal devices, such as smartwatch and other IoT devices in the store. Besides interaction and preferences, additional data such as the exact shelf, locations and type of product picked were identified. Therefore, it is possible for retailers to build a scalable item picking model with a minimal loss in accuracy [41].

Queue/Billing For retailers, IoT payment devices can be of great utility, since consumer are able to make purchases more conveniently with voice-activated devices like smart home appliances or Amazon Echo, thus, transforming the way to reach consumers [42]. Yet, in order to decrease waiting time or avoid standing in long queues, Shankar et al. [43] propose a smart trolley based on IoT and attached with raspberry-pi, barcodes scanner and LCD display with the information of the purchased items and the total bill which is charged in cart saving valuable time at the charging counter. With the use of NodeMCU [43] and RFID [44], similar projects have been developed, proposing the use of smart carts supported by barcodes scanning systems, weight sensors, LCD display and mobile applications [45]. The products placed in the cart share through sensors the equivalent weight in both devices, generating QR codes for future payment. The systems also share purchases history information and personalized customer offers based on selected products, thus benefiting consumers by optimizing purchase budget and their activities in the purchase process [44, 45].

Out-store Kaur and Kaur [46] present a design of a retail business model using sensor-based measurement containers (SBMCs) and a mobile application. SBMC mitigates the pressure related to a cloud IOT-enabled grocery management system (CE-GMS) for the management of household grocery needs, by sensing their quantities using an ultrasonic sensor. Four automate entities are considered in the system (units produced, warehouses, customer sites and stores), in which the information flow from customers to the nearest store supported with GPS and GSM modules and the supply processes starts by sensing the demand.

4.2 Blockchain in Grocery Retail

With the potential to streamline processes and automate trust across many sectors, Blockchain has gained significant traction in business [47]. In the retail sector, this 'distributed ledger' is used to record and verify faster and cheaper transactions processes, allowing cryptocurrencies being used as decentralized payment systems [48]. Thus, it is expected to foster greater trust and confidence in online ecommerce for instance, contributing to the security of personal data and economic growth [49].

Supply Chain Blockchain can provide important data to be recorded, maintained and shared by private and public communities [50]. In grocery retail, traceability and transparency have emerged as two reliable enhancers in terms of their application in processes related to the supply chain management. Traceability during the entire lifecycle of products from raw materials to its disposability on the shelves, thus ensuring efficiency and risk reduction [10, 51]. On the other hand, transparency offers clarity in the perception, exchange and disclosure of information and knowledge in every transaction in the supply chain process [52].

In grocery retail, Blockchain has started to be adopted mainly in food supply chain, where products are digitally tracked along every step of the process from an ecosystem of suppliers to stores, and consequently to consumer, offering valuable information such as farms details, track codes, processing data, storage temperatures, shipping details and expiration dates [53]. In order to gain global competitive advantages, world major food retailers such as Walmart are already trialing this disruptive technology. In 2017, the company announced a partnership with IBM and its platform IBM Food Trust, and with collaboration of giants of retail industry such as Dole, McCormick, Nestlé, Unilever and Tyson Foods, by September 2018 was possible to trace over 25 products including fruits and vegetables, dairy products, salads and meat and poultry [47, 54]. In this way, companies stand out for improving the speed, traceability and trackability, saving time and cost and mitigating risks [53].

Contracts and Payments process Nuseir [10] redacts in his research the potential impacts of Blockchain on business practices in grocery stores. Focused on payments and contracts between stakeholders, the author mentions the advantages of offering funding via tokens between suppliers and retailers or store and customers, processes that might guaranty secure transactions and management of stocks and the inventory available in the store thus improving the after-sales services.

4.3 *Artificial Intelligence (AI) and Machine Learning (ML) in Grocery Retail*

Focused on the development of intelligent machines, those technologies refer to systems that based upon data that is observed, collected and analyzed, are able to change behaviors without being programmed. AI includes different disruptive technologies such as deep learning, machine leaning and computer vision [3]. Such technologies are yet available for indoor shopping in grocery retails, to improve consumer's shopping experiences and better understand their interests and preferences [54].

Customer Relationship Management (CRM) In 2017, AI were identified as the newest trend in CRM, since AI-based bots will become essential tools based on ML and predictive algorithms that can help take the guesswork out of marketing decision [55]. For example, simply through the analysis of the consumer profile, AI allows recommendation received by consumers to be as personalizes as possible, or even the development of chatbots, customers can obtain immediate and accurate information when necessary [56].

Payment's process Through the use of AI, mobile platforms have become capable to offer smart solutions supported by ML. According with Jakhiya et al. [42], in counties like India those solutions are noticeable by improving peer-to-peer (P2P) payment platforms and reducing frauds in high-risk mobile payments. With the aid of text messages pre-ordering, speech recognition and payment by bots, customers can complete the purchase process using the store's apps [42]. Artificial Intelligence (AI) and smart solutions are noticeable in the growth of mobile payments in India in dual ways, firstly in reducing fraud in high-risk mobile payments and secondly in improved Peer to Peer (P2P) Payment Platforms. Mobile payments through AI solutions is just one illustration of technology that is broadly used today, and that has already started advancing the FinTech industry forward for the Fourth Industrial Revolution.

Forecasting
Stock and inventory. Based on customer segmentation, the development of forecasting models supported by AI and ML start to be analyzed through simulation to improve the performance of inventories in retail stores. Capturing the knowledge of the different segments, the research of Bala [57] proposed a data mining model for demand forecasting with inventory replenishment system able to improve the inventory performance and profitability of operation by reducing reaching days and the increase in customer services levels simultaneously. Out-of-Stock (OOS) problems have also been analyzed within the parameters of AI and ML automating the process of detecting anomalies in the transactional data provided by retailers [58]. This is possible with the use of point-of-sale (POS) data share by retailers to suppliers, measuring and quantifying new predictor variables and comparing ML algorithms performance in a real-world scenario [59].

Customer decision making. Consumer Purchasing Behavior (CPB) have been widely analyses in the last years [60], and with the support of AI and ML models, several methods have been proposed and suggested by authors. This, in order to better understand consumer behavior, enabling the development of applications to engage them with personalized experiences based on behavioral patterns resulting from their grocery retail transactions [61]. For example, Kulkarni et al. [62] propose an algorithm relating consumer psychology and traversal paths in on-line and physical shops based on visited sub-paths and prominent nodes, mapping consumer behavior and establishing their buying patterns.

Table 3 shown diverse algorithms based on ML and AI being applied in many fields in grocery retail to compare and identify their performance and accuracy in forecast futures shopping transactions and consuming patterns [63, 64].

Table 3 ML and AI algorithms applied in grocery retail

Algorithms	Fields
Artificial Neuronal Networks (ANN) k-Nearest Neighbors (k-NN), Generalized Linear Model (GLM), Support Vector Regration (SVR), Autoregressive Integrated Moving Average (ARIMA), Support Vector Machines (SVM), Bayes Classifier, Recurrent Neural Network (RNN), Long-Short Term Memory (LSTM), t-distributed Stochastic Neighbor Embedding (t-SNE), Skyp-Gram (SG), Memory-level Parallelism (MLP-Boost), Convolutional Neural Net-work (CNN)	conservation of products [60], runtimes, costs, convenience, generalization abilities [65], sales predictive analytics [66], multi-task predictions [67], aisle and shelf layouts [68], products purchased, returned products, amount of discount and delivery delay [63], classification of products [18].

In-store Shopping Nowadays, customer is able to purchase products without using self-checkout station or cashier counters. Retail stores like Amazon Go are equipped with just walk out (JWO) technologies that heavily depends on deployment of multiple sensors based on computer vision and deep learning algorithms that can accurately predict the type of products and their quantities [69, 70]. In the field of image classification, CNN have been successfully used in computing vision task due to the advancement in hardware issues and well-annotated datasets (i.e., collection of data) [71].

4.4 Unmanned Aerial Vehicles (UAV) in Grocery Retail

By using Drone Delivery Systems (DDS), a type of UAV for individual orders, retailers obtain logistical benefits offering unprecedented delivery speed and adaptable delivery lead times cost-effectively [16]. Postal service companies worldwide such as UPS from United States, Swiss Post from Switzerland, DHL from Germany and retail delivery companies such as JD.com from China, Wing from USA, Rakuten from Japan and DDC from Canada have already been operating in delivery goods in the first and last mile, increasing delivery with point-to-point and automated operations [72, 73].

With drones operating at progressively faster speeds, retailers' logistics network will progressively become more decentralized whit more last-mile warehouses for instance, reducing operational and overhead cost and minimizing shipping times [74]. Amazon is already testing delivery packages in less than 30 minutes and under 5 pounds [75].

*Along with Sidewalk Autonomous Delivery Robots (SADRs) and Road Autonomous Delivery Robots (RADRs), the UAVs are being analyzed due to furthe*r benefits. In a matter of sustainability, this disruptive technology brings to the retail sector a significant reduction of CO2 emissions, by taking out off the road motorized vehicles usually used for delivery services [76]. On the other hand, such deliveries can accelerate the growth of e-commerce due potential cost saving that driverless technologies might bring [77].

4.5 Virtual Reality (VR) in Grocery Retail

VR in retail or v-commerce how is stated to be identified for some authors [7, 78], have begun to lay its foundations in delivering exceptional online experiences and yet influencing the consumer's shopping behaviors. Currently, the challenge remains in understanding how VR can impact on retail determining the use of this technology compared to visiting physical shops, specifically on the grocery sector [79]. In United States for example, the grocery sector is the largest single retail category, however, it is still one of the sectors that remains predominantly in-store, not online [24]. Compared to e-commerce, tool widely used by large grocery retail companies, VR increase sensory depth by transmitting detailed 3D images instead of two-dimensional static images [78] providing more engaging and realistic experiences and enhanced consumer trust [7, 80].

Pilot studies have been developing to analyses aspects regarding to the implementation of virtual shopping tasks, its acceptance through diverse customer's characteristics and purchase decisions [81]. According with Weerasinghe et al. [82] with well-designed virtual stores or supermarkets based on VR and Augmented Reality (AR) technologies, consumers will be allowed to do their shopping from their own home in the near future.

On the other hand, the focus on products has also been tested through VR, thus verifying the acceptability of products that are difficult to commercialize such as perishable foods. Verhulst et al. [80] present VR as a tool to investigate how customers interact with non-standard (misshaped) fruit and vegetables and evaluate consumers perceptions into a virtual supermarket in which trade-off between aspect of the products and quality were successfully verified.

4.6 Augmented Reality (AI) in Grocery Retail

In the retail sector, comparing with VR, AI is more appropriate for physical stores by overlapping of virtual elements in the physical world, allowing consumers to test products without trying them on [83]. Based on that, it was able to find in literature, the application of AI specifically in grocery retail, identifying alternatives to manage grocery lists on mobile application that enable consumers to add, delete or cross items find in nearby groceries stores. The solution proposed by Firoz and Ratnayaka [84] aims to reduce the level of complexity faced in grocery list management and improve customers' shopping process.

4.7 Big Data in Grocery Retail

In recent years, the need to extract the most valuable information that flows between suppliers, retailers and customers has made of Big Data, one of the most used technologies in retail. This is in order to relay updated information about products, customers experiences and patronage behaviors, location and demand allocation from the existing and growing datasets [8, 85]. According with Wothington and Fear [86], CRM has been one of the primary areas for innovation using Big Data across retail sectors involving data analytics to improve the customer experience through personalized product and services. This technology offers though customer profiling and segmentation, trend analysis and predictive modelling, as well for the evaluation of supplier activities, pricing new products lines and in-store advertising, mainly in grocery retail where loyalty programmed has gain greater relevance [86].

For this research, two papers were identified that link the grocery retail business with the integration of crowdsourcing Big Data sources integrated with conventional statistics and survey datasets. During the COVID-19 pandemic, Brandtner et al. [87] evaluates the impact of the situation on grocery retailers from the viewpoint of consumer satisfaction, applying data acquisition techniques and service-quality assessment model to analyze the sentiments of consumers. On the other hand, Ye et al. [88] identify the grocery retail demand in a small urban area. Thus, it was able to indicates the potential with higher demands to be used as an additional input to grocery retail location modelling tools.

4.8 Robotics Services in Grocery Retail

According with Rivera et al. [89], as new business models emerge, innovative strategies allow the application of disruptive technologies in retail processes such Robotics, a technology that endure continuous transformation due its short innovation cycles. Some of these models based on robotic services emphasize the service and support for people with reduced mobility or with some type of disability. For example, the use of autonomous and self-driven shopping carts [90] or teleoperated robots that allow handling objects or identifying the location of specific products [91, 92]. For the visually impaired, Gharpure and Kulyukin propose the addition of shopping assistance robots capable of providing spatial cognition and information about selected products [93]. On the other hand, as mentioned in Sect. 4.4, Sidewalks Autonomous Delivery Robots (SADRs) and Road Autonomous Delivery Robots (RADRs) are able to travel on roadways, sidewalks or pedestrian paths, optimizing delivery services without a driver, supported by sensors and navigation technologies [77].

4.9 Cloud Computing in Grocery Retail

The use of Cloud Computing converge with other technologies such as IoT, deep learning, computing vision and 4G and 5G networks, has brought a revolution on smart retail market [46], the joint work of these technologies has allowed the creation of structures for the assistance of consumers through the recognition of products based on the exchange and processing of images from open sources, resulting in useful correspondences in the purchase process, mainly for customers with visual difficulties [91].

5 Challenges and Limitations

The diverse studies, projects and prototypes found in this paper, highlight the forceful impact of the application of disruptive technologies on grocery retail business. In addition to enhance the importance of using data based on I4.0 for selection of products, places of purchase and sale and payment methods in this sector, the results also show how much these technologies have surpassed an environment properly physical to integrative models based on the rational an emotional interaction of customers and other stakeholders.

This change has made it possible to challenge the psychological motives of consumption and strengthen the link between man-machine relationships and thus, improve business performance and the customer journey through personalized experiences. This constitutes a competitive differentiation strategy, managing interactions with customers to influence the necessary perceptions in order to guarantee their loyalty.

On the other hand, it is important to highlight that the transition from manual controls to the implementation of digital systems can generate enormous demands for logistical work, storage, specialized engineering, etc. [4]. However, the process is hampered by many challenges such as difficulty in adapting sound logistics, lack of adequate skills at different levels of the workforce, complexity in the technologies to be implemented and understanding the dynamic capabilities of disruptive technologies, many of which are yet to be discovered [53]. Such challenges can be solved by creating awareness and educational programs about the benefits of those disruptive technologies, since the leading cause of changes in the grocery retail is digitalization [92].

6 Conclusions

The far-reaching impact of those disruptive technologies on grocery retail have been analyzed in this research. The authors aimed to contribute by filling the literature gap in this sector and establish the starting point for future research. Causality mechanisms for grocery retail affected by such technologies were investigated, presenting a summary of projects in progress and implementation of each of the technologies mentioned in this research: IoT, Blockchain, AI e ML, UAV, VR, AR, Big Data, Robotics Services and Cloud Computing. In addition, it was supposed to obtain information from 3D technologies applied to grocery retail; however, the results referred to the process of transforming images from 2D to 3D, characteristic of augmented reality technologies, and did not refer to 3D printing or manufacturing additive, also recognized as a type of disruptive technology.

Perhaps more results would had been obtained when exchanging search terms regarding the variants that identify grocery retail, for example, some authors identify the term Grocery Retail as Bricks and Mortar (B&M) [10], Grocery Industry [70], Convenience stores [70], proximity retail [93] and neighborhood groceries [94].

It is important to note that complementary technologies, such as digital monitoring, wireless networks, use of sensors, mobile application, automated data processing and the technological infrastructure of stores, play an essential role in the use of those disruptive technologies, thus developing a technological ecosystem and creating new resources for customers, retailers, and suppliers [95, 96].

In future works, it is intended to identify, through Key Performance Indicators (KPI) the projected evolution of retail grocery stores in order to assess changes in the commercial ecosystem during the application of disruptive technologies in this market niche. As well as estimating the socio-technical behavior of users through the Technology Acceptance Model (TAM) in order to evaluate the interaction between consumers and the use of disruptive technologies.

Acknowledgments The authors gratefully acknowledge the financial support from the research Unit of Governance, Competitiveness and Public Policies (GOVCOPP) from Aveiro University – Portugal.

References

1. Parry, M.E., Kawakami, T.: The encroachment speed of potentially disruptive innovations with indirect network externalities: the case of E-readers. J. Prod. Innov. Manag. **34**(2), 141–158 (2017). https://doi.org/10.1111/jpim.12333
2. Christensen, C.M.: The innovator's Dilemma, Rev edn. Harvard Business School Press, Boston (1997)
3. World Customs Organization.: Study Report on Disruptive Technologies (2019, Jun)
4. Majumdar, D., Banerji, P.K., Chakrabarti, S.: Disruptive technology and disruptive innovation: ignore at your peril! Technol. Anal. Strateg. Manag. **30**(11), 1247–1255 (2018). https://doi.org/10.1080/09537325.2018.1523384
5. Blitz, A.: Surviving disruption: the grocer's tale. J. Bus. Strateg. **42**(1), 13–22 (2021). https://doi.org/10.1108/JBS-07-2019-0152
6. Pantano, E., Priporas, C.V., Dennis, C.: A new approach to retailing for successful competition in the new smart scenario. Int. J. Retail Distrib. Manag. **46**(3), 264–282 (2018). https://doi.org/10.1108/IJRDM-04-2017-0080
7. Xue, L., Parker, C.J., Hart, C.: How to design fashion retail's virtual reality platforms. Int. J. Retail Distrib. Manag. **48**(10), 1057–1076 (2020). https://doi.org/10.1108/IJRDM-11-2019-0382

8. Sikos, T.T., Kozák, T., Kovács, A.: New retail models in online and offline space. J. Region. Dev. Tour. **11**, 2019
9. Christensen, C.M.: The opportunity and threat of disruptive technologies. MRS Bull. **27**(4), 278–282 (2002). https://doi.org/10.1557/mrs2002.81
10. Nuseir, M.T.: Potential impacts of blockchain technology on business practices of bricks and mortar (B&M) grocery stores. Bus. Process. Manag. J. **27**(4), 1256–1274 (2020). https://doi.org/10.1108/BPMJ-06-2020-0267
11. Bower, J.L., Christensen, C.M.: Disruptive technologies: catching the wave. Harv. Bus. Rev. **73**(1), 43–53 (1995)
12. McKinsey Global Institute.: Notes from the AI frontier: Applications and value of deep learning (2018, Apr 17) https://www.mckinsey.com/featured-insights/artificial-intelligence/notes-from-the-ai-frontier-applications-and-value-of-deep-learning
13. Kim, S., Connerton, T.P., Park, C.: Exploring the impact of technological disruptions in the automotive retail: a futures studies and systems thinking approach based on causal layered analysis and causal loop diagram. Technol. Forecast. Soc. Chang. **172** (2021). https://doi.org/10.1016/j.techfore.2021.121024
14. Das, S.: An IoT business model for public sector retail oil outlets. Inf. Technol. People. (2021). https://doi.org/10.1108/ITP-08-2020-0570
15. Salhieh, L., Shehadeh, M., Abushaikha, I., Towers, N.: Integrating vehicle tracking and routing systems in retail distribution management. Int. J. Retail Distrib. Manag. **49**(8), 1154–1177 (2021). https://doi.org/10.1108/IJRDM-12-2019-0400
16. Perera, S., Dawande, M., Janakiraman, G., Mookerjee, V.: Retail deliveries by drones: how will logistics networks change? Prod. Oper. Manag. **29**(9), 2019–2034 (2020). https://doi.org/10.1111/poms.13217
17. Coetzee, J.: Strategic implications of fintech on South African retail banks. S. Afr. J. Econ. Manag. Sci. **21**(1) (2018). https://doi.org/10.4102/sajems.v21i1.2455
18. Hafez, M.M., Vilas, A.F., Redondo, R.P.D., Pazó, H.O.: Classification of retail products: from probabilistic ranking to neural networks. Appl. Sci. (Switzerland). **11**(9) (2021). https://doi.org/10.3390/app11094117
19. Chang, R.M., Kauffman, R.J., Kim, K.: How strong are the effects of technological disruption? Smartphones' impacts on Internet and cable TV services consumption. In: Proceedings of the Annual Hawaii International Conference on System Sciences, pp. 2891–2899 (2013). https://doi.org/10.1109/HICSS.2013.252
20. Jesson, J., Matheson, L., Lacey, F.M.: Doing your Literature Review : Traditional and Systematic Techniques. SAGE Publication Inc., Los Angeles (2011)
21. Elsevier B.V.: What is Scopus about? (2022) https://service.elsevier.com/app/answers/detail/a_id/15100/supporthub/scopus/
22. Reis, J., Melão, N., Salvadorinho, J., Soares, B., Rosete, A.: Service robots in the hospitality industry: the case of Henn-na hotel, Japan. Technol. Soc. **63**, 101423 (2020). https://doi.org/10.1016/j.techsoc.2020.101423
23. Kalyani, D.: Supermarkets & Grocery Stores in the US, Illinois, (2019, Apr)
24. Hyland, R.: Online Grocery Sales in the US (2018, Oct)
25. Deloitte University Press.: The retail transformation Cultivating choice, experience, and trust (2015)
26. McKinsey Global Institute.: Disruptive technologies: Advances that will transform life, business, and the global economy (2013) [Online]. Available: www.mckinsey.com/mgi
27. Digital McKinsey.: Insights Winning in digital ecosystems (2018)
28. Lans, L., van der Voordt, D.: Descriptive research. In: de Jong, T.M., van der Voordt, D.J.M. (eds.) Ways to Study and Research Urban, Architectural and Technical Design, pp. 53–60. DUP Science (2002)
29. Koh, E.T., Owen, W.L.: Descriptive research and qualitative research. In: Introduction to Nutrition and Health Research, pp. 219–248. Springer US, Boston, MA (2000). https://doi.org/10.1007/978-1-4615-1401-5_12

30. Liberati, A., et al.: The PRISMA statement for reporting systematic reviews and meta-analyses of studies that evaluate health care interventions: explanation and elaboration. PLoS Med. **6**(7), e1000100 (2009). https://doi.org/10.1371/journal.pmed.1000100
31. Prisma statement.: Welcome to PRISMA (2021) http://www.prisma-statement.org/
32. Pitula, M., Spangenberg, M.: Disruption Report Online Grocery Shopping (2021)
33. Federation of Wholesale Distributors.: The Future of UK Food and Dring Wholesaling, Kent (2019, Jun) [Online]. Available: www.futurestudies.co.uk
34. Jongen, W.: Big data + Internet of Things = smart retail economy (2018, Apr) https://retail-loyalty.org/en/news/big-data-internet-of-things-smart-retail-economy/
35. Ng, C.K., Wu, C.H., Yung, K.L., Ip, W.H., Cheung, T.: A semantic similarity analysis of Internet of Things. Enterprise Inf. Syst. **12**(7), 820–855 (2018). https://doi.org/10.1080/17517575.2018.1464666
36. Fagerstrøm, A., Eriksson, N., Siguresson, V.: What's the 'thing' in Internet of Things in Grocery shopping? A customer approach. Proc. Comput. Sci. **121**, 384–388 (2017). https://doi.org/10.1016/j.procs.2017.11.052
37. Thilagavathi, S., Nivethitha, K.S., Preeti, P., Vikram, D.T.: IoT based smart retail system with social distancing for Covid19 outbreak. J. Phys. Conf. Ser. **1917**(1) (2021). https://doi.org/10.1088/1742-6596/1917/1/012030
38. Fagerstrøm, A., Eriksson, N., Sigurdsson, V.: Investigating the impact of Internet of Things services from a smartphone app on grocery shopping. J. Retail. Consum. Serv. **52** (2020). https://doi.org/10.1016/j.jretconser.2019.101927
39. Pal, I., Basak, R., Bandyopadayay, A., Ghosh, T., Sarkar, M.: Design of a User-Friendly Market System by Introducing Motorised Shelves Automated Billing System (2020). https://doi.org/10.1109/IEMENTech51367.2020.9270123
40. Raad, O., Makdessi, M., Mohamad, Y., Damaj, I.: SysMART indoor services: a system of smart and connected supermarkets. In: Canadian Conference on Electrical and Computer Engineering, Aug. 2018 (2018, May). https://doi.org/10.1109/CCECE.2018.8447626
41. Radhakrishnan, M., Sen, S., Vigneshwaran, S., Misra, A., Balan, R.: IoT+Small Data: Transforming in-Store Shopping Analytics & Services (2016). https://doi.org/10.1109/COMSNETS.2016.7439946
42. Jakhiya, M., Mittal Bishnoi, M., Purohit, H.: Emergence and growth of mobile money in modern India: a study on the effect of mobile money. In: 2020 Advances in Science and Engineering Technology International Conferences, ASET 2020 (2020). https://doi.org/10.1109/ASET48392.2020.9118375
43. Shankar, S.K., Balasubramani, S., Basha, S.A., Ariz Ahamed, S., Kumar Reddy, N.S.: Smart trolley for smart shopping with an advance billing system using IoT. In: Proceedings - 5th International Conference on Computing Methodologies and Communication, ICCMC 2021, Apr, pp. 390–394 (2021). https://doi.org/10.1109/ICCMC51019.2021.9418348
44. Paul, C., Sabu, S., Angelin, R., Pardeshi, A.: Smart shopping application using IoT and recommendation system: an effective mobile assisted software application for grocery shopping. In: 2021 7th International Conference on Advanced Computing and Communication Systems, ICACCS 2021, Mar, pp. 522–526 (2021). https://doi.org/10.1109/ICACCS51430.2021.9441762
45. Mekruksavanich, S.: The smart shopping basket based on IoT applications. In: 2019 IEEE 10th International Conference on Software Engineering and Service Science (ICSESS), Oct 2019, pp. 714–717 (2019). https://doi.org/10.1109/ICSESS47205.2019.9040750
46. Kaur, J., Kaur, P.D.: CE-GMS: a cloud IoT-enabled grocery management system. Electron. Commer. Res. Appl. **28**, 63–72 (2018). https://doi.org/10.1016/j.elerap.2018.01.005
47. Yacoub, G., Castillo, M.: Blockchain in your grocery basket: trust and traceability as a strategy. J. Bus. Strateg. (2021). https://doi.org/10.1108/JBS-02-2021-0032
48. Antonopoulos, A.M.: Mastering Bitcoin: Unlocking Digital Cryptocurrencies, 1st edn, (2014)
49. Ally, M., Gardiner, M., Lane, M.: The Potential Impact of Digital Currencies on the Australian Economy. ACIS 2015 Proceedings, vol. 136, 2015

50. Magas, J.: Immutability in Doubt: Do We Need to Protect Blockchain Data? (2018, Jul 05). https://cointelegraph.com/news/immutability-in-doubt-do-we-need-to-protect-blockchain-data
51. Awaysheh, A., Klassen, R.D.: The impact of supply chain structure on the use of supplier socially responsible practices. Int. J. Oper. Prod. Manag. **30**(12), 1246–1268 (2010). https://doi.org/10.1108/01443571011094253
52. Tucker, C., Catalini, C.: What Blockchain Can't Do (2018, Jun 28) https://hbr.org/2018/06/what-blockchain-cant-do#:~:text=Blockchain%20technology%20can%20track%20which,digital%20identifier%20requires%20offline%20verification
53. Čirjevskis, A.: Exploring Dynamic Managerial Capabilities for Digital Transformation in Global Grocery Retail: Walmart's Alliance with IBM (2020)
54. Rogojanu, I., Suciu, G., Ditu, M.C., Pasat, A.: Smart shopping technologies for indoor markets. In: Proceedings - 21st IEEE International Conference on Computational Science and Engineering, CSE 2018, Dec, pp. 99–103 (2018). https://doi.org/10.1109/CSE.2018.00020
55. Yakuel, P.: 7 Top CRM Trends for 2017: A Look Ahead (2017, Jan 3) https://www.cmswire.com/customer-experience/7-top-crm-trends-for-2017-a-look-ahead/
56. Marín-García, A., Gil-Saura, I., Ruiz-Molina, M.E., Berenguer-Contrí, G.: The moderating effect of store format on the relationships between ICT, innovation and sustainability in retailing. Front. Psychol. **12** (2021). https://doi.org/10.3389/fpsyg.2021.678991
57. Bala, P.K.: Improving inventory performance with clustering based demand forecasts. J. Model. Manag. **7**(1), 23–37 (2012). https://doi.org/10.1108/17465661211208794
58. Papakiriakopoulos, D.: Developing a mechanism to support decisions for products missing from the shelf. J. Decis. Syst. **20**(4), 417–441 (2011). https://doi.org/10.3166/JDS.20.417-441
59. Andaur, J.M.R., Ruz, G.A., Goycoolea, M.: Predicting out-of-stock using machine learning: an application in a retail packaged foods manufacturing company. Electronics (Switzerland). **10**(22) (2021). https://doi.org/10.3390/electronics10222787
60. Vázquez-Martínez, U.J., Morales-Mediano, J., Leal-Rodríguez, A.L.: The impact of the COVID-19 crisis on consumer purchasing motivation and behavior. Eur. Res. Manag. Bus. Econ. **27**(3), 100166 (2021). https://doi.org/10.1016/j.iedeen.2021.100166
61. T. Lang and M. Rettenmeier, "Understanding Consumer Behavior with Recurrent Neural Networks," 2017
62. Kulkarni, H., Patil, P., Menon, R.: Multi-agent system for customer behavior tracking using shoppers' path or traversal. In: 2019 IEEE International Conference on Electrical, Computer and Communication Technologies (ICECCT), Feb 2019, pp. 1–5 (2019). https://doi.org/10.1109/ICECCT.2019.8869132
63. Zuo, Y., Yada, K., Ali, S.: Prediction of consumer purchasing in a Grocery store using machine learning techniques; prediction of consumer purchasing in a Grocery store using machine learning techniques. In: 2016 3rd Asia-Pacific World Congress on Computer Science and Engineering (APWC on CSE) (2016). https://doi.org/10.1109/APWC.on.CSE.2016.13
64. Muqodas, A.U., Kusuma, G.P.: Promotion scenario based sales prediction on E-retail groceries using data mining. Int. J. Emerg Technol. Adv. Eng. **11**(6), 9–18 (2021). https://doi.org/10.46338/IJETAE0621_02
65. Kaunchi, P., Jadhav, T., Dandawate, Y., Marathe, P.: Future Sales Prediction for Indian Products Using Convolutional Neural Network-Long Short Term Memory (2021). https://doi.org/10.1109/GCAT52182.2021.9587668
66. Cirqueira, D., Helfert, M., Bezbradica, M.: Towards Preprocessing Guidelines for Neural Network Embedding of Customer Behavior in Digital Retail (2019). https://doi.org/10.1145/3386164.3389092
67. Gabel, S., Guhl, D., Klapper, D.: P2V-MAP: mapping market structures for large retail assortments. J. Mark. Res. **56**(4), 557–580 (2019). https://doi.org/10.1177/0022243719833631
68. Khodabandehlou, S., NikNafs, A.A.: Improving customer loyalty evaluation methods in the grocery retail industry: a data mining approach. Int. J. Electron. Customer Relationship Manag. **10**(2/3/4), 158 (2016). https://doi.org/10.1504/ijecrm.2016.10003155

69. Wingfield, N.: Amazon Moves to Cut Checkout Line, Promoting a Grab-and-Go Experience (2016, Dec 05) https://www.nytimes.com/2016/12/05/technology/amazon-moves-to-cut-checkout-line-promoting-a-grab-and-go-experience.html
70. Ives, B., Cossick, K., Adams, D.: Amazon go: disrupting retail? J. Inf. Technol. Teach. Cases. **9**(1), 2–12 (2019). https://doi.org/10.1177/2043886918819092
71. Liu, C., Wang, X., Ni, J., Cao, Y., Liu, B.: An edge computing visual system for vegetable categorization. In: Proceedings - 18th IEEE International Conference on Machine Learning and Applications, ICMLA 2019, Dec, pp. 625–632 (2019). https://doi.org/10.1109/ICMLA.2019.00115
72. DHL Global.: UNMANNED AERIAL VEHICLES Ready for Take-off? (2022). https://www.dhl.com/global-en/home/insights-and-innovation/thought-leadership/trend-reports/unmanned-aerial-vehicles.html
73. Corrigan, F.: Drones For Deliveries From Medicine To Post, Packages And Pizza (2020, Jul 02). https://www.dronezon.com/drones-for-good/drone-parcel-pizza-delivery-service/
74. DDC.: Reimagining the Way You Deliver, https://dronedeliverycanada.com (2022)
75. D'Onfro, J.: Amazon's New Delivery Drone Will Start Shipping Packages 'In A Matter Of Months (2019) https://www.forbes.com/sites/jilliandonfro/2019/06/05/amazon-new-delivery-drone-remars-warehouse-robots-alexa-prediction/?sh=6de0367b145f
76. Wing Aviation LLC.: Celebrating Earth Month with a glimpse inside our office, and a passion for sustainability that inspires our team (2020, May 07) https://blog.wing.com/2020/05/celebrating-earth-month-with-glimpse.html
77. Figliozzi, M.A.: Carbon emissions reductions in last mile and grocery deliveries utilizing air and ground autonomous vehicles. Transp. Res. Part D: Transp. Environ. **85** (2020). https://doi.org/10.1016/j.trd.2020.102443
78. Klein, L.R.: Creating virtual product experiences: the role of telepresence. J. Interact. Mark. **17**(1), 41–55 (2003). https://doi.org/10.1002/dir.10046
79. Chaffey, D.: The reasons why consumers shop online instead of in stores (2017, Oct 19). https://www.smartinsights.com/ecommerce/ecommerce-strategy/the-reasons-why-consumers-shop-online-instead-of-in-stores/.
80. Verhulst, A., Normand, J.-M., Lombart, C., Moreau, G.: A study on the use of an immersive virtual reality store to investigate consumer perceptions and purchase behavior toward non-standard fruits and vegetables. In: 2017 IEEE Virtual Reality (VR), pp. 55–63 (2017). https://doi.org/10.1109/VR.2017.7892231
81. Plechata, A., Fajnerova, I., Hejtmanek, L., Sahula, V.: Development of a virtual supermarket shopping task for cognitive remediation of memory and executive functions in schizophrenia. In: 2017 International Conference on Virtual Rehabilitation (ICVR), pp. 1–2 (2017, Jun). https://doi.org/10.1109/ICVR.2017.8007514
82. Weerasinghe, N., Jayawardena, S., Mahawatta, D., Navaratne, H., Sriyaratna, D., Gamage, I.: Mixed reality supermarket: a modern approach into day - to - day grocery shopping. In: 20th International Conference on Advances in ICT for Emerging Regions, ICTer 2020 - Proceedings, pp. 184–189 (2020, Nov). https://doi.org/10.1109/ICTer51097.2020.9325501
83. Caboni, F., Hagberg, J.: Augmented reality in retailing: a review of features, applications and value. Int. J. Retail Distrib. Manag. **47**(11), 1125–1140 (2019). https://doi.org/10.1108/IJRDM-12-2018-0263
84. Firoz, A., Ratnayaka, G.: ShopLister - A Grocery List Management Application (2020). https://doi.org/10.1109/ICIP48927.2020.9367367
85. Suhara, Y., Bahrami, M., Bozkaya, B., Pentland, A.S.: Validating gravity-based market share models using large-scale transactional data (2019) [Online]. Available: http://arxiv.org/abs/1902.03488
86. Worthington, S., Fear, J.: The Hidden Side of Loyaltycard Programs. The Australian Centre For Retail Studies (2009)
87. Brandtner, P., Darbanian, F., Falatouri, T., Udokwu, C.: Impact of COVID-19 on the customer end of retail supply chains: a big data analysis of consumer satisfaction. Sustainability (Switzerland). **13**(3), 1–18 (2021). https://doi.org/10.3390/su13031464

88. Ye, Z., Clarke, G., Newing, A.: Estimating small-area demand of urban tourist for groceries: the case of Greater London. J. Retail. Consum. Serv. **58** (2021). https://doi.org/10.1016/j.jretconser.2020.102263
89. Rivera, R., Amorim, M., Reis, J.: Technological evolution in Grocery retail: a systematic literature review. In: 2021 16th Iberian Conference on Information Systems and Technologies (CISTI), pp. 1–8 (2021). https://doi.org/10.23919/CISTI52073.2021.9476598
90. Gross, H.-M., et al.: TOOMAS: Interactive Shopping Guide robots in everyday use - final implementation and experiences from long-term field trials. In: 2009 IEEE/RSJ International Conference on Intelligent Robots and Systems, pp. 2005–2012 (2009). https://doi.org/10.1109/IROS.2009.5354497
91. Kulyukin, V., Zaman, T., Andhavarapu, A., Kutiyanawala, A.: Eyesight sharing in blind grocery shopping: remote P2P caregiving through cloud computing. Lect. Notes Comput. Sci. (including subseries Lecture Notes in Artificial Intelligence and Lecture Notes in Bioinformatics). **7383 LNCS**(PART 2), 75–82 (2012). https://doi.org/10.1007/978-3-642-31534-3_12
92. Agnihotri, A.: Can Brick-and-Mortar retailers successfully become multichannel retailers? J. Mark. Channels. **22**(1), 62–73 (2015). https://doi.org/10.1080/1046669X.2015.978702
93. Lenglet, F., Mencarelli, R.: Proximity in marketing: an integrating theoretical framework and research agenda. Recherche et Applications en Marketing (English Edition). **35**(4), 99–124 (2020). https://doi.org/10.1177/2051570720957152
94. Ananias, E., Gaspar, P.D., Soares, V.N.G.J., Caldeira, J.M.L.P.: Artificial intelligence decision support system based on artificial neural networks to predict the commercialization time by the evolution of peach quality. Electronics (Switzerland). **10**(19) (2021). https://doi.org/10.3390/electronics10192394
95. Marin-Hernandez, A., Hoyos-Rivera, G.D.J., García-Arroyo, M., Marin-Urias, L.F.: Conception and implementation of a supermarket shopping assistant system. In: Proceedings of Special Session - Revised Papers, 11th Mexican International Conference on Artificial Intelligence 2012: Advances in Artificial Intelligence and Applications, MICAI 2012, pp. 26–31 (2012). https://doi.org/10.1109/MICAI.2012.21
96. Rivera, R., Amorim, M., Reis, J.: Robotic services in smart cities: an exploratory literature review. In: 15th Iberian Conference on Information Systems and Technologies (CISTI). IEEE. pp. 1–7 (2020). https://doi.org/10.23919/CISTI49556.2020.9141099

A Network Modeling and Analysis of COVID-19 Hospital Patient Data

Filipe Loyola Lopes, Ronaldo Lopes Inocêncio Júnior, Victor Castro Nacif de Faria, Filipe Alves Neto Verri, and Ana Carolina Lorena

Abstract There are currently large amounts of public databases on COVID-19 patients. Among them the FAPESP COVID-19 Data Sharing/BR has gathered laboratory tests and hospitalization data from large health centers in the São Paulo metropolitan area. This paper uses part of such a repository to assemble a set of networks of positive COVID-19 patients according to the similarity of their age and laboratory tests results. Next, popular complex network metrics such as clustering coefficient and average path length are extracted from such networks and compared to the expected values observed for classical networks. Similarities of the clustering coefficient values with those of Watts-Strogatz networks were observed, although there are no sustainable characteristics of Small World networks. There are also similarities to scale-free networks, such as high-degree variance and the presence of hubs of nodes. In addition, a partition of the networks based on the modularity measure using the Fast Greedy algorithm is obtained and analyzed. An structure with four clusters and modularity values greater than zero was found, indicating that the network has some community structure.

Keywords Complex networks · COVID-19 networks · Patient network

1 Introduction

COVID-19 is the infectious respiratory disease caused by the SARS-COV-2 virus. The first case of the disease was reported in December 2019, at the Chinese province

F. L. Lopes (✉) · V. C. N. de Faria
Aeronautics Institute of Technology, Sao Jose dos Campos, Brazil

Federal University of Sao Paulo, Sao Jose dos Campos, Brazil
e-mail: filipe.loyola@unifesp.br

R. L. Inocêncio Júnior · F. A. N. Verri · A. C. Lorena
Aeronautics Institute of Technology, Sao Jose dos Campos, Brazil

V. M. López Sánchez et al. (eds.), *Industrial Engineering and Operations Management*, Springer Proceedings in Mathematics & Statistics 400,
https://doi.org/10.1007/978-3-031-14763-0_26

of Wuhan [1]. On March 2020, the World Health Organization (WHO) declared a pandemic status for the disease. Two years later, in March 2022, 440 million cases and almost 6 million deaths were already recorded worldwide [2]. The pandemic drew attention of scientists from diverse areas around the world by studying disease mechanisms, prevention, treatments and the development of vaccines [3, 4].

Various recent studies are dedicated to COVID-19 data analysis and the usage of statistics and Artificial Intelligence techniques for prediction, diagnosis and decision-making [3, 5–10]. In order to encourage these scientific researches, COVID-19 open data sharing platforms and repositories were created [11, 12]. One commendable initiative in Brazil is the FAPESP COVID-19 Data Sharing/BR, which stores data on thousands of patients diagnosed with COVID-19 from referral hospitals located at the São Paulo metropolitan area, the most populated region of Brazil [13].

In this paper we take data from one of these health centers, namely Sírio Libanês Hospital, and assemble a set of networks of positive COVID-19 patients, based on the similarity of their laboratory test results. Complex networks are graphs that have a non-trivial topological structure, composed by a set of nodes connected by edges [14]. They allow to represent relationships between observations in a natural way and to reveal interesting aspects concerning data structure and organization. Popular since the late 1990s, the study of complex networks has evolved and established itself as a research topic with many contributions to applied domains such as social data analysis and Bioinformatics [15]. There are some previous work in the literature employing a complex network modelling in the analysis of COVID-19 data. Nonetheless, most of them are focused on modelling the dynamics of the disease spread and pandemic waves [16–18].

The networks built in this paper have positive COVID-19 patients attended at the Sírio Libanês Hospital as nodes and edges are added between patients who have similar age and laboratory tests results. Different degrees of similarity are considered for building these networks. Next, properties of the networks are analyzed, such as the average degree of the nodes, their clustering coefficient and average path length [19]. These values are compared to those obtained for classical network models from the literature, such as Watts-Strogatz and scale-free networks [19]. Modularity optimization algorithms were used to discover evidence of community structure [20]. Through these analyses, we can observe an assortative behavior, that is, similar nodes tend to be very connected to each other. This behavior can generate insights about possible clusters division. In addition, it is possible to observe the presence of community structure, which may allow future studies about community analysis.

This paper is structured as follows: Sect. 2 presents the materials and methods employed in this work. Section 3 presents and discusses the results achieved. Section 4 concludes this paper.

2 Materials and Methods

This section presents the materials and methods employed in this paper. All codes were implemented in the *Python* language using the integrated *Google Colaboratory* environment, with support of libraries such as *Igraph* for graph generation, *Numpy* for managing mathematical functions, *Pandas* for structuring complex data sets and *Matplotlib* for plotting graphs.[1]

2.1 Dataset

The raw database used in this study comes from Sírio Libanês Hospital and is available at the FAPESP COVID-19 Data Sharing/BR repository [13]. It contains data from 8971 suspicious or confirmed COVID-19 patients who were attended by the hospital from March to October, 2020. More than 950 types of laboratory tests are recorded in this database, which has a high rate of missing data (on average 96.55% $\pm$12.37 of missing values by laboratory test). Age and sex demographic features are also available.

Some filters were applied to this database, resulting in a dataset with 4320 patients and 33 input attributes (the preprocessed dataset is available at reference [21]). The inclusion criteria were: patients with COVID-19 positive result up to 15 days after attendance, for keeping only certified COVID-19 positive patients; laboratory test results from up to the first three days of hospital attendance and only from emergency room, for screening the conditions of the patients upon hospital admission. And as exclusion criteria we have: removal of laboratory tests with more than 50% of missing values, in order to avoid missing data imputation biases, and removal laboratory tests with non numerical results.

Patients were labeled according to the origin of laboratory tests, into four groups: 79% have been attended at the emergency room only (Group 0); 12% have also been attended at the infirmary (Group 1); 2% have been attended at the emergency room and progressed straight to intensive care (ICU beds) (Group 2); and 7% have been attended at emergency room, infirmary and ICU (Group 3). The latter two groups of individuals can be regarded as severe COVID-19 cases, whilst the first two are considered mild or non severe cases [21].

[1] The developed code is available at https://doi.org/10.5281/zenodo.6399915.

2.2 Assembling the Patients' Networks

From the dataset described in the previous section, two additional pre-processing steps were applied. The first was feature selection, where ten attributes with highest correlation to the patients' clinical condition (severe or non severe) were maintained [21]. The attributes kept after this selection were: Age, Alanine Aminotransferase (ALT), C-reactive Protein, Aspartate Aminotransferase (AST), Creatinine, Urea, Red Cell Distribution Width (RDW), Potassium, Erythrocytes and Neutrophils. All chosen features were normalized in the [0,1] interval.

The networks were then built based on the similarity of the values of the previous attributes. Each patient is one node of the network. Edges are added between patients with at least six similar attribute values, a number which was empirically determined. The similarity of values is calculated based on the standard deviation of the corresponding variable, as:

$$x_{im} \text{ is similar to } x_{jm} \iff \left|x_{im} - x_{jm}\right| < \frac{\sigma_m}{c}, \tag{1}$$

where x_{im} and x_{jm} represent the m-th attribute values for two distinct patients (i and j). σ_m is the standard deviation of the m-th attribute and c is a constant dividing the standard deviation, which was varied in our study from 1 to 10 with increments of one. Therefore, for a given pair of patients (indexed as i and j) and a c value, an edge ij exists if the number of similar features between patients is greater than six. As c increases, the similarity ranges are smaller and more concentrated around the attribute values of the i-th patient. For each value of c, the connections between vertices were determined and samples of 1000 unique and random vertices were created with those connections, in a undirected and unweighted manner, in order to analyze them as instances of a network model.

2.3 Generation of Synthetic Networks

In the study of complex networks, it is necessary to use classical models, because they follow a rule of creation which can be used for comparison purposes. These artificial networks models may or may not present similarities with the instances of the patients networks studied. If the networks analyzed have similar properties to those of classical networks, the latter can be used as reference for the analysis of statistical properties and the isolation of some factors to be studied, in order to facilitate causal association analyzes [15].

For comparing the characteristics of the patients' networks to those of well-known networks from the literature, some synthetic networks were generated. Each network instance was built with 1000 non-repeated vertices and 100 instances per scenario are considered. Only the giant component of these networks were kept for analysis, since there is no guarantee of connectivity of all vertices in the sample.

Next, the following metrics were extracted from the networks: clustering coefficient (CC), average path length (APL) and the average and variance of degrees [19]. For a given node i, the CC accounts for the ratio between the number of neighbor nodes i has, divided by the maximum number of edges that can be formed between those neighbors. The average of the CC values of each node is taken for a given network. As more nodes are connected in the network the CC values tend to be larger. APL is an arithmetic average of the minimum distance between all possible pairs of vertices in the network. In networks with a low APL value, information spreads much faster than in structured networks with the same number of nodes and the same average degree. In this research, the CC and APL values were calculated to compare with classical networks. The degree of a node is given by the number of connected neighbors it has and is closely related to CC.

The classical networks considered in this research were Erdös-Renyi (ER) or random networks, Watts-Strogatz (WS) or Small World, and Barabási-Albert (BA), which is a particular instance of the "scale-free network models".

The ER model is the simplest complex network model [14]. In this model, each edge is connected with the others based on a probability value (p). When the number of nodes (n) tends to infinity, the computation of the average degree (Z) can be obtained as a function of (n) and (p):

$$Z = pn. \tag{2}$$

In ER networks, when Z grows faster than $log(n)$, the sequence of graphs has a high probability of being fully connected [20]. On the other hand, when the order of growth is smaller than the logarithmic function, the network has a high probability of being disconnected, comprised of smaller components. A connectivity phase transition occurs at $Z = 1$ [22]. In this paper, the degree value was controlled by varying the value of the probability p in the range from 3.50% to 40.93%.

The WS network model, also known as Small World network, is generated from the rewiring of links in a fully structured (lattice) network with a value of rewiring probability and has a typical small distance between any pair of nodes (geodesic distance). This type of network has three special characteristics: high clustering coefficient, low average distance between vertices and sparseness [23]. Its generation is also governed by a rewiring probability value p. When $p = 0$, the network is completely regular (as a lattice network) and, when $p = 1$, the network is completely random, similar to an ER network model. So the characteristic Small World of the model lies between regularity and randomness. The average degree can be calculated by:

$$Z = 2d, \tag{3}$$

where d represents the number of closest vertices to the right and to the left of each vertex in the lattice during the creation of the WS network [24]. In this work, the d parameter was used to control the average degree and was varied between 18 and

205. In this case, the probability value was adjusted in order to ensure the presence of the small world characteristic values ranging from 3.50% to 40.93%.

In the BA classical model, a specific type of scaling-free network, few nodes have a high number of connections, becoming hubs in the network [14]. In the BA networks, the average degree value was controlled by varying a numerical constant m, which represents the average number of edges added to the network at each time step during its creation.

2.4 Community Detection

Modularity is popularly used to detect similar communities in a graph [2]. In this paper, we also managed to find communities in some of the patient's networks, namely for those networks with the c value equal to four (Eq. 1). The Fast Greedy algorithm was used to detect communities in the networks, guided towards maximizing modularity [25]. The idea is to verify if consistent groups of patients with similar attributes can be identified in the data.

3 Results and Discussion

The number of edges and diameter of the patients' networks built, as a function of c, can be seen in Table 1. All 10 networks have 4320 vertices, which represent the number of patients within the processed dataset. As the network diameter measures the shortest distance between the farthest pair of nodes in a network, it also means, in the context of patients' network, how much different the most different patients are from each other.

From Table 1, it is possible to observe that the number of edges decreases as the constant c increases. This occurs because the higher the c value, the more restricted

Table 1 Characteristics of the patients' networks constructed as function of the c variation in order to control the level of similarity between the attributes for the formation of edges

c	♯ Edges	Diameter
1	3,814,460	4
2	1,110,353	5
3	505,705	7
4	317,658	8
5	266,109	9
6	239,452	12
7	220,523	17
8	209,647	18
9	205,548	24
10	202,660	23

Table 2 Instances description obtained from the patients' network for $c = 4$. The averages and standard deviation values are computed from 100 instances

Metric	Mean	*Std*
Number of nodes	288.97	19.78
Average degree	74.92	7.52
CC	0.67	0.03
Degree variance	3987.51	542.26

is the degree of similarity required between the attributes for a connection to be established. Therefore, the networks are more connected for lower c values. In contrast, it is possible to observe from Table 1 that in general the smaller the number of edges (or higher c values), the greater the diameter of the network. This ought to be expected because, the less the patients are similar to each other, the harder it is to make a series of connections between any pair of vertices. One exception is observed for c between 9 and 10, where there is a slight decrease in the network diameter.

Next we generated instances of each of the synthetic networks described in Sect. 2. Hundred networks for each of the hyper-parameter variations and for each model are produced. Although each network instance has 1000 nodes, only the giant (largest) connected component of each instance was considered in order to guarantee a connected network. There is a direct mapping between the generated networks' average characteristics and the c values of the patients' networks, in order to allow a comparison between them. A summary of the main characteristics of the networks produced for c equal 4 can be seen in Table 2.

From Table 2, it is possible to observe that the giant component had about 289 vertices on average. As each instance was produced with 1000 vertices, the giant component's average size was 28.9% of each instance. The average degree also has a low standard deviation, suggesting that all instances had similar average degrees with each other. The same phenomenon is observed in relation to the standard deviation of the number of vertices and the CC. On the other hand, the degree variance mean value and its corresponding standard deviation have higher values. The high degree variance may be indicative of the existence of hubs, which is typical for BA networks, however, other analyses are still necessary.

3.1 Comparison with Classical Networks

For a fair comparison between the patients' networks instances and the instances of classical models, the number of nodes and their average degree were made equivalent for all networks. Figure 1 shows a comparison between the average degree of all network models.

In Fig. 1 it is possible to observe a drop effect of the average degree for values of c lower than or equal to 5. However, for values greater than 5, there is an increase in the average degree, contrary to the trends of the previous values. This phenomenon

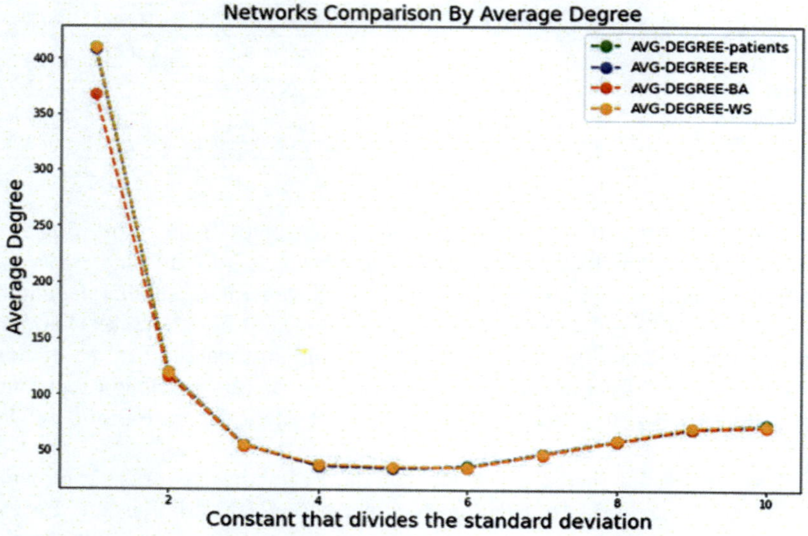

Fig. 1 Average degree comparison of networks

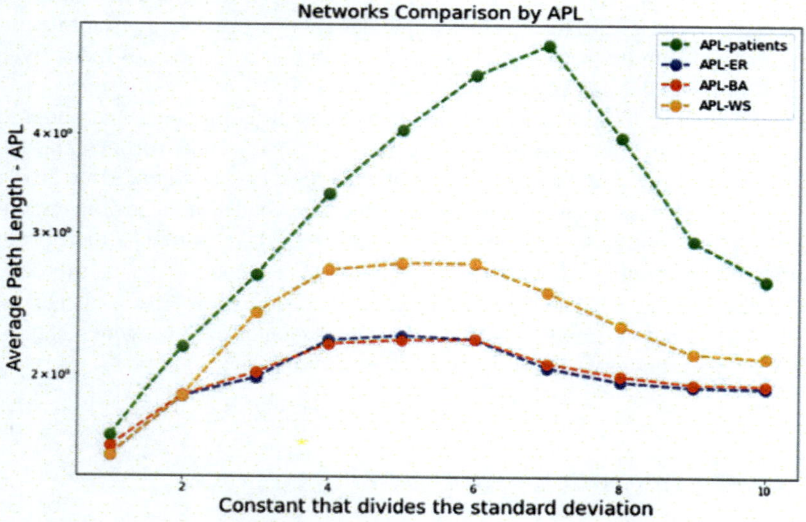

Fig. 2 Average Path Lengths (APL) of the networks. The APL value axes are in a logarithmic scale

can be caused by the degree distribution homogenization of the network instances' giant components. Notably, the same behavior is observed for all network models. Figures 2 and 3 show a comparison between the networks' APL and CC values.

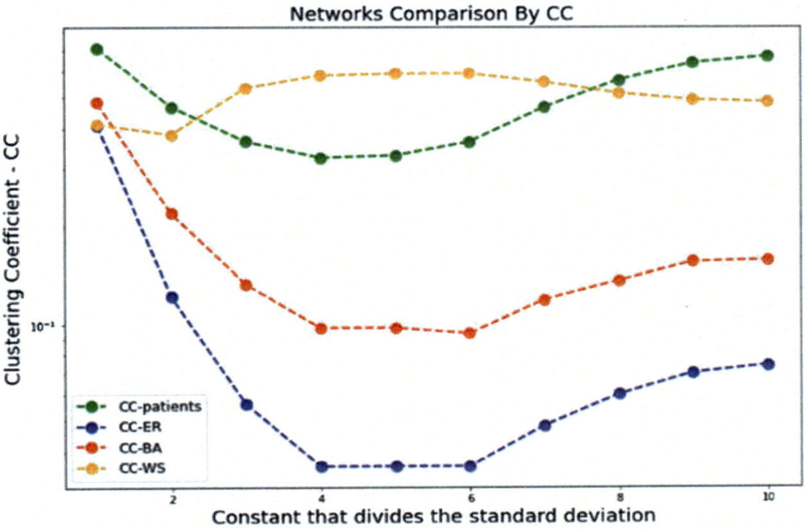

Fig. 3 Clustering coefficient (CC) of the networks. The CC value axes are in a logarithmic scale

In Fig. 2, the APL of the patient network is greater than the corresponding values of the other networks. This includes the WS networks, making it possible to infer that the patients' networks instances do not show the small world phenomenon. In other words, it is not possible to generate an instance of the patients' network based on a WS network model, because of the great average distance between vertices.

In Fig. 3, is possible to observe that there is an increase in CC in the patient networks approximately at $c = 4$, despite a decline in the number of connections. This indicates a trend of neighboring patients who are very connected to each other, discarding any indication of sparsity, being instead an indication of network assortativity [26].

In Fig. 4, a variance analysis of the degree of the networks was performed, as a way to assess the homogeneity of the networks. It is possible to verify that the degree variance of the patient network tends to stabilize as the value of c increases. The stability of the network degree variance value for the same range of values of c is another indication to support a hypothesis of homogenization of the networks. In a more detailed analysis, it is possible to observe that up to the value of c equal to 4 there is a variance decrease, followed by an uninterrupted increase. With the greater restriction of connections between vertices, there is a tendency for the neighbors of a node to be more connected to each other and less connected to nodes with little to no similarity with it.

Still on the variance of the degree distribution, it is also worth assessing the high variance of the patients network when compared with the WS and ER network models. There is a decrease in variance up to a value of c equal to 3, followed by a level of increase to an apparent stability from values of c greater than 8. Even though

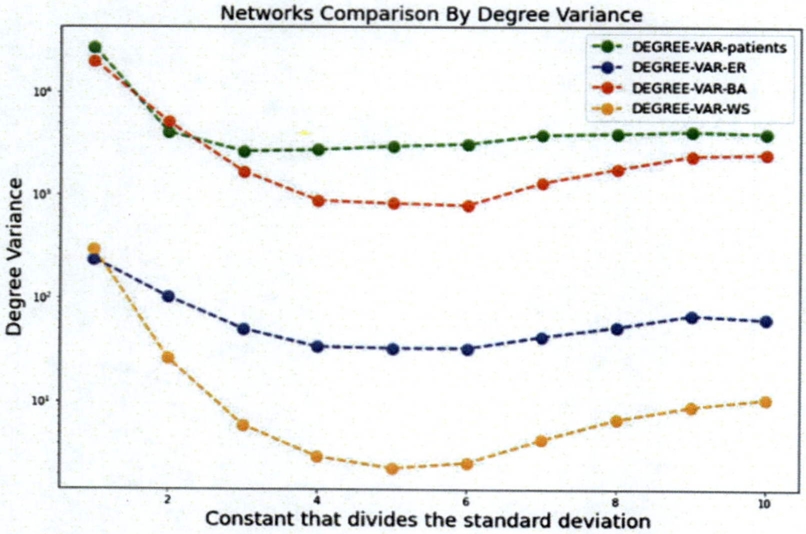

Fig. 4 Degree variance comparison of the networks. The variance value axis is on a logarithmic scale

the values remain comparable to those of the BA model, such stability indicates an effect of degree distribution homogenization, for a greater similarity constraint between laboratory test values.

Summarizing, in the comparison with the classical network models performed in this work, we verified no Small World characteristics, due to the high APL compared to similar WS networks. Despite this, the CC values of the patient's instances are closer to the WS network model, or a Lattice network, than the other network models discussed.

Another hypothesis about the patients' network structure can be inferred from the variance of the degree and CC. As the network has instances with similar degree variance to BA networks (typically with high degree variance), it is possible to deduce the presence of hubs of nodes in the network, that is, patients highly connected to others. Nodes with lower degree are also further away from the hub nodes, despite having strong clustering with their own neighbors. Therefore, it is possible to suggest an assortative structure for the patient network. There is a large number of patients with a high degree of exams similarity, while patients with a low degree of similarity are further away from the hubs.

Table 3 Modularity values for patients' networks with $c = 4$ obtained from five instances by the Fast Greedy algorithm. Instances were selected at random

♯ Vertices	♯ Groups	Modularity
928	2	0.32494
917	5	0.35391
927	4	0.31119
927	3	0.36873
911	4	0.28167

3.2 *Modularity Analysis*

Table 3 shows some of the results obtained in the modularity analysis of network instances with c equal to 4. This value of c was chosen empirically to ensure the generated networks would have sparsity, as well as connectivity between nodes, thus showing characteristics of community formation.

The Fast Greedy algorithm presented different recommendations of numbers of clusters. On average, it recommended four clusters with a standard deviation equal to 1.27. And, in general, the modularity obtained by the Fast Greedy algorithm was greater than zero. Modularity values greater than zero indicate that the networks have some kind of community structure. In a future work, we plan to provide statistics of these groups to a domain specialist in order to validate the clustering structures found.

4 Conclusion

This work modelled a dataset composed of data from 4320 COVID-19 patients from a large hospital trough network structures. Each vertex of the networks represented a patient and pairs of patients are connected when their attribute values (age and laboratory test exams) are similar for at least six attributes. Each network was instantiated 100 times, enabling probabilistic analysis of APL, CC, degree variance and average degree distribution. The instantiated networks were compared with the classic networks ER, WS and BA. In addition, a community detection algorithm guided by modularity was employed to find clusters of patients in the network.

In the comparison with classic networks, the patients network did not present Small World characteristics due to the higher APL, despite having a high CC (like a lattice or WS network). The high level of CC and the degree variance stability denoted an homogeneity effect of the instances, related to degree distribution and the tendency of similar patients to be very connected with each other. The homogeneity of the networks is an indication of assortativity characteristic. That is, there is a concentration of patients with a high degree of similarity, while patients with a low degree of similarity are further away from the hub nodes. The assortative behavior of the network may suggest hypotheses such as: "are nodes far from the hubs linked to the severity of the disease?" or "do communities with low degree nodes represent rare yet similar cases while communities with high degree nodes the common ones?" These hypotheses will be addressed in future studies. The presence of hubs is a typical characteristic of BA networks. In terms of modularity, the obtainment of cluster structures with modularity values greater than zero indicate that the networks have some community structure.

As future work we plan to use other network construction methods that allow better analysis of variations in network density in the original space, as is the case with the k-nearest neighbor graph. Furthermore, we plan to apply such analysis to data from other health centers and compare their network structures. Other community detection algorithms can also be employed in the modularity analysis. And a domain expert will be involved in analyzing and validating the clustering results.

Acknowledgments This study was financed in part by the Coordenação de Aperfeiçoamento de Pessoal de Nível Superior - Brasil (CAPES) - Finance Code 001 and the project "Data Science fighting outbreaks, epidemics and pandemics in hospitals" (grant 88887.507037/2020-00). The authors also thank the Brazilian research agency FAPESP for providing data under the FAPESP COVID-19 Data Sharing/BR.

References

1. Guan, W.J., et al.: Clinical characteristics of coronavirus disease 2019 in China. N. Engl. J. Med. **382**, 1708 (2020)
2. Worldometers: Covid-19 coronavirus pandemic. https://www.worldometers.info/coronavirus/ (March 2022)
3. Sibemberg, L.S., et al.: Classificação de risco em redes complexas: o caso da covid-19 no rio grande do sul. Ciência e Natura, v. 43 (2021)
4. Sohrabi, C., et al.: World health organization declares global emergency: A review of the 2019 novel coronavirus (covid-19). Int. J. Surg. **76**, 71 (2020)
5. Alimadadi, A., et al.: Artificial intelligence and machine learning to fight covid-19. Physiol. Genom. **52**, 200–202 (2020)
6. Lalmuanawma, S., et al.: Applications of machine learning and artificial intelligence for covid-19 (sars-cov-2) pandemic: A review. Chaos Solitons Fractals **139**(2020), 110059 (2020)
7. Nabi, K.N.: Forecasting covid-19 pandemic: A data-driven analysis. Chaos Solitons Fractals **139**, 11004 (2020)
8. Randhawa, G.S., et al.: Machine learning using intrinsic genomic signatures for rapid classification of novel pathogens: Covid-19 case study. PLoS One **15**, 4 (2020)
9. Rodriguez-Morales, A.J., et al.: Clinical, laboratory and imaging features of covid-19: A systematic review and meta-analysis. Travel Med. Infect. Dis. **34**, 101623 (2020)
10. Tuli, S., et al.: Predicting the growth and trend of covid-19 pandemic using machine learning and cloud computing. Internet Things **11**, 100222 (2020)
11. Bertsimas, D., et al.: An aggregated dataset of clinical outcomes for covid-19 patients. https://www.covidanalytics.io (November 2021)
12. Mello, L.E., et al.: Opening Brazilian covid-19 patient data to support world research on pandemics. Zenodo (2020)
13. FAPESP: FAPESP COVID-19 DataSharing/BR. https://repositoriodatasharingfapesp.uspdigital.usp.br (November 2021)
14. Metz, J., et al.: Redes complexas: conceitos e aplicações (2007)

15. Ronqui, J.R.F.: Estudo e comparação da topologia de redes de interação de proteínas. Tese (Doutorado em Ciências) - Instituto de Física de São Carlos, Universidade de São Paulo, São Carlos (2018)
16. Demertzis, K., Tsiotas, D., Magafas, L.: Modeling and forecasting the covid-19 temporal spread in Greece: An exploratory approach based on complex network defined splines. Int. J. Environ. Res. Public Health **17**(13), 4693 (2020)
17. Tsiotas, D., Magafas, L.: The effect of anti-covid-19 policies on the evolution of the disease: A complex network analysis of the successful case of Greece. Physics **2**(2), 325–339 (2020)
18. Silva, C.J., Cantin, G., Cruz, C., Fonseca-Pinto, R., Passadouro, R., Dos Santos, E.S., Torres, D.F.: Complex network model for covid-19: human behavior, pseudo-periodic solutions and multiple epidemic waves. J. Math. Anal. Appl. **514**, 125171 (2021)
19. Vega-Redondo, F.: Complex Social Networks. Cambridge University Press, Cambridge UK (2007)
20. Newman, M.E.: Modularity and community structure in networks. Proc. Natl. Acad. Sci. **103**(23), 8577–8582 (2006)
21. Lopes, F.L., Lorena, A.C.: Filipeloyola/laboratory-tests-preprocessing: Initial Release (Mar 2022). https://doi.org/10.5281/zenodo.6392307
22. Bedia, E.C.: Conectividade do grafo aleatório de Erdös-Rényi e uma variante com conexões locais. Tese (Dissertação de mestrado ICMC/USP), Universidade de São Paulo (2021). https://repositorio.ufscar.br/handle/ufscar/7493
23. Vosgerau, R.A.: Análise da Entropia em Redes Complexas. Tese de Doutorado da Universidade Estadual de Ponta Grossa (2016)
24. Suzuki, H.T., Ribeiro, C.H.C.: Estudo empírico do fenômeno small world em redes sociais. Anais do 13o Encontro de Iniciação Científica e Pós-graduação do ITA-XIII ENCITA, São José dos Campos (2007)
25. Lancichinetti, A., Fortunato, S.: Community detection algorithms: a comparative analysis. Phys. Rev. E **80**(5), 056117 (2009)
26. Barbosa, L.M., Attux, R., Godoy, A.: Uma análise de assortatividade e similaridade para Artigos Científicos. Simpósio Brasileiro de Sistemas Colaborativos (2014)

University-Industry Collaboration from the Perspective of a Spin-off: A Case Study

Laís de Sousa Leite Nascimento, Letícia Lemos Martins, and Lillian do Nascimento Gambi

Abstract This study aims to describe the perceptions and perspectives regarding the University-Industry collaboration of a spin-off developed by the technology park of a Federal University in Brazil. Initially, a literature review was conducted to identify facilitating aspects and barriers in UIC, and then a case study was conducted. This way, by means of an interview supported by a semistructured questionnaire, the facilitating aspects and barriers that they observe and experience during this process of development, collaboration, and technology transfer were addressed, as well as the positive and negative points of this process. In addition, the spin-offs perspectives related to the future of the contributions between universities and firms were analyzed. As a result of the case study, the need to further study and deepen the perspective of a new developing company, with the purpose of creating ways to eliminate obstacles so that the collaboration process occurs in a more fluid manner, thus providing greater technological development between the parties.

Keywords University-industry collaboration · Technology transfer · Innovation · Spin-off

1 Introduction

Collaboration between universities and industries (UIC) has become increasingly significant in recent years and scientists study facilitating factors and barriers that determine the success and failure of such collaborations [1]. This collaboration occurs through technology transfer (TT) between these two organizations, which happens through the creation of patents and contracts, by the knowledge of academics applied directly in the areas of research and development in industry, research projects, publications, among other forms [2]. Moreover, according to [3],

L. de Sousa Leite Nascimento (✉) · L. L. Martins · L. do Nascimento Gambi
Federal University of Viçosa, Campus Rio Paranaíba, Rio Paranaíba, Brazil
e-mail: lais.leite@ufv.br

V. M. López Sánchez et al. (eds.), *Industrial Engineering and Operations Management*, Springer Proceedings in Mathematics & Statistics 400,
https://doi.org/10.1007/978-3-031-14763-0_27

improving the degree of UIC has become a key policy objective of much of the world economy.

In this manner, some differences between developed and emerging countries can be highlighted in relation to this policy of the development of UIC. In relation to emerging countries, the exchange consists mostly of literature in the form of publications and reports [4, 5], and when focusing on the scope of entrepreneurship at the university, there are business incubator services and entrepreneurial education, in addition to internships for undergraduate/postgraduate students in companies that need a specific vision. On the other hand, what occurs in developed countries is private participation in graduate programs, long-term research partnerships, research consortia, entrepreneurship education, patent licensing, and spin-off companies [6].

Nevertheless, the importance of the study of technology transfer between companies and universities is due to the gains that may occur with this exchange of information. According to [7], when there is this connection, companies profit from highly qualified professionals who begin to contribute to the development of the company. In addition, students have access to technological resources of high performance and enjoy knowledge of the business environment [8] and benefit from additional funding that can be subsidized by companies, access to industrial equipment and patents [9].

It is worth noting, however, that there are barriers that hinder the development and transfer of technologies, so this becomes a challenge in the implementation and expansion of the use of innovations in the market. Thus, according to [10], it is important to be aware of possible barriers before making decisions related to technology development to avoid unforeseen problems while conducting research. It is known that there are methods for reducing the number of barriers, however, it is necessary to develop more efficient methods that meet the demands of TT processes.

Thus, for [11], it is of utmost importance that UICs have successful management since this generates benefits for all involved. Therefore, to ensure this management, it is important that this topic be discussed and explored in many different areas. Since the collaboration between universities and industries is of substantial relevance for the development of innovations, for the dissemination of knowledge, and for favoring the economy, it is essential that the scientific literature addresses this theme in a continuous and in-depth way, given the various factors that may influence it. In view of this, this study is justified to address the facilitating and inhibiting aspects, as well as the strategies used to manage the technology transfer process by the spin-off in question.

2 Theoretical Framework

2.1 University-Industry Collaboration

Technology transfer has as a characteristic the tangibility of knowledge, which promotes changes and means for information dissemination [12]. From the analysis of a historical context, that universities are the main sources of knowledge dissemination through research, being this a fundamental institution not only for the development of numerous industries, consolidation of technologies and construction of technical know-how but also of an entire society and improvements of its welfare [13].

In addition to universities, several other institutes collaborate for the creation and dissemination of technology, since this enables the significant growth of sectors relevant to society, such as hospitals, public sector entities, and industry [14]. Thus, there is a strong relationship between investment and support to research institutes and the competitiveness of a country, i.e., a developed country certainly invests a lot in research [15]. Other aspects that influence TT, according to [16], are the pretension of greater productivity and competitive advantage seeking survivability in the market.

According to [17], for TT to occur there must be some similarity between the technologies of the institutions, while [18] require a relationship between geographic and temporary cognitive proximity, the latter being a coparticipated knowledge base [19]. Other factors that influence this process are the culture of the institution, entrepreneurship, the environment and its respective infrastructure and many other organizational issues [20]. It is possible to list several sectors that can benefit from technology transfer, especially when the knowledge used is multidisciplinary [1].

Thus, it is possible to transform this exchange into social and economic transformation, either for the creators of a given innovation or for those who enjoy it [21]. Thus, it can be concluded that the incentive to science and research is an essential public policy for the social cohesion and construction of a more just state, thus ensuring a positive and relevant impact on the growth and development of a population [22]. In addition, it is considered a strategic factor providing innovation, economic strengthening and competitiveness [23].

Despite the various advantages offered in this type of knowledge exchange, technology transfer has some barriers that hinder the effective operation of this exchange. According to [24] knowledge transfer can be unsuccessful if the participants in the process do not collaborate or are reluctant about the changes involved. When looking from the perspective of the link between the university and the company, it is possible to analyze the presence of people with very different mentalities, making clear the difference between the characteristics of different generations and the context of the environment in which they operate, being necessary to break cultural paradigms to execute an exchange project with quality [25, 26].

2.2 Barrier Aspects and Facilitators in UIC

The TT process between U-I involves several points that must be analyzed with the aim of understanding the complexities and singularities of this exchange. For [27], the main points of analysis and attention in relation to this partnership are the distinction of the nature of the institutions; the culture and the specific objective of each sector; the lack of an entrepreneurial vision at certain times, both on the part of students and entrepreneurs; the conflicts generated by the relationship of two groups with different convictions; the results; the knowledge about legislation regulating these partnerships and, finally, the concern about the management of the process.

From the point of view of scholars in the area regarding the barriers involved in TT, several classifications have been proposed as shown in Table 1.

It is worth noting that in the classification made by [30], the author divided the barriers into twenty-six categories; thus, in the previous table, only the main classifications were addressed. Moreover, another consideration is that [10] emphasize that the classification of barriers elaborated by the authors can be observed at different levels: strategic, tactical, and operational.

When dealing with the point of view of Technological Innovation Centers (TICs), the classification of barriers to TT between universities and industries undergoes changes. Thus, in a study conducted by [33], the classifications of four different cores located in the state of São Paulo were analyzed, in which the authors were able to identify several barriers of different natures. However, there were also some points of congruence that allowed for an effective classification, these being legal, governmental political, relational, economic, human resources, operational, technological, and marketing.

Many authors address facilitating aspects for U-I TT, and they can also be categorized. Thus, according to the literature, access to adequate infrastructure [34], qualified human resources [7], and technical equipment [35] play a crucial role in

Table 1 Classification of barriers in UIC according to different authors

Year	Author(s)	Classification
1972	Creighton [28]	Formal (procedural) and informal (behavioral).
1974	Jervis and Sinclair [29]	Political barriers and institutional barriers
1974	Mock [30]	Financial, competence, communication, and market-related barriers.
1983	Sharif [31]	Organization-ware, information-ware, technique-ware and human-ware.
2011	Mojaveri, Nosratabadi, and Farzad [32]	Technical, behavioral, cultural and market barriers.
2017	Mazurkiewicz and Poteralska [10]	Technical barriers, organizational-economic barriers, and system barriers to technology transfer.

collaborations. The use of appropriate language within the field of work [36], good relationships [37], constant feedback [11] and mutual trust between the institutions [38].

Additionally, other authors show different relevant factors to facilitate this process, especially in regard to the organization of the partnership and aspects concerning the environment and external influences, such as project management [39], the formal job description to differentiate the performance of each team within the collaboration [40], government policies, such as tax incentives [41], government networks [42] or public funding [43, 44], in addition to geographical distance, since this can improve access to adequate infrastructure and qualified human resources [7].

2.3 *TT and UIC in Brazil*

According to [6] universities currently have three missions: the first two classic missions related to teaching and research and the third and most recent missions related to the commercialization of academic knowledge. This exchange can occur through technology transfer offices, technology/science parks, incubators and patents; thus, such modes help and corroborate the existence of UIC [45, 46]. Another point to be raised is that the literature highlights a difference in incentives by the government between developed and developing countries when dealing with UIC [47].

In developed countries such as Malaysia, a country that belongs to the High Performing Asian Economies [48], the government encourages universities to generate strategic collaborations for the development and commercialization of such innovations and technologies through the UIC to promote positive implications for the national economy [49]. An example of such an incentive is the proposed transformation of the University of Malaysia, which is intended to delegate education and research for resource and profit generation [50]. Similar incentives occur in countries such as the United States, Japan and China and other major world powers [51, 52].

However, from a view related to emerging countries, which are in transition of income levels, the same incentive and search for UIC declines [53]. An example is Pakistan, which according to [54], has hampered technology transfer between U-Is due to a lack of coordination between the government, industries, and universities. According to the same authors, in emerging country situations, the success of collaborative relationships is subject to the quality and type of institutions involved and the proximity between them. Another example is Africa, which according to [55], African universities in particular lack the competence to engage more frequently with business due to research and innovation with low qualification and sophistication.

When narrowing this scope and focusing on Brazil's innovation base, it is known that it is a small base, which counts on a few research centers classified as excellent and these centers are mostly located in the southern region of the country [56]. Furthermore, it is worth noting that Brazil is considered a country that comprises companies with limitations in terms of innovation capacity [57–59], focusing on meeting the demands of low and medium technology industries [41]. Thus, the role of universities as promoters becomes even more relevant.

As seen earlier, technology transfer between U-I can take place through the patenting of technologies, and according to the Resident Depositor Ranking – 2020, the public and federal universities in Brazil accounted for a total of 71.44% of invention patents filed that year [60]. This reveals that despite low investments in recent years in public higher education, public universities remain ahead of other institutions in relation to promoting education and innovations [61, 62]. In 2017, only 516 companies benefited from funding for Research and Development and Technological Innovation projects in partnerships with universities or research institutes [63].

On the website of the Government of Brazil, there is a place where it is possible to make the request for partnerships, that is, collaborations with federal universities in the country, promoting assistance in the formalization of collaborations between educational institutions and companies, directed to innovation, development of processes and/or products, technology transfer and others [64]. According to [65], Brazil has medium-high technological competencies in certain industries, such as development in agriculture and the study of biomass (ethanol) and aeronautics, however, the industries of electronic equipment, or information technology instruments and chemicals or pharmaceuticals still need to break through innovation barriers.

3 Methods

The case study was conducted through a semistructured interview with a spin-off that originated from a Federal University located in the state of Minas Gerais. The spin-off originated from a doctoral thesis project related to the development of recombinant proteins. Through this research, outside companies started to look for the doctoral student, and this is how researcher partner saw the opportunity to enter this market. Thus, they transformed the spin-off into a startup that had its maturation process through the incubation processes of the technology park of this same university. Thus, the present work refers to the deepening of the perspectives of this startup in relation to the processes of technology transfer and collaboration with the university.

Having said this, the work was divided into stages, namely the deepening of the content through academic literature, the development of a semistructured questionnaire, the contact with the company, the interview, the analysis of the results, and future projections about the theme. As far as the knowledge acquired by the literature is concerned, it can be seen in the theoretical framework. The development of the semistructured questionnaire was based on the authors [66, 67], and thus, by means of their elaboration, it was possible to conduct the interview with the company. The company was determined by means of contacts provided by the technology park and filtered based on its level of evolution within the incubation process. Finally, the results obtained in the interview were described in the Results section and subsequently analyzed, and suggestions were made for future research to be conducted in relation to the perspective of spin-offs and startups.

4 Results

To understand the relationship between the Federal University (FU) and the company under study, it is necessary to initially understand the bond that the two institutions have. Thus, among the classifications of types of collaborations proposed by [67], the company reported three types of contracts with the university: technology licensing (which refers to the codevelopment of the product); the partnership agreement; and the permission to use a laboratory for research and testing. It is also known that there are several factors that influence technology transfer, such as geographical distance, government support, and effective communication, among other factors published by [11]. Thus, for didactic purposes, the findings were divided into barriers and facilitators of this process, as well as the positive and negative aspects from the perspective of the company.

When dealing with the barriers in the TT process, the company reported relevant points in relation to the performance of the most diverse entities along the university. Thus, among the various challenges faced in this process, explained by [2], those that refer to the document processing time within some instances, bureaucratization, and the lack of innovation policies stand out. In this way, they explained that the Technological Innovation Center (TIC) involved plays a very efficient and solid role regarding the necessary procedures. However, it is necessary the participation of other instances, do not have such an agile methodology or such a refined knowledge regarding innovation as for TIC, demanding more time for the processing and release of documents. The company reports that the university in question, as well as the entire process of technological innovation in the country, is in a process of maturation, there are public universities that stand out and others that are on the way to progress.

Another point is the issue of bureaucratization, that even though they are time consuming processes, the company understands and emphasizes the need for these, and states that they could be improved and streamlined, allied to the thought of [68], which states that any university regulation that leads to excessive bureaucracy or that hinders the interaction between researchers and companies should be removed. It also has a major challenge raised by [2], regarding the internal management challenges of the collaboration processes within universities. That can be confirmed by the author [69] who portrays that a good part of the universities act in an amateurish way in the collaboration process.

In research conducted by [14] and by issues addressed by the company as favorable aspects to TT, the cost reduction was raised, since by being associated with a university, all the paperwork and access to resources are facilitated; for example, the price of rent is lower and laboratory facilities are more accessible, given that startups/spin-offs are seeking to generate their prototypes and test their products in the market, not counting, most of the time, with a high capital for the construction of these facilities. Moreover, the company believes that among the relevant factors in this process are the access to technology developed by the university, contact with highly renowned researchers, cost reduction, and access to inputs made available by the FU. Furthermore, it highlights the importance of this process for the strengthening of technology, the generation of national products, which consequently reach the consumer at a more viable price, the generation principle by the competition itself, the strengthening of science and technology in the country, and the opening of new markets to encourage the trade of technologies other than commodities.

In relation to the existing negative points in this TT process, they portray that there are no negative points, because all involved come out benefited from this collaboration. However, they again emphasized the barrier issues in this process and issues that can be improved in this process. Such as a clear and defined innovation policy for all actors, both the student institution and the company, as well as ways to speed up the processes and contracts.

Finally, conniving with [14] considerations about the importance of the topic under discussion for the progress and performance of a state, the company believes that the future of innovation lies in the interaction between university-industry since the main research environment of a country are the public universities, and that this should be increasingly facilitated and deepened. Another aspect is in relation to master's and PhD professionals, who spend years developing their research and have the opportunity to insert it into the market, being able to impact others with their contributions. Thus, he considers that public universities already create professionals of excellence it is now up to the market to strengthen itself to receive these professionals and make the innovation and technology area in the country become a reference.

5 Conclusion

As for the knowledge acquired throughout this research and the understanding of the spin-off's perspectives and visions, it was possible to understand the relevance of studying technology transfer by means of collaboration between universities and industries to propose improvements and ways to optimize this exchange of knowledge. Thus, it is expected to contribute to the development of new technologies and the improvement of existing technologies, with the purpose of making the country less dependent on external technologies, impacting the economy, research, scientific knowledge generation, and the expansion of the economic matrix.

Thus, for future studies, it is suggested that the researchers delve into issues related to the internal processes of the university for the signing of agreements, documentation and the like, to map the possible bottlenecks and, thus, propose specific improvements for each situation to streamline the demands. For public universities, it is suggested the creation of a training manual so that, as far as possible, there is a standardization of the sectors that are involved in bureaucratic processes to accelerate administrative measures and avoid this being an obstacle or hindrance in the achievement of partnerships between educational institutions and business organizations.

References

1. Falleiros, V.B.: Transferência de tecnologia do meio acadêmico para o setor produtivo: uma abordagem funcional (Dissertation). Universidade de São Paulo (2008)
2. Fernandes, R.F., Antenor, M.C., Andrade, J.S., Barros Filho, M.M.L., Araújo, A.: Práticas de transferência de tecnologia: uma análise multicasos. Cadernos de Prospecção. **11**(5), 1342 (2018)
3. Banal-Estañol, A., Jofre-Bonet, M., Lawson, C.: The double-edged sword of industry collaboration: evidence from engineering academics in the UK. Res. Policy. **44**(6), 1160–1175 (2015)
4. Mgonja, C.: Enhancing the university-industry collaboration in developing countries through best practices. Int. J. Eng. Trends Technol. **50**(4), 216–225 (2017)
5. Sá, C.M.: Perspective of industry's engagement with African universities. Draft report for the Association of African Universities (2015)
6. Perkmann, M., Walsh, K.: University–industry relationships and open innovation: towards a research agenda. Int. J. Manag. Rev. **9**(4), 259–280 (2007)
7. Myoken, Y.: The role of geographical proximity in university and industry collaboration: case study of Japanese companies in the UK. Int. J. Technol. Transf. Commer. **12**(1–3), 43–61 (2013)
8. Ankrah, S., Omar, A.T.: Universities–industry collaboration: a systematic review. Scand. J. Manag. **31**(3), 387–408 (2015)
9. Pereira, R.D.A.B.: Gestão de benefícios em programas de projetos de I&D em colaboração Universidade-Indústria (Doctoral dissertation) (2016)
10. Mazurkiewicz, A., Poteralska, B.: Technology transfer barriers and challenges faced by R&D organisations. Procedia Eng. **182**, 457–465 (2017)
11. Rybnicek, R., Königsgruber, R.: What makes industry–university collaboration succeed? A systematic review of the literature. J. Bus. Econ. **89**(2), 221–250 (2019)
12. Gopalakrishnan, S., Santoro, M.D.: Distinguishing between knowledge transfer and technology transfer activities: the role of key organizational factors. IEEE Trans. Eng. Manag. **51**(1), 57–69 (2004)
13. Hou, B., Hong, J., Chen, Q., Shi, X., Zhou, Y.: Do academia-industry R&D collaborations necessarily facilitate industrial innovation in China? The role of technology transfer institutions. Eur. J. Innov. Manag. **22**, 717–746 (2019)
14. Marques, J.L.: Desempenho dos Núcleos de Inovação Tecnológica do Brasil no período de 2006 a 2016: da implementação à transferência de tecnologia (Master's thesis). Universidade Federal de Pernambuco (2018)
15. Terán-Bustamante, A., Colla-De-Robertis, E.: Vinculando el talento de investigadores y emprendedores para la innovación. Revista mexicana de economía y finanzas. **13**(4), 547–569 (2018)
16. Zammar, G.: Interação universidade-indústria: um modelo para transferência de tecnologia (2017)

17. Woerter, M.: Technology proximity between firms and universities and technology transfer. J. Technol. Transf. **37**(6), 828–866 (2012)
18. Mathisen, L., Jørgensen, E.J.: The significance of knowledge readiness for co creation in university industry collaborations. Innovations. **23**(4), 534–551 (2021)
19. Motta, K., Luna, M., Santos, J., Romero, F.: Uma revisão da literatura sobre transferência de tecnologia. HOLOS. **8**, 255–272 (2016)
20. Aldridge, T.T., Audretsch, D.: The Bayh-Dole act and scientist entrepreneurship. In: Universities and the entrepreneurial ecosystem. Edward Elgar Publishing, Cheltenham (2017)
21. Mendes, C.I.C.: Transferência de tecnologia da Embrapa: rumo à inovação. Embrapa Informática Agropecuária-Tese/dissertação (ALICE) (2015)
22. Bendul, J., Rosca, E., Hoffmann, T.: Sustainable technology transfer for poverty alleviation: a unified framework for challenges and transdisciplinary solution approaches. WIT Trans. Ecol. Environ. **193**, 823–834 (2015)
23. Rast, S., Khabiri, N., Senin, A.A.: Evaluation framework for assessing university industry collaborative research and technological initiative. Procedia Soc. Behav. Sci. **40**, 410–416 (2012)
24. Pasaribu, B.I., Afrianti, A., Gumilar, G.G., Rizanti, H.P., Rohajawati, S.: Knowledge transfer: a conceptual model and facilitating feature in start-up business. Procedia Comput. Sci. **116**, 259–266 (2017)
25. Rikap, C.: La vinculación de la universidad con el sector productivo. Transferencia tecnológica: technological transference. Ecos de Economía. **16**(34), 127–149 (2012)
26. Dutrénit, G., Núñez, J.: Vinculación universidad-sector productivo para fortalecer los sistemas nacionales de innovación: experiencias de Cuba, México y Costa Rica. Editorial UH 432, La Habana (2017)
27. da Silva, C.V.: Processo de transferência de conhecimento na interação universidade-empresa: programas de incubação do distrito federal (2010)
28. Creighton, J.W.: Enhancement of Research and Development Output Utilization Efficiencies. Linker Concept Methodology in the Technology Transfer Process. Defense Technical Information Center, Ft. Belvoir (1972)
29. Jervis, P., Sinclair, T.C.: Conditons for successfull technology transfer and innovation in the U.K. In: Davidson, H.F., et al. (eds.) Proceedings of the NATO Advanced Study of Institute on Technology Transfer. Nordhoff International Publishing, The Netherland (1974)
30. Mock, J.E.: Barriers and stimulants to the transfer of public technology, technology transfer. In: Davidson, H.F., et al. (eds.) Proceedings of the NATO Advanced Study of Institute on Technology Transfer. Nordhoff International Publishing, The Netherland (1974)
31. Sharif, M.N.: Management of Technology Transfer and Development. UNESCAP: Regional Centre for Technology Transfer, Bangkok (1983)
32. Mojaveri, H.S., Nosratabadi, H.E., Farzad, H.: A new model for overcoming technology transfer barriers in Iranian health system. Int. J. Trade Econ. Financ. **2**(4), 280 (2011)
33. Pakes, P.R., Borrás, M.Á.A., Torkomian, A.L.V., Gomes, A.O., Silva, B.B.: A percepção dos núcleos de inovação tecnológica do estado de São Paulo quanto as barreiras à transferência de tecnologia universidade-empresa. Tecno-Lógica. **22**(2), 120–127 (2018)
34. Boardman, C., Bozeman, B.: Academic faculty as intellectual property in university industry research alliances. Econ. Innov. New Technol. **24**(5), 403–420 (2015)
35. Arvanitis, S., Kubli, U., Woerter, M.: University-industry knowledge and technology transfer in Switzerland: what university scientists think about co-operation with private enterprises. Res. Policy. **37**(10), 1865–1883 (2008)
36. Baba, Y., Yarime, M., Shichijo, N.: Sources of success in advanced materials innovation: the role of "core researchers" in university–industry collaboration in Japan. Int. J. Innov. Manag. **14**(02), 201–219 (2010)
37. Collier, A., Gray, B.J., Ahn, M.J.: Enablers and barriers to university and high technology SME partnerships. Small Enterp. Res. **18**(1), 2–18 (2011)
38. Attia, A.M.: National innovation systems in developing countries: barriers to university–industry collaboration in Egypt. Int. J. Technol. Manag. Sustain. Dev. **14**(2), 113–124 (2015)

39. Rajalo, S., Vadi, M.: University-industry innovation collaboration: reconceptualization. Technovation. **62**, 42–54 (2017)
40. Canhoto, A.I., Quinton, S., Jackson, P., Dibb, S.: The co-production of value in digital, university–industry R&D collaborative projects. Ind. Mark. Manag. **56**, 86–96 (2016)
41. Freitas, I.M.B., Marques, R.A., Silva, E.M.D.P.: University–industry collaboration and innovation in emergent and mature industries in new industrialized countries. Res. Policy. **42**(2), 443–453 (2013)
42. Rampersad, G.C.: Developing university-business cooperation through work integrated learning. Int. J. Technol. Manag. **68**(3–4), 203–227 (2015)
43. Flores, M., Boër, C., Huber, C., Plüss, A., Schoch, R., Pouly, M.: Universities as key enablers to develop new collaborative environments for innovation: successful experiences from Switzerland and India. Int. J. Prod. Res. **47**(17), 4935–4953 (2009)
44. Piva, E., Rossi-Lamastra, C.: Systems of indicators to evaluate the performance of university-industry alliances: a review of the literature and directions for future research. Meas. Bus. Excell. **17**, 40–54 (2013)
45. Marhl, M., Pausits, A.: Third mission indicators for new ranking methodologies. Eval Higher Educ. **5**(1), 43–64 (2011)
46. Perkmann, M., Tartari, V., McKelvey, M., Autio, E., Broström, A., D'este, P., et al.: Academic engagement and commercialisation: a review of the literature on university–industry relations. Res. Policy. **42**(2), 423–442 (2013)
47. Terán-Bustamante, A., Martínez-Velasco, A., López-Fernández, A.M.: University– industry collaboration: a sustainable technology transfer model. Adm Sci. **11**(4), 142 (2021)
48. Page, J.M.: The East Asian miracle: an introduction. World Dev. **22**(4), 615–625 (1994)
49. Salleh, M.S., Omar, M.Z.: University-industry collaboration models in Malaysia. Procedia Soc. Behav. Sci. **102**, 654–664 (2013)
50. Yaakub, N.I., Hussain, W.M.H.W., Rahman, M.N.A., Zainol, Z.A., Mujani, W.K., Jamsari, E.A., et al.: Challenges for commercialization of university research for agricultural based invention. World Appl. Sci. J. **12**(2), 132–138 (2011)
51. Hanel, P., St-Pierre, M.: Industry–university collaboration by Canadian manufacturing firms. J. Technol. Transf. **31**(4), 485–499 (2006)
52. Perkmann, M., Neely, A., Walsh, K.: How should firms evaluate success in university–industry alliances? A performance measurement system. R&D Manag. **41**(2), 202–216 (2011)
53. Stojčić, N.: Collaborative innovation in emerging innovation systems: evidence from Central and Eastern Europe. J. Technol. Transf. **46**(2), 531–562 (2021)
54. Malik, K., Bashir, T., Ali, T.M.: University-industry collaborations in Pakistan: current challenges and future opportunities. Foresight. **23**, 496–508 (2021)
55. Adegbile, A.S., Sarpong, D., Cao, D.: Industry–university collaborations in emerging economies: a legitimacy perspective. IEEE Trans. Eng. Manag., 1–13 (2021)
56. Garcia, R., Araujo, V., Mascarini, S., Gomes Santos, E., Costa, A.: Looking at both sides: how specific characteristics of academic research groups and firms affect the geographical distance of university–industry linkages. Reg. Stud. Reg. Sci. **2**(1), 518–534 (2015)
57. Mazzoleni, R., Nelson, R.R.: The roles of research at universities and public labs in economic catch-up (No. 2006/01). LEM Working Paper Series (2005)
58. Ryan, M.P.: Patent incentives, technology markets, and public–private bio-medical innovation networks in Brazil. World Dev. **38**(8), 1082–1093 (2010)
59. Fischer, B.B., Schaeffer, P.R., Vonortas, N.S., Queiroz, S.: Quality comes first: university-industry collaboration as a source of academic entrepreneurship in a developing country. J. Technol. Transf. **43**(2), 263–284 (2018)
60. INPI: https://www.gov.br/inpi/pt-br/central-de-conteudo/estatisticas-e-estudos economicos/arquivos/estatisticas-preliminares/rankdepositantesresidentes-2020.pdf. Last accessed 2022/03/07
61. PROIFES: https://www.proifes.org.br/infografico-interativo-impacto-dos-cortes-deinvestimentos-nas-universidades-federais/. Last accessed 2022/03/15

62. RUF: https://ruf.folha.uol.com.br/2019/ranking-de-universidades/pesquisa/. Last accessed 2022/03/07
63. IBGE: https://www.ibge.gov.br/estatisticas/multidominio/ciencia-tecnologia-einovacao/9141-pesquisa-de-inovacao.html?=&t=destaques. Last accessed 2022/03/12
64. Governo do Brasil: https://www.gov.br/pt-br/servicos/solicitar-auxilio-naformalizacao-de-parcerias-entre-a-ufv-e-empresas. Last accessed 2022/03/12
65. Dahlman, C.J., Frischtak, C.: Os Desafios para o Brasil da Economia do Conhecimento: educação e inovação em um mundo crescentemente competitivo. Reforma Política e Economia do Conhecimento: dois projetos nacionais. José Olympio, Rio de Janeiro (2005)
66. Siegel, D.S., Waldman, D., Link, A.: Assessing the impact of organizational practices on the relative productivity of university technology transfer offices: an exploratory study. Res. Policy. **32**(1), 27–48 (2003)
67. Bekkers, R., Freitas, I.M.B.: Analysing knowledge transfer channels between universities and industry: to what degree do sectors also matter? Res. Policy. **37**(10), 1837–1853 (2008)
68. Guimón, J.: Promoting university-industry collaboration in developing countries. World Bank. **3**, 12–48 (2013)
69. Reynolds, E.B., Schneider, B.R., Zylberberg, E. (eds.): Innovation in Brazil: Advancing Development in the 21st Century. Routledge, London (2019)

GOAL-OKR: A Framework for Strategic Focus Using Objectives & Key Results and Theory of Constraints

Tobias Leonardo Kunrath, Mariana Paim Machado, and Daniel Pacheco Lacerda

Abstract Business environments are becoming increasingly complex and uncertain, and this requires companies to formulate strategies to adapt to changes. In this sense, it is necessary for organizations to use tools that allow them to develop and monitor the strategy, ensuring that the organization's goal is achieved. Considering this context, this article aims to present a framework to integrate throughput accounting and the focusing process of the Theory of Constraints in the planning and monitoring of an organization's Objectives and Key Results. To develop this research, an adaptation of Design Science Research was used since the research method is oriented towards the development of artifacts. As a result, the GOAL-OKR is presented, and each step of the framework is detailed. In addition, the GOAL-OKR is evaluated in a business environment through application in a technology company and its results are presented. Finally, an overview of the framework is explained, demonstrating future opportunities and limitations of this work.

Keywords Theory of constraints · Objective and key results · Strategy development

1 Introduction

The companies are creating strategies for the dynamic market, as business environments have become more complex and uncertain [1]. These strategies need to allow organizations to adapt to changes as they occur, and this can be materialized through strategic planning, as it is a mechanism where decision alternatives are generated and considered from opportunities or threats that arise in the environment [2]. In

T. L. Kunrath (✉) · M. P. Machado · D. P. Lacerda
Research Group on Modeling for Learning – GMAP | UNISINOS, São Leopoldo, RS, Brazil

Universidade do Vale do Rio dos Sinos, São Leopoldo, RS, Brazil

V. M. López Sánchez et al. (eds.), *Industrial Engineering and Operations Management*, Springer Proceedings in Mathematics & Statistics 400,
https://doi.org/10.1007/978-3-031-14763-0_28

this sense, reassessing changes in the environment and adjustments are normal and necessary actions in the strategic management process [3].

The performance management should be combined with strategic management to ensure that strategic planning is put into practice and that it can adequately react to the context changes caused by business [3]. Performance management is responsible for meeting strategic objectives, and this is associated with the positive result for the organization, and a good result is characterized by the organization's ability to generate resources economically and use these efficient form resources [4].

One of the consequences of strategic planning is the degree of flexibility that the organization has concerning changes, and this presents itself as an important mediator between strategic planning and performance [2]. In this regard, Objective & Key Results (OKR) presents itself as an alternative to developing and monitoring the organization's strategy and dealing with continuous changes [5]. Therefore, OKR is a tool used to assist in the development of strategic planning and the execution of the strategy by unfolding business objectives and key results, contributing to the focus of the strategy [5].

However, it is necessary for the company to clearly define the organization's overall goal [6]. In this sense, the company's goal is to make money today and, in the future [6]. However, directly measuring this goal is a difficult task and for this, it is necessary to create indicators to determine whether management actions are guiding the company to the goal.

This article presents a framework to integrate the throughput accounting and the process of focusing on the Theory of Constraints in the planning and monitoring of Objective & Key Results. The purpose of the framework is to provide a structured method that (a) establishes a logical validation sequence of Objective & Key Results with throughput accounting, (b) assists in the strategic focus process of organizations, and (c) reduces the gap between the company's goal and strategy. This article has been supported by the Design Science Research method that aims to develop artifacts for real problems [7]. The framework created has been evaluated in the business environment through an application in an information technology company.

This article is organized into six sections. Next, the theoretical framework used for the development of the research will be presented. In the third section are the methodological procedures used in the research. The fourth section offers the construction of the artifact and after that, the application of the artifact is submitted. Finally, the main conclusions and opportunities for future work are made explicit.

2 Theoretical Framework

This section will present the concepts that led to the development of this study, being *Objectives and Key Results* and the Theory of Constraints.

2.1 Objectives and Key Results (OKR)

OKRs are defined as "a management methodology that helps to ensure that the company focuses efforts on important issues across the organization" [8] and "a critical thinking structure and ongoing discipline that aims to ensure that employees work together, concentrating their efforts to make measurable contributions that drive the company forward" [5]. The objectives are clear and an accurate statement that describes a qualitative goal and provides the organization with a driver for what it wants to achieve [5]. These objectives should be meaningful, concrete, action-oriented, and time-limited [8]. A didactic example related to the development of the book "Objectives and Key Results" is presented, in which the authors are creating a strategy to disseminate OKRs and one of the objectives of this strategy is "to design an attractive website that attracts people to OKRs" [5].

The Key Result is a quantitative statement that acts as a mediator between achieving the goal and the current state of the organization [5, 8]. In addition, Key Results must be measurable and verifiable, in addition to being specific, aggressive, and realistic [8]. Exemplifying the process of relating the objective with the key result, two key results for the example presented earlier are presented, being: (i) 20% of visitors should return to the site after a week and (ii) 10% of visitors should ask about our training and consulting services [5].

The process of implementing OKRs is organized in two phases: (i) planning and (ii) development [5]. Since the planning phase is responsible: (i) for ensuring the sponsorship of executives with OKR, (ii) determining where the OKRs will be implemented, and (iii) creating the implementation plan [5, 8]. The development phase is responsible for: (i) conceptually leveling employees about the tool, (ii) developing or confirming the mission, vision, and strategy of the organization, (iii) developing OKRs for the chosen environment, (iv) presenting the OKR to the company, (v) monitoring the OKRs and (vi) reporting the results [5, 8].

2.2 Theory of Constraints (TOC)

The TOC is presented as a structuring and troubleshooting method applicable to various areas of operations, such as production, projects, services, accounting, and finance [9]. TOC considers that the company is a system, composed of a set of elements that are connected and that are responsible for the overall performance of the whole, where an element can limit the ability to obtain the best result of the system [10, 11]. In addition, TOC presents five steps of the focusing process, where the goal is to achieve continuous improvement and decision support [6, 12].

The five steps of the focusing process are: (1) Identify the system restriction; (2) decide how to exploit the restriction to the maximum; (3) Make everything else subject to the previous decision; (4) Raise the system restriction; (5) If, in step 4, the restriction is eliminated, return to the first step, but do not allow inertia to become the system constraint [13]. In this sense, it is necessary to combat the increase in the efficiencies of resources and isolated sectors since the sum of local optima is not equal to global optima [12].

2.3 TOC Performance Indicators

TOC assumes that the company's goal is to make money today and, in the future, and for this must meet three basic conditions of the strategy: (i) increase in profits today and in the future, (ii) satisfy markets today and in the future and (iii) employee safety and satisfaction today and in the future [6]. For to achieve it, three global indicators to evaluate the unfolding of the target are associated: Net Income, Return on Investment, and Cash Flow [6]. Net income is an absolute measure of the result, return on investment is a relative measure obtained through the division of net income and investment [6]. Cash flow is presented as an indicator to evaluate the unfolding of the target as a function of the company's survival [6]. Furthermore, three operational indicators are presented to connect management decisions with global indicators, these operational indicators are Throughput, Inventory/Investment, and Operating Expenses of operations [12].

The throughput is understood as the rate of money generation through operational activities and is considered the first place on the scale of importance of TOC [6]. Inventory/Investment is considered all the money spent by the system to generate revenue, in addition, this indicator determines the company's ability to compete in its markets [6]. Finally, the operating expenses of the operations are all the capital spent by the system to transform inventory/investment into throughput [6].

Using the three global performance indicators it is possible to evaluate the impact of any action on the system, and for an action to be viable it should increase the throughput and/or decrease investment/inventory and operating expenditure [12]. In this sense, organizations should evaluate the decisions taken through three questions: (i) will the decision increase the throughput? If so, how much? (ii) will the decision decrease investment/inventory? If so, how much? (iii) will the decision reduce operating expenses? If so, how much? [14].

3 Method

This research has been conducted using an adaptation of the Design Science Research proposed (DSR), which is a research method oriented to the development of something new, being responsible for creating artifacts and solutions to real problems [7]. A research method was developed, presented in Fig. 1. The first step of the method is focused on identifying the problem and it aims to respond to the problem to be studied. In the case of this research, the problem to be studied arose

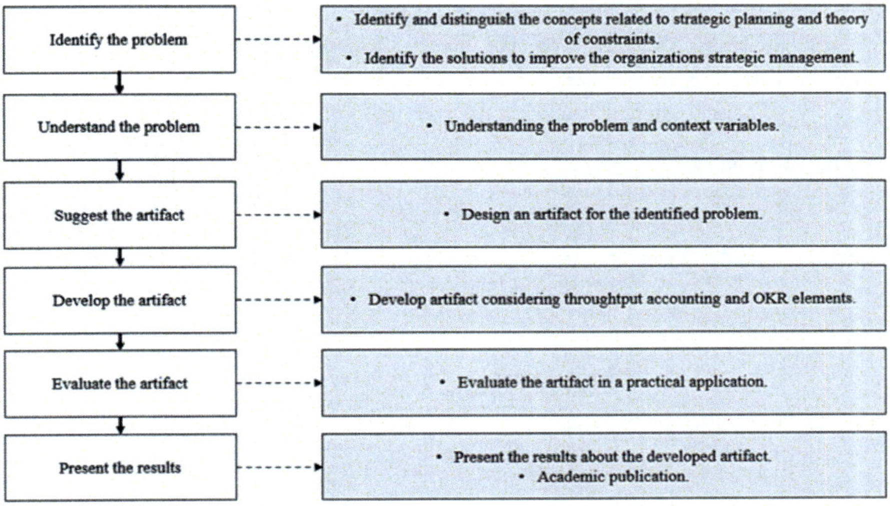

Fig. 1 Research method

from the researcher's interest in finding a practical solution for the development and managing of the company's strategy considering the assumptions of the TOC.

An artifact for the identified problem will then be suggested, considering the internal characteristics and application context. The development of the artifact will begin by considering the throughput accounting and the elements that form the OKRs. After the development, the artifact will be evaluated in a practical application, identifying contributions to strategic management. Finally, the results of the artifact developed will be communicated.

4 Framework GOAL-OKR

GOAL-OKR consists of a framework based on the five steps of the focusing process proposed by Goldratt, contemplating the logic development of Objectives & Key Results and the performance indicators proposed by TOC. This framework is represented in Fig. 2.

GOAL-OKR considers integrating throughput accounting assumptions with objectives & key results. To implement the GOAL-OKR cycle, it is considered that it is necessary to connect the OKRs vertically and horizontally in the organization. Thus, the cycle starts from the development of the company's OKRs, followed by the business units, after the teams that work in the business units, and ultimately the individuals. Each of the GOAL-OKR steps will be detailed below.

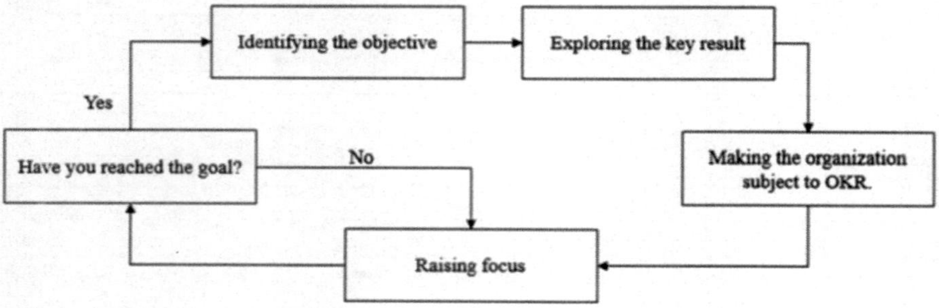

Fig. 2 Framework GOAL-OKR

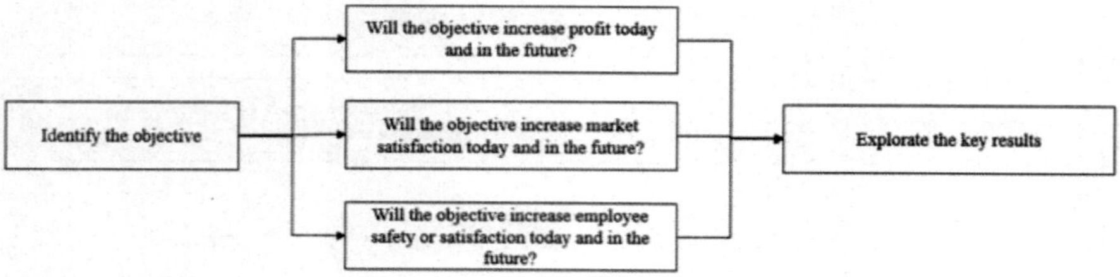

Fig. 3 Goal validation

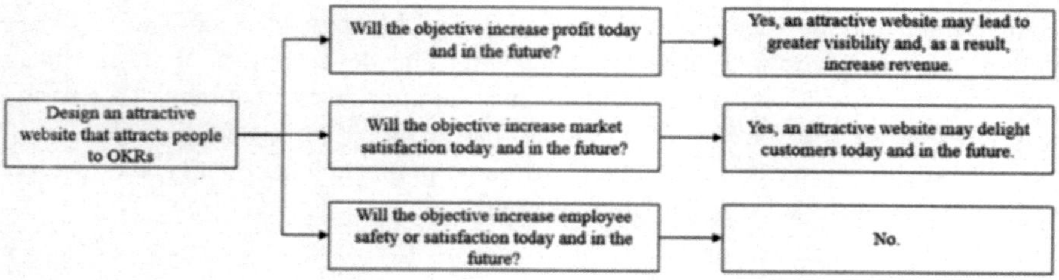

Fig. 4 Goal consistency test

4.1 Identifying the Objective

GOAL-OKR performs a consistency test of the objective suggested by the organization with the basic conditions of strategy, as shown in Fig. 3.

The consistency test questions the objective regarding the basic condition of the strategy and expects to receive at least one yes in the test responses. The Fig. 4 exemplify the goal consistency test.

In Fig. 4, the goal was submitted to the questions and outlined an answer for each of them. This thought process aims to give us clarity about the objective being proposed and mainly validate the objective with the company's goal.

4.2 Exploring the Key Result

From the validation of the objective, the key results are explored. On this point, the key result is a quantitative statement that measures the achievement of a given goal [5, 8]. However, using the operational indicators of performance Throughput, Investment/Inventory, and Operational Expenditure it is possible to evaluate the performance of any action on the system [12]. In this sense, GOAL-OKR performs a second consistency test about the key result, according to Fig. 5.

The validation of key results tests the consistency of the key result about the global indicators of the TOC: Net Income, Return on Investment, and Cash Flow. At this stage, all responses to operational indicators should be yes, as these operational indicators reflect on the overall indicators. This ensures that the key result directs the actions to the company's goal. As an example, in Fig. 6 we will use the key result described earlier "20% of visitors should return to the site after a week".

When performing the result consistency test, we found that the key result met the operational indicators of throughput increase and inventory/investment reduction, but the reduction in operating expenses does not apply in this case. Once the reduction in operating expenses, in this example, is not related to the increase or reduction of wages or rent.

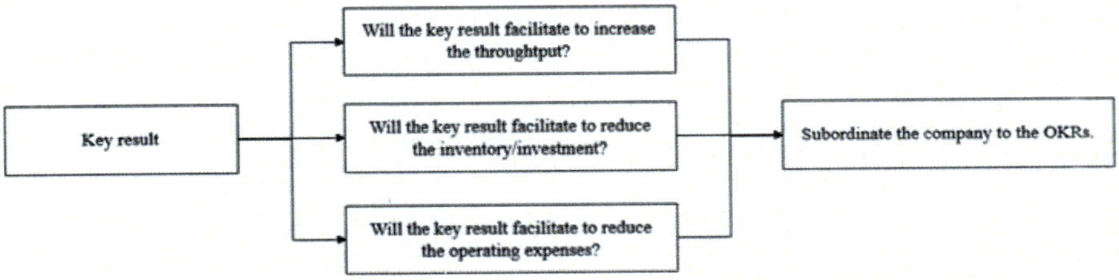

Fig. 5 Validation of key results

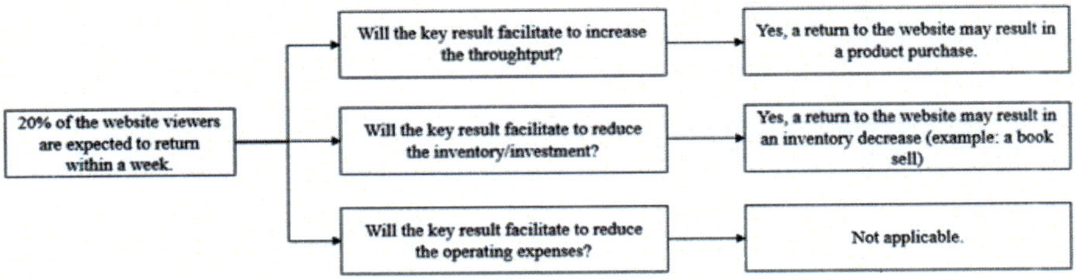

Fig. 6 Key result consistency test

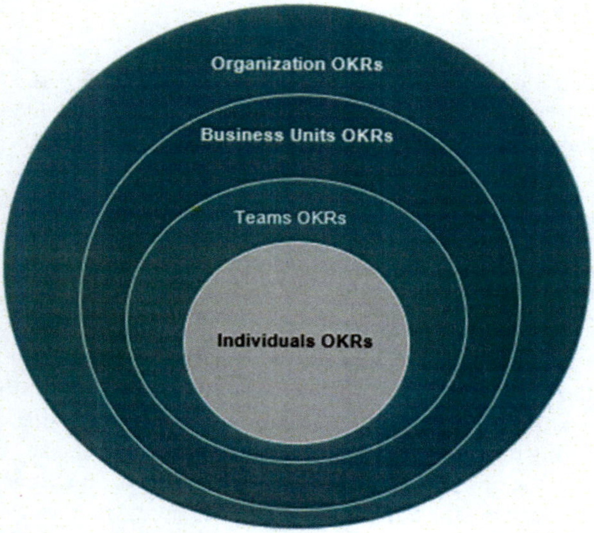

Fig. 7 Connection of OKRs

4.3 *Making the Organization Subject to OKRs*

After linking the objective to Key Results, you must subordinate the organization to OKRs. For this, it is suggested that the OKRs developed are coupled to the superior unit of analysis. For example, when developing a set of OKRs for an organization's sales sector, it is important to link these OKRs at the enterprise level of the enterprise. It is important to emphasize that the first development cycle will not be coupled with any other, since the cycle represents the initial development of the company. The process of connecting the OKRs needs to happen vertically and horizontally, as shown in Fig. 7.

Subordination is the mechanism of coupling these OKRs throughout the development cycles since the ORGANIZATION's OKRs are unfolding in business units, teams, and individuals. It is in this step that the OKR verification built for the level analysis unit above takes place.

4.4 *Raising Focus*

The stage called how to raise the focus aims to reduce the distractions that divert attention to what is important, in this sense, raising the focus means promoting constant analyses of the results of OKRs. For this, we used the approach composed of three fronts: Monday meetings, mid-quarter check-ins, and end-of-quarter reviews

[5]. Using these approaches, it is expected to collect information from the current states of OKRs and transform this knowledge into actions. The following will be detailed on each of these fronts.

The purpose of Monday meetings is to evaluate weekly the progress of OKRs and identify potential problems before they occur [5] Mid-quarter check-ins are control points used to conduct a more formal review of the progress of OKRs to align expectations, require changes, raise results or even disregard OKRs [5]. The end-of-quarter reviews evaluate the performance of OKRs with what it proposes, and what were the failures and learnings during the quarter in the conduct of OKRs [5].

4.5 *Have You Reached the Goal?*

The last step aims to ensure that strategic focus is conducted continuously. After checking the objective result in the end-of-quarter reviews, the OKR may have been completed, so new OKRs should be updated using the GOAL-OKR steps. On the other hand, OKR may not have succeeded in the quarter and should have its focus maintained. To do this, approaches to raising the focus of the OKR should be repeated.

5 Application of GOAL-OKR in a Technology Company

GOAL-OKR has been tested at an information technology company developing an ERP software for the leather industry. The company in question has the system of OKRs deployed at the corporate level. The use of GOAL-OKR will aim at verifying the consistency of OKRs about the company's goal (steps 1 and 2) and to perform approaches to raise focus (step 4). The step of subordinating the organization to OKRs was disregarded because the company operates only with OKRs at the corporate level. The data used were extracted from the first quarter of 2021.

From the data extracted from the first semester, each of the objectives was analyzed considering the following conditions: (i) increase in profits today and in the future, (ii) satisfying markets today and in the future, and (iii) safety and employee satisfaction today and in the future. Table 1 represents the objectives belonging to the corporate level and the consistency test.

Through the consistency test, it was possible to verify that the objectives were adhered to at least one of the basic conditions of the strategy. Objective number 1 in Table 1 will help to increase profit today and, in the future, as it contributes directly to the company's bottom line. However, this objective needs to be tested against its key outcome to ensure that efforts to increase revenue are oriented to operational indicators of the TOC.

Table 1 Corporate level objectives

Goal	Will the objective increase profit today and in the future?	Will the objective increase market satisfaction today and in the future?	Will the objective increase employee safety or satisfaction today and in the future?
1. Increasing the company's revenue stemming from the sale of new projects	Yes	No	No
2. Ensuring quality in deliveries and customer satisfaction	No	Yes	No
3. Ensure delivery of projects and change requests within the predetermined deadline with the client	No	Yes	No
4. Reducing the lead time of delivery of solutions	Yes	Yes	No
5. Develop research to innovate products/services	Yes	Yes	No
6. Develop in teams the focus on collaborative growth strategy, including the organizational human aspect	Yes	Yes	Yes
7. Ensure the company's financial sustainability	No	No	Yes

Objective number 2 will help to increase market satisfaction today and, in the future, as it contributes to the development of quality deliveries focused on the end customer. Objective number 3 allows the organization to focus on actions to meet the deadline and as a result, the goal will increase customer satisfaction today and in the future. However, the key results of objective number 3 need to be validated about the operational indicators of the TOC, to avoid actions that go against the company's goal.

Objective number 4 contributes directly to the increase in profit today and in the future, since the reduction of lead time makes the return on investment happen faster and increases market satisfaction. Objective number 5 can help to increase profits from the moment new products/services are launched and sales are effective, and the market may be satisfied with these new products/services.

Objective number 6 will help to meet all the basic conditions of the strategy, because the development of human resources focusing on collaboration will help to establish relational gains in the organization, and this can contribute to increased profits, market satisfaction, and employee satisfaction. Finally, objective number 7 can help employee satisfaction today and, in the future, once the sustainability of the company financial will generate a stable environment with less risk to employees.

After verifying the objectives concerning meeting the basic conditions of the strategy, each of the key results was analyzed. Key results are expected to ensure increased throughput, reduced inventory/investment, and reduced operating expenses. However, it is worth mentioning that some of the key results contribute directly or indirectly to the operational indicators of the theory of restrictions, therefore, it is up to a systematic analysis of how the result behaves in the organization.

Some of the key results did not pass the consistency test with the operational indicators of the theory of restrictions, being: 5.4 Promote the participation of at least 20% of employees in research and innovation projects in 2021, 6.1 Achieve the participation of at least 40% of employees in the bank of ideas and 6.2 Ensure the conversion of at least 10% of the ideas of the thought bank by December/2021. Result 5.4 is associated with objective number 5 in Table 1 and results in 6.1 and 6.2 relate to objective number 6 in Table 1.

Key result 5.4 will not increase the throughput, reduce inventory/investment, and reduce operational expenditure, as increasing employee participation in research projects should be a premise for the development of research and innovation projects. The 6.1 results also did not contribute to the increase in throughput, reduction of inventory/investment, and reduction of operating expenses and can generate an undesirable effect that is the generation of ideas by obligation. An alternative to measure this key result could be to define a percentage of employees in the use of troubleshooting tools. The 6.2 results also do not contribute to the increase in throughput, reduction of investment, and reduction of inventory since converting the ideas of a bank of ideas can generate a behavior of implementing the idea without evaluating the economic impact of the suggestion.

Of the approaches suggested in the step of raising the focus, the company in question used mid-quarter check-ins and end-of-quarter reviews. Through mid-quarter check-ins, it was verified that several key results would not be achieved if the current performance was maintained. To this end, the company's board has developed an action plan to direct efforts to achieve the expected results. In the end-of-quarter reviews, the main difficulties throughout the quarter and which OKRs were completed were verified.

By the end of the first quarter of 2021, the company had not reached any of the proposed OKRs. In this sense, the company chose to keep the OKRs for the second quarter of 2021 and monitor the action plan originated in the step of raising the focus to achieve the expected result.

5.1 Discussions of the Results of the Application

Among the results of the practical application of the GOAL-OKR framework, there are some considerations relevant to strategic management. The process of verifying objectives about the basic conditions of strategies provide a holistic view of how objectives connect with the company's strategy, allowing those involved to conduct the process of building OKRs in a structured way. In addition, verifying key results against the operational indicators of the theory of constraints allows those involved to see whether the company is moving toward the company's goal.

This check helps you to rethink key results that are not adhering to the company's goal and allows you to create key results based on operational indicators. GOAL-OKR also allows organizations to focus efforts to raise the focus on their OKRs, as the proposed approaches are simple to be implemented in various contexts of organizations.

6 Conclusions

GOAL-OKR collaborates for the development and monitoring of OKRs in the organization through a sequence of steps with structured activities. Through the logics of throughput accounting, Objectives & Key Results are validated against the company's goal, reducing the gap between the company's strategy and goal. In addition, through the framework, it is possible to carry out the development and monitoring process in a focused and structured way, thus contributing to the translation of the strategy in organizations.

Through the conceptual development of the framework, it was noticed that the stage of subordinating the organization to OKRs ensures that the process occurs in a systematized manner and that each of the OKRs is coupled at a higher level, and that this coupling helps organizations to unfold the strategy throughout their organizational structure. In addition, the framework enables the step of raising focus and adapting the context changes that are required in the business environment, since the approaches mitigate the risk of surprise about the expected key result.

However, GOAL-OKR presents some limitations concerning the traditional process of implementation of OKRs. GOAL-OKR considers that the sponsorship of executives with OKR is an action to be carried out a priori by the implementation team and that the conceptual leveling of the tool and the development or confirmation of the mission, vision, and strategy of the organization are fundamental requirements for the development of OKRs in organizations.

Another limitation of GOAL-OKR is that no interviews were conducted with experts to validate the framework. These interviews would help capture framework limitations and establish contingency heuristics for the artifact. In this sense, suggestions are presented for the development of future work, both for the advancement of the framework and application of GOAL-OKR in an organizational environment that does not have the OKRs system.

To advance in the construction of the framework, it is suggested that interviews with specialists should be conducted. Another possible advance in the construction of the framework is to carry out focus groups with companies that use the OKRs system to verify the acceptance concerning the consistency tests based on the throughput accounting proposed in TOC. In addition, these future studies could provide a comprehensive view of how the framework can be used in different organizations and what are the main difficulties related to deploying the OKRs system.

References

1. Aracıoğlu, B., Zalluhoğlu, A.E., Candemir, C.: Measuring and evaluating performance within the strategic management perspective: a study on performance measurement of a seafood company. Procedia. Soc. Behav. Sci. **99**, 1026–1034 (2013)
2. Rudd, J.M., Greenley, G.E., Beatson, A.T., Lings, I.N.: Strategic planning and performance: extending the debate. J. Bus. Res. **61**, 99–108 (2008)
3. Cao, Y., Zhao, K., Yang, J., Xiong, W.: Constructing the integrated strategic performance indicator system for manufacturing companies. Int. J. Prod. Res. **53**, 4102–4116 (2015)
4. Vasile, E., Croitoru, I.: Management by objectives – factor of growth in organizational performance. Intern. Audit. Risk Manag. **52**, 44–53 (2018)
5. Niven, P.R., Lamorte, B.: Objectives and Key Results (2016). https://doi.org/10.1002/9781119255543
6. Goldratt, E.M., Cox, J.F.: The Goal. North River Press, Croton-on-Hudson (1984)
7. Dresch, A., Lacerda, D.P., Antunes Jr., J.A.V.: Design science research: a method for science and technology advancement. Springer, New York (2015)
8. Doerr, J.: Measure What Matters: How Google, Bono, and the Gates Foundation Rock the World with OKRs, 2nd edn. Penguin Publishing Group, New York (2018)
9. Kim, S., Roscoe Davis, K., Cox, J.F.: An investigation of output flow control, bottleneck flow control and dynamic flow control mechanisms in various simple lines scenarios. Prod. Plan. Control. **14**, 15–32 (2003)
10. Cox III, J.F., Schleir, J.G.: Handbook da Teoria das Restrições. Bookman, Porto Alegre (2013)
11. Goldratt, E.M.: Computerized shop floor scheduling. Int. J. Prod. Res. **26**, 443–455 (1988)
12. Goldratt, E.M.: The Haystack Syndrome: Setting Information Out of the Ocean. North River Press, Croton-on-Hudson (1990)
13. Schragenheim, E.: From DBR to simplified-DBR for make-to-order. In: Theory of Constraints Handbook, 1st edn, pp. 217–245. The McGraw-Hill Companies, New York (2010)
14. Barnard, A.: Continuous improvement and auditing. In: Theory of Constraints Handbook, 1st edn, pp. 455–454. The Philosophy of Science Association, New York (2010)

Influence of the Covid-19 Pandemic on Purchasing Behavior and the Perception of Shipping Price and Delivery Time by Brazilian E-shoppers

Igor Luiz de Almeida Sousa, Kivia Mota Nascimento, and Roberta Alves

Abstract The covid-19 pandemic accelerated many digital transformations in Brazil and the world, and one of them was online shopping, which already had an upward trend in revenue and grew even more from 2020 onwards due to the social isolation actions as a way as how to contain the spread of coronavirus. This above-expected increase directly impacted the value and time delivery of online purchases. It is relevant, in this moment of advanced vaccination and return of presidential economic activities, to study whether the dynamics of online shopping tend to remain constant. To characterize the online purchase behavior of Brazilian consumers throughout the pandemic and their intentions of online shopping in the post-pandemic future, as well as their perceptions about the shipping prices and delivery time, questionnaires, were applied to e-shoppers from Brazil. Information was collected regarding the profile of the sample, the isolation situation in the face of the pandemic, the online purchase behavior during the period, and their perceptions about the freight changes. The results indicate that part of the changes in online shopping that occurred during the period of the covid-19 pandemic should remain in the post-pandemic scenario. They also indicate that there is a perception that the shipping prices has increased and that the delivery time has decreased during this period.

Keywords E-commerce purchase behavior · Covid-19 pandemic · Changes in freight

---

I. L. de Almeida Sousa · K. M. Nascimento (✉)
Universidade Federal de São João del-Rei, Campus Santo Antônio, Sao Joao del Rei, Brazil
e-mail: kiviamn@ufsj.edu.br

R. Alves
Universidade Federal de Lavras, Campus Paraíso, Lavras, Minas Gerais, Brazil

V. M. López Sánchez et al. (eds.), *Industrial Engineering and Operations Management*, Springer Proceedings in Mathematics & Statistics 400,
https://doi.org/10.1007/978-3-031-14763-0_29

1 Introduction

The Covid-19 pandemic, decreed by the World Health Organization (WHO) in March 2020, affected purchasing relationships, due to the social isolation necessary to contain the contagion of the virus [1]. To favor social distancing actions, several companies considered "non-essential", such as bars, stores, and gyms, were forced to keep their doors closed and, in the face of this scenario, entrepreneurs had to adapt quickly to keep their businesses competitive [2–4].

With the need to maintain their operations, physical stores began to see the digital market as an opportunity [5]. In line with this, consumers' shopping needs substantially changed purchasing patterns and influenced a higher-than-expected increase in e-commerce sales [6]. In Brazil, in 2020, there was a 68% growth in e-commerce sales, totaling more than 300 million online purchases and a record value of US$ 43.6 billion [7]. In addition, it is estimated that 20 million Brazilians made their first online purchase in 2020 and 150,000 stores began to sell their products online during this period [7].

The advance in vaccination, with more than 70% of the population fully vaccinated [8], and the consequent reduction in the number of deaths caused by Covid-19, allowed sectors of the economy to resume their face-to-face activities. Despite the growth in the economic activities, the recovery was uneven across sectors, and traditional retail in December 2021 was still 2.3% below the pre-pandemic level, with an emphasis on the book, newspaper, magazine, and office segment, which was in a sales level of 34.4% lower than in February 2020 [9].

In this sense, it is necessary to study what changes that occurred in the dynamics of online shopping during the pandemic tend to remain in a post-pandemic scenario, and which should return to patterns closer to those observed in the pre-pandemic period. Thus, this research aims to characterize the change in purchasing behavior in Brazilian e-commerce and consumer perceptions regarding shipping prices and delivery time during the pandemic.

2 Research Method

The present research has an applied nature characterized by its practical interest and qualitative-quantitative approach [10]. This research is classified as exploratory since there is not much accumulated and systematized knowledge about the topic to be addressed [11]. This research fits into the survey technique regarding the methodological procedures since the data will be collected from a sample taken from a certain population that we wanted to know.

A survey must be performed according to the following steps [12]:

1. Specification of objectives
2. Operationalization of concepts and variables
3. Development of the data collection instrument

4. Instrument pre-test
5. Sample selection, data collection, and verification
6. Analysis and interpretation of data
7. Presentation of results

Thus, based on the objective established for this study, which is to characterize the change in purchasing behavior in Brazilian e-commerce and e-shoppers' perceptions regarding deadlines and shipping prices during the pandemic, the main information collected refers to the profile of the respondents, the situation of isolation in the face of the Covid-19 pandemic, the behavior of online shopping during the period and the perceptions about the change in the regarding shipping prices and delivery time of online shopping. Then, the concepts and variables to be measured were defined, such as demographic data, information on isolation, the frequency of purchases, types of products purchased, etc. From there, the data collection instrument was elaborated containing 27 questions to collect all the necessary information. A pre-test was carried out with the researchers involved to check if there were any errors. After validation, the online questionnaire was used as an instrument to collect data and information from participating e-shoppers. A total of 388 responses were sent via e-mail between August and November 2021. E-shoppers from all regions of Brazil responded to the questionnaire, most of them residing in the Southeast region. The collected data were analyzed and interpreted in Microsoft Office Excel® and RStudio® software, to prepare the report.

3 Results and Discussion

3.1 Personal Attributes

Table 1 shows the characterization of the sample that responded to the survey.

The total number of respondents to this survey is 388 people, 60% women and 39% men. Most of the sample, 65.5%, is concentrated in the Southeast Region of Brazil, 18.6% resides in the South Region, and the other 16% is divided between the Northeast, North, and Center-West regions. Regarding marital status, 42.2% of the sample is married, 38.7% are single and 20.1% are divorced, widowed, or in a stable relationship. As for age, 2/3 of the sample is between 18 and 45 years old and 1/3 is 46 years old or older. The predominant schooling in the sample is complete postgraduate education, which corresponds to 76.3% of respondents, while the other 23.4% have completed high school or higher education.

Most respondents, 68.8%, are public sector employees and in terms of income, 68.6% of the sample have an average monthly family income between 5 and 20 minimum wages. Regarding the situation of social isolation, 54.1% of respondents said they were only going out a few times and only 13.1% said they were not in isolation. Regarding the work situation, 64.4% said they were working from home,

Table 1 Statistical characteristics of the sample

Variable	Category	Frequency	Relative frequency
Gender	Female	233	60.10%
	Male	153	39.40%
	Other	1	0.30%
	Rather not answer	1	0.30%
Region of Brazil	Southeast	254	65.50%
	South	72	18.60%
	Northeast	38	9.80%
	North	18	4.60%
	Midwest	6	1.50%
Marital status	Married	160	41.20%
	Single	150	38.70%
	Divorced	30	7.70%
	Widower	3	0.80%
	Stable Union	45	11.60%
Age	18–25 years	51	13.10%
	26–35 years	94	24.20%
	36–45 years	112	28.90%
	46–55 years	79	20.40%
	Over 55 years	52	13.40%
Education	Complete high school	47	12.10%
	Complete higher education	45	11.60%
	Complete postgraduate	296	76.30%
Household income (in Brazilian minimum wage[a])	Up to 4	87	22.40%
	5–10	123	31.70%
	11–20	143	36.90%
	Above 20	35	9.00%
Primary activity	Public employee	267	68.80%
	Private sector employee	20	5.20%
	Student	66	17.00%
	Freelancer	10	2.60%
	Unemployed	11	2.80%
	Retired	3	0.80%
	Businessperson	3	0.80%
	Others	8	2.10%
Social isolation situation	Go out sometimes	210	54.12%
	Be in isolation	103	26.55%
	Not be in isolation	51	13.14%
	Just go out for work	24	6.19%

(continued)

Table 1 (continued)

Variable	Category	Frequency	Relative frequency
Work situation	Is working from home-office	250	64.43%
	Not working	53	13.66%
	Is working in person in a sector not considered "essential"	49	12.63%
	Is working in person in a sector considered "essential"	36	9.28%

[a]Brazilian monthly minimum wage in 2021: BRL 1100.00; US$ 213.17 [13]

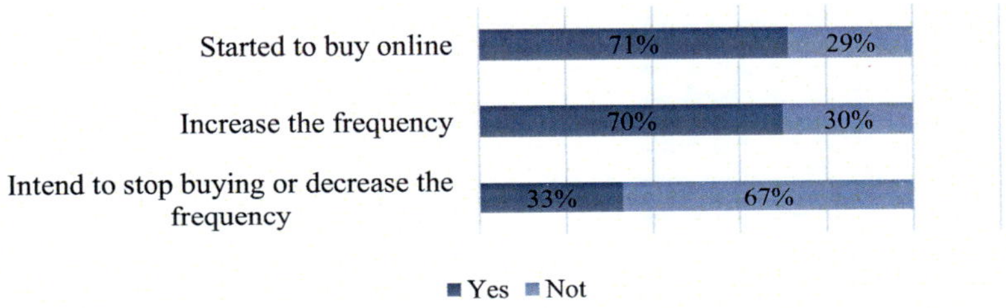

Fig. 1 Distribution of respondents who changed their purchasing behavior

12.6% said they were working in person in a sector considered non-essential and 9.3% worked in person in a sector considered essential.

3.2 Change in Purchasing Behavior

Regarding online shopping behavior, respondents were asked if there was any category of products that were not purchased online and started to be purchased through this means after the beginning of the pandemic and if there was any category that they already bought through e-commerce, but who started to buy more often that way. In addition, they were asked if they intend to stop using e-commerce to buy a category of products after the end of the pandemic. The distribution of responses is shown in Fig. 1.

When asked if there was any category that was not purchased online before the covid-19 pandemic, and started to be purchased online, 71.2% of the sample answered in the affirmative. A similar case occurred when the question was whether there was any category that was already purchased through e-commerce before the start of the health crisis, but which from the beginning of the pandemic began to be purchased online more frequently, in what 70.1% of respondents said yes.

When asked if there was any category that the respondent bought online during the pandemic but intended to stop buying or to buy less frequently online, only 33% of the sample answered yes, while 67% answered that they did not intend to

Table 2 Distribution of e-commerce behavior changes by category

Category	Increased the frequency	Started to buy online	Intend to stop buying or decrease the frequency
Ready-to-eat food and drinks	122	8	36
Food Products	84	55	29
Books and Magazines	77	38	14
Supermarket shopping	76	112	53
Health/Cosmetics/Perfumery Items	63	100	18
Clothing	62	100	36
Housewares	47	73	11
Computer Items	40	56	7
Home Appliances	40	55	11
Electronics	39	41	8
House and Decoration	36	57	7
Stationery and Office	29	35	10
Pet Products	25	39	8
Sports and Entertainment	24	41	5
Toys	18	27	5
Baby Items	8	14	3
Construction	7	14	1
Telephony	4	8	1
Others	2	1	0

stop buying or decrease the frequency of online shopping. This may indicate that the increase in the use of e-commerce as a means of purchase during the pandemic tends to continue, even after the end of the restrictions imposed during this period.

The distribution of the categories of products that respondents started to buy or increased the frequency of online purchases, and those that intend to stop buying or decrease their frequency in e-commerce, is shown in Table 2, which is classified in descending order by the column "Increased the frequency".

It is possible to observe that, when it comes to the increase in purchase frequency, the most cited category is the ready-to-eat food and drinks, which is one of the least cited categories in the item started to buy after the beginning of the pandemic. This can be explained by the fact that the delivery business was already widespread before the pandemic, but it became the only way to buy ready-to-eat food and drinks during trade restrictions in the most acute stages of contagion. Among the 130 people who answered that they increased their purchase frequency or started to buy in this category, only 36 (27.7%) intend to stop buying or reduce the frequency of purchase of ready-to-eat foods and drinks.

The food products and books and magazines segments complete the top 3 of the most cited when asked about the increase in online shopping frequency, with 84 and 77 responses respectively. The category of books and magazines was also mentioned by 38 respondents, who said they started to buy these products online after the beginning of the pandemic, but only 14 people said they intend to stop buying or reduce the frequency of online purchases. In line with this, we have the book, newspaper, magazine, and office segments, which were the retail segment that in December 2021 was furthest from the pre-pandemic level, 34.4% below [9]. This may point to a trend that the increase in online purchases in this category should continue and even cause a possible reduction in sales in the physical market.

Among the categories that respondents started to buy online after the beginning of the health crisis, supermarket shopping was the most cited, with 112 responses. This was also one of the most cited categories when it came to increasing online shopping frequency. With 53 responses, the supermarket shopping category is also the most cited among those in which respondents intend to stop buying or reduce the frequency of purchases on the internet, but even so, it represents only 28.2% of those who have increased the frequency or started to buy this category online during the pandemic.

The categories of items for health/cosmetics/perfumery and clothing are in second place among the most cited by respondents who started to buy a category of products on the internet, with 100 responses each. These two categories were also cited similarly about the segments that respondents claim to have increased their online shopping frequency, with 63 for the category of items for health/cosmetics/perfumery and 62 responses for the clothing category. However, only 18 people responded that they intend to stop shopping or reduce the frequency of online purchases in the first category, compared to 36 in the second.

3.3 *Change in Shipping Perception*

It was asked if the sample noticed any change in the shipping prices during the pandemic and the answers were stratified by region of Brazil (see Figs. 2 and 3).

In this sense, 37.1% noticed an increase in the shipping prices, 10.3% noticed that the value decreased, 33.5% did not perceive any change in the value and 19.1% did not know how to respond. When we segment the sample by region, this scenario remains, with the proportions remaining similar in all regions of the

Fig. 2 Distribution of responses about change in the shipping prices

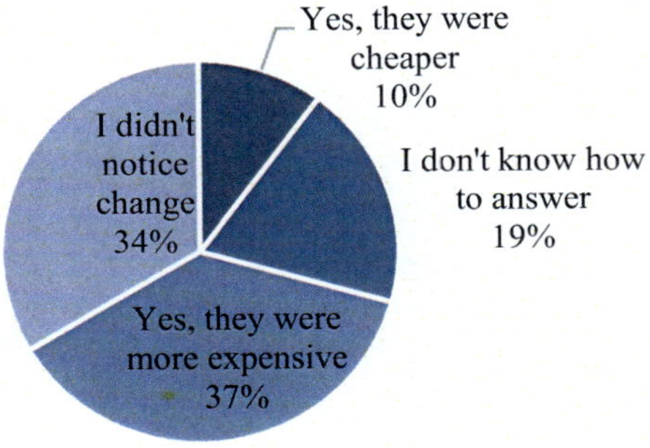

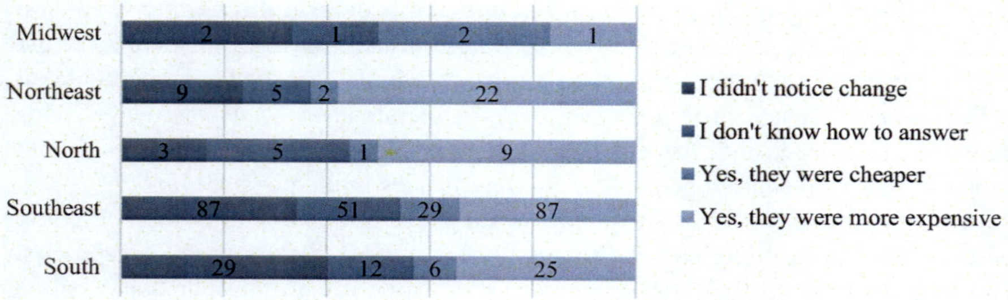

Fig. 3 Distribution of responses about change in the shipping prices by region

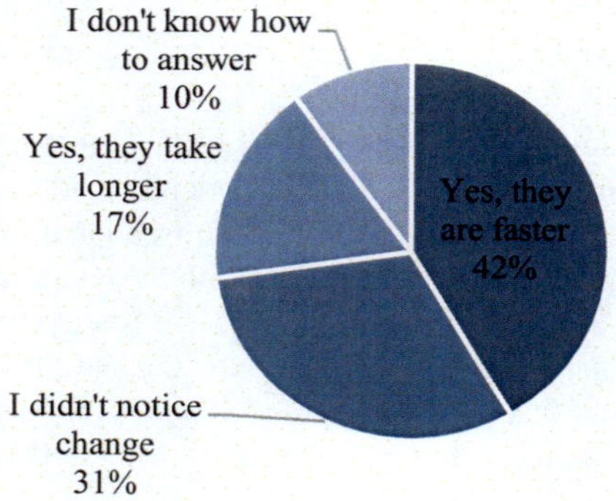

Fig. 4 Distribution of responses about changes in the delivery time

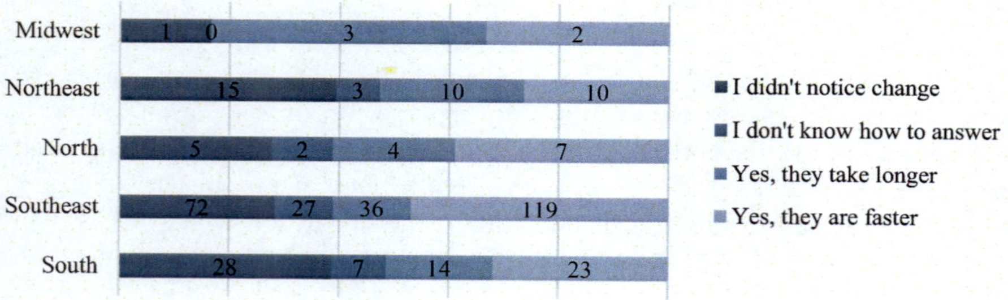

Fig. 5 Distribution of responses about changes in the delivery time by region

country, showing that this increase in value was similarly perceived throughout Brazil. However, a survey carried out by the Promo-bit shows that the shipping prices for the same product costs up to 10 times more in the Northeast region than in the Southeast region [14].

Respondents were also asked if they noticed any changes in the delivery time for purchases made through e-commerce during the pandemic and the answers were stratified by Brazilian regions (see Figs. 4 and 5).

Regarding the delivery time of purchases made through e-commerce, 41.5% of respondents noticed that deliveries are arriving faster, 17.3% of those deliveries are taking longer, 31.2% did not notice any change and 10.1% could not answer. This difference was even more pronounced in the Southeast region, where 47.0% of respondents noticed that delivery time are faster. In the Northeast region, perceptions are more divided, 26.3% realized that deliveries are faster, the same percentage realized that deliveries are taking longer, and 39.5% did not perceive any change in values. In fact, the delivery time for the Northeast region can be up to twice as long when compared to the delivery time for the Southeast region [14].

4 Conclusion

Based on the answers obtained, it is possible to assume that the purchasing behavior in e-commerce underwent significant changes during the pandemic and that, apparently, many of these changes that occurred in this period should remain after the end of the health crisis, and the e-commerce should gain more in lives of Brazilians.

The scenario for the management of shipping these orders is difficult, as with the increase in demand, shipping prices also seems to have increased. On the other hand, the general perception is that during the pandemic the delivery time for e-commerce was faster in Brazil, which indicates that the sector apparently has improved during the crisis. In order for e-commerce to continue growing in Brazil, it is necessary to think of solutions that increase the efficiency of urban logistics systems, reducing costs and, at the same time, further reducing the delivery time.

References

1. Fabris, D.C., et al.: Mapa do Comportamento do Consumidor a Partir da Covid-19: Uma Análise das Reações do Consumidor Frente às Mudanças Enfrentadas pela Pandemia. In: Clav (2020)
2. Farias, H.S.: O Avanço da Covid-19 e o Isolamento Social como Estratégia para Redução da Vulnerabilidade. Espaço e Economia. Revista Brasileira de Geografia Econômica (2020)
3. Silva, G.L.: Varejo e Covid-19: Uma Análise da Covid 19 e seus Impactos no Varejo de Varginha. In: TCC – Comunicação Social com Habilitação em Jornalismo, p. 40. Centro Universitário do Sul de Minas, Varginha (2020)
4. Montemurro, N.: The emotional impact of covid-19: from medical staff to common people. Brain Behav. Immun. **87**, 23–24 (2020)
5. Do Nascimento, A.C., Do Prado, N.B., Da Cunha, C.F.: Covid-19 e modelos de gestão nas micro e pequenas empresas: qual a melhor saída? Rev. Expectativa. **20**, 50–72 (2021)
6. González-Varona, J.M., Villafáñez, F., Acebes, F., Redondo, A., Poza, D.: Reusing newspaper kiosks for last-mile delivery in urban areas. Sustainability. **12**, 9770 (2020)
7. Cesar, C.: O Crescimento dos Marketplaces em 2021, https://abcomm.org/noticias/o-crescimento-dos-marketplaces-em-2021/, last accessed 22 Mar 2022

8. Brasil, Ministério da Saúde: Vacinação Contra a Covid-19 no Brasil – #PÁTRIAVACINADA, https://www.gov.br/saude/pt-br/vacinacao/, last accessed 15 Mar 2022
9. Alvarenga, D.: Retomada da Economia: Veja Quais Atividades já Retomaram o Patamar Pré-Pandemia e Quais Ainda Não, https://g1.globo.com/economia/noticia/2022/02/11/retomada-da-economia-veja-quais-atividades-ja-retomaram-o-patamar-pre-pandemia-e-quais-ainda-nao.ghtml, last accessed 22 Mar 2022
10. Turrioni, J.B., Mello, C.H.P.: Metodologia de Pesquisa em Engenharia de Produção - Estratégias, Métodos e Técnicas para Condução De Pesquisas Quantitativas E Qualitativas. In: Programa De Pós-Graduação Em Engenharia de Produção da Universidade Federal de Itajubá – UNIFEI (2012)
11. Longaray, A.A., Beuren, I.M.: Caracterização da pesquisa em contabilidade. Como elaborar trabalhos monográficos em contabilidade: teoria e prática. **2**, 21–45 (2009)
12. Gil, A.C.: Como Elaborar Projetos de Pesquisa, 6th edn. Atlas, São Paulo (2017)
13. CNN: Salário Mínimo do Brasil é Menor que da Bolívia e Paraguai; na Venezuela é US$ 1, https://www.cnnbrasil.com.br/business/salario-minimo-do-brasil-e-menor-que-da-bolivia-e-paraguai-na-venezuela-e-us-1/, last accessed 29 Mar 2022
14. Filgueiras, I.: Frete até 10 Vezes Mais Caro para Algumas Regiões é Empecilho para 'Viver' de Compras On-line, https://valorinveste.globo.com/objetivo/gastar-bem/noticia/2020/06/18/frete-ate-10-vezes-mais-caro-para-algumas-regioes-e-empecilho-para-viver-de-compras-on-line.ghtml, last accessed 22 Mar 2022

The Impacts of the COVID-19 Pandemic on an Educational Institution's Administrative Process, a Process Mining Approach

Mirley Bitencourt Ferreira, Isabelle Martins Santos, Rafael Garcia Barbastefano, and Diego Carvalho

Abstract The COVID-19 pandemic has affected virtually every human activity over the past 2 years. This paper examines how the COVID-19 pandemic interfered with the business processes in Brazil's public vocational and higher education institution. Throughout the pandemic, the Organization forced the enactment of the paper-recorded processes in a virtual implementation. To unveil how the referred paper-recorded processes subset got executed during the pandemic, we conduct a process mining on the company's information system. The process mining data shows various indications of task merging, precluding, and duration modifications. The analysis of 4231 instances of administrative processes between 2019 and 2021 showed a reduction in duration times and the number of tasks.

Keywords Process mining · Pandemic · COVID-19

1 Introduction

The pandemic of COVID-19 has affected virtually every human activity over the past 2 years, and the crucial social distancing procedures preconized by the World Health Organization have affected people's lives, organizations, and the economy. In this context, Businesses and Organizations have had to readapt their business model, management, routines, and processes [1].

This paper examines how the COVID-19 pandemic interfered with the functioning of the enterprises by looking at their core: the business processes. To cope with the non-pharmacological interventions implemented to diminish the virus spreading, the enterprise digitized its operation in the shortest time possible and motivated this research: how did the business processes come to behave during the pandemic?

M. B. Ferreira (✉) · I. M. Santos · R. G. Barbastefano · D. Carvalho
CEFET/RJ – Production & Systems Engineering Program, Rio de Janeiro, Brazil

V. M. López Sánchez et al. (eds.), *Industrial Engineering and Operations Management*, Springer Proceedings in Mathematics & Statistics 400,
https://doi.org/10.1007/978-3-031-14763-0_30

To answer this question, we investigate a public vocational and higher education institution in Brazil to achieve this goal since a subset of its internal processes is performed with conventional paper-recorded administrative processes due to law restrictions. This subset got directly impaired by the pandemic condition because home office labor was the only viable way to operate due to non-pharmacological interventions.

During the pandemic, the Organization forced the enactment of the paper-recorded processes in a virtual implementation. To unveil how the referred paper-recorded processes subset got executed during the pandemic, we conduct a process mining on the company's information system. The process mining's primary purpose is to discover, monitor, and improve actual processes by extracting knowledge from event logs registered in information systems, linking processes and data and models and processes [2]. Amongst the main drivers behind the development and increasing use of process mining techniques is the need for companies to learn more about how their processes operate in the real world [3]. These process mining techniques can be applied to event logs containing data related to processing execution to discover business processes and their behavior, thereby improving decision support [4]. The process mining data shows various indications of task merging, precluding, and duration modifications.

This analysis is essential to evaluate the effectiveness of the administrative processes' management in a public agency during the pandemic, contributing to better management in public education. Researchers point out that there is a gap in applying the techniques and use of process mining in organizations [3, 13]. The majority of works in the literature focus on the development of tools and algorithms [13], and, in many of them, the data used for testing the tools are artificial, implying a lack of practical application, not generating new knowledge and improvements for management. Moreover, the studies analyze a few departments such as finance, health, and industry in most cases, and many of them generally reuse the same data files [13].

Besides this introduction, the current work presents a theoretical background on process mining in Sect. 2. Section 3 describes our method, explaining how we extracted, preprocessed, and mined the information from the information system database. We analyze the number of tasks and time durations distributions of the most common processes performed before and after the pandemic in Sect. 4. Finally, Sect. 5 offers our conclusions.

2 Public Administration and Administrative Processes

Public administration is formed by a set of agencies, services, State agents, and public entities that ensure through laws the satisfaction of the collective needs of security, culture, and economic and social well-being of a people [12]. It is fundamental to have plans based on general guidelines to meet these needs. To ensure that this planning is carried out, it is necessary to control and monitor the results achieved.

Although a public institution is always subordinated to the public interest, administrative management is not different from managing a company. Four main activities are necessary to achieve good administration: planning, organization, direction, and control [12].

The first activity is planning, in which the objectives and results to be achieved are defined. This phase will define the action plans [12]. Government planning must promote the country's social-economic development and national security, and it is necessary to build **udm** of the general government plan and general, sectoral, and regional programs [12].

The second activity, organization, aims to structure the operations, technical and human resources, and their responsibilities in the institution, ensuring the integration between these resources and departments and the coordination of activities [12].

The next activity is directing, which seeks to translate the plans into actions, giving guidance to people on how to execute them, guaranteeing the objectives planned in the previous phases. Unlike other administrative functions, this function constitutes an interpersonal process between the administrator and his collaborators [12].

Finally, control activity comes to monitor and verify the achieved results [12]. The attainment of the objectives described in the first phase is checked, and, in case they are not adhered to, action plans are created [12].

The importance of having efficient process management within an institution is undeniable, whether in private or public organizations, of education or another category, so that there is a correct functioning of its operations, transparency, and ethics of those in government.

3 Process Mining

Process mining involves the discovery, monitoring, and improvement of actual processes by extracting knowledge from event logs recorded in information systems, establishing linkage between processes and data, and models and processes [2]. These techniques offer a means to verify information more rigorously about an organization's core processes [2] by looking for relatively simple patterns in large data sets [5]. Researchers and Practitioners discover these patterns using data mining methods such as association rules, decision trees, clusters, and frequent and similar sequences [5]. Moreover, they also use neural networks to explore and learn the intrinsic structure of the enterprise's business processes [7, 11].

Machine learning methods use recorded database events rather than hand-made models [5], avoiding bias and errors introduced by those performing the process mapping. Process mining approaches focus on performance and control flow issues [6]. However, they are not limited to the analysis of data collected over time and can assist in operational support, allowing predictions and guidance about the analyzed process; nor are they restricted only to discovery, compliance, and improvement techniques. When possible, extra information is used - such as resources, time of occurrence of the event, and agent, not limited only to the process control flow [2].

Process mining researchers focus on the flow control perspective, where the process is analyzed by how the activities are triggered. Besides, Enterprise Resource Planning's event logs can contain a massive amount of information related to other perspectives, such as the resources spent in performing a task and its relationship with resources [7]. Moreover, this analysis, combined with external organizational data, creates a scenario where the process mining helps understand organizational and social structure [8], focusing on learning more about people, machines, organizational functions, work distribution, and work patterns [9]. Identifying such behavioral patterns of resources can help improve individual employee and team performance, directly impacting overall process and organizational performance [10].

The organizational process mining techniques are derived according to the three existing process mining classes – discovery, enhancement, and compliance. In discovery, the most relevant models are (1) organizational model discovery, (2) social network discovery, (3) personnel allocation rule discovery (how functions or units take responsibility for a task), and originator (who is authorized to do which tasks) [8]. The enhancement class analyzes the processes, and the compliance focuses on how the disclosed log data map onto an existing formal processes model [8].

4 Methods

The Federal Centre for Engineering Studies and Technological Education (Cefet/RJ) is Brazil's public vocational and higher education institution. It has a distinct characteristic: two categories of business processes. The first category represents the academic practices overseen by information systems, and the second category depicts administrative procedures regulated by a robust law framework that aims at accountability. Every process in this second category runs in a traditional paper form, and Cefet/RJ's General Electronic Protocol System (GEPS) records id, description, owner, dates, movements, location, and organizational unit during the process' life.

The institution enforced a home office policy on March 15th, 2020. Figure 1 shows the two 278 days periods analyzed in this work. The target pandemic period (CD) comprises GEPS data from the start of the home office policy until the very last available data on the system (December 18th, 2020). Moreover, the non-pandemic period comprises GEPS data from December 13th, 2018, until September 17th, 2019, which is arbitrarily 6 months before the home office policy.

We performed the data cleaning by removing all records with missing values and standard normalization by removing special characters from the description since it is recorded in Portuguese. The next step on data cleaning removed any process with

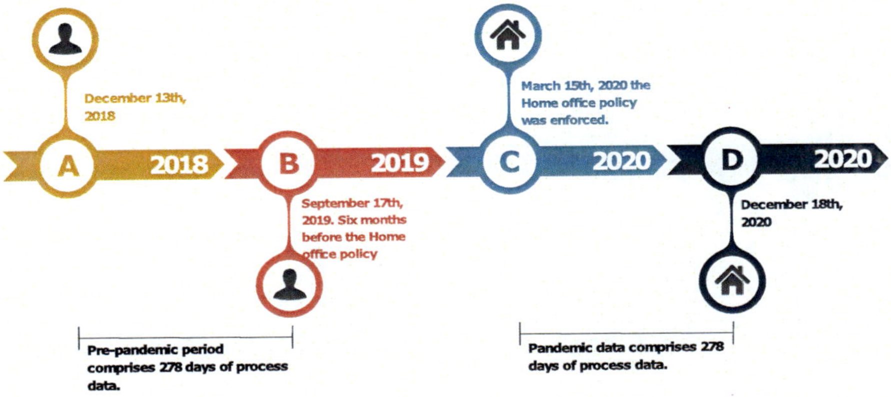

Fig. 1 Important dates on the data collection

more than a year of lifespan since, due to the current system logic, the process is waiting for archiving. The removal of the lengthy process does not compromise the analysis since each removed process started before the target period.

We also reduced data by removing unnecessary columns and keeping only the description, date, and organization unit for every process movement. This step was followed by removing noisy data where any process with loops or repetitions got verified and validated.

To group every process into a category, we performed the grouping with the Open Refine Software [https://openrefine.org/] on the description field. The final preprocessing step was to record both periods on a different database.

Using the database created in the beforementioned steps, we compute the distribution for the number of tasks and task duration for every process category.

5 Results

The analysis encompasses 3144 and 1177 processes in the pre-pandemic and pandemic periods, and in a bird's eye, the business process changed drastically. Figure 2 displays the histograms of the process span, where the process tends to be shorter on average after the pandemic.

In addition, there are concrete indications that the workers implemented a new process structure during the pandemic, as shown in Fig. 3. Systematically, the process presents a shorter number of tasks.

Before the pandemic, an organizational unit oversaw physically moving and enregistering processes in the system. During the home office policy, the workers electronically transact the prior paper-recorded process. The hypothesis for the dramatic change in duration and number of tasks per process is the sub notification

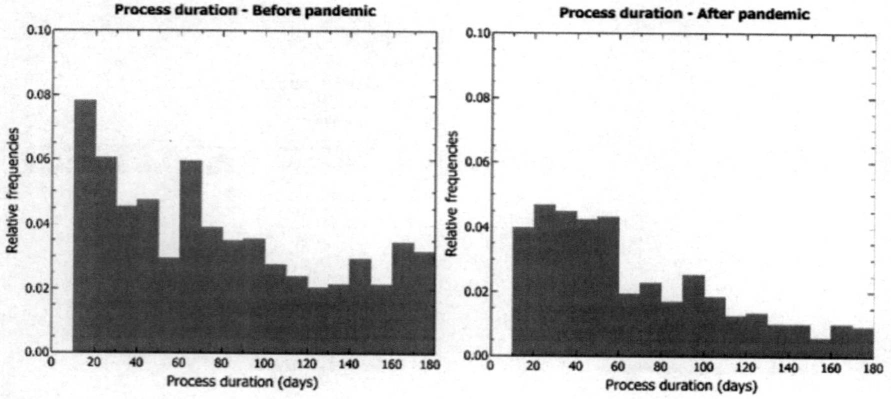

Fig. 2 Histogram of process duration in days

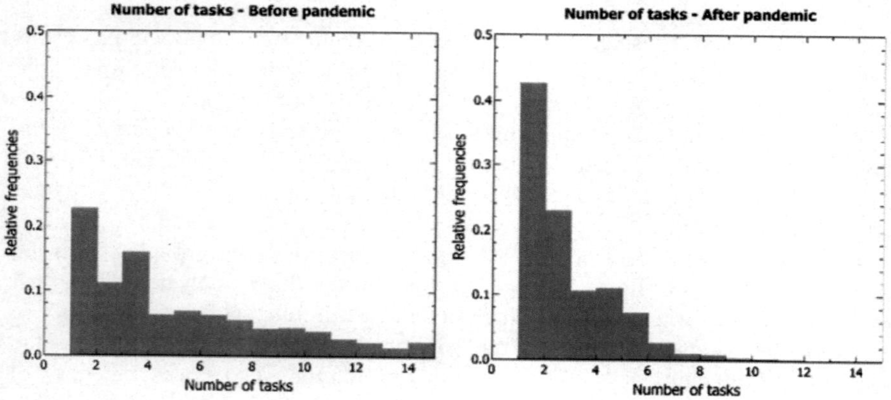

Fig. 3 Histogram of the number of tasks per process

supposition, where the worker precludes recording processes that change the system.

The most common process groups were first defined to get an overview of the processes. Typical classes in the periods before and after the pandemic are shown in Tables 1 and 2. Twenty groups were identified as having the highest occurrence before the pandemic. These represent 80.92% of the total (3144 macro processes). In this table, there are two columns about the position.

The first column shows the ranking, and the Diploma group represents 17.56% of the total processes. Text is the Faculty Career Management and TAE Career Management. These last two are about the career evolution of both the professors and the administrative-technical employees.

To order the most common classes in the post-pandemic period, we used the Pareto Law, which showed 18 macro processes with the highest occurrence. These

Table 1 Most common process groups before the pandemic

Ranking	Process group	Quantity	Pareto
1	Diploma	552	17.56%
2	Faculty Career Management	361	29.04%
3	TAE Career Management	213	35.81%
4	Agreement	213	42.59%
5	Absence	167	47.90%
6	Reimbursement and general payment	128	51.97%
7	Scholarship	105	55.31%
8	Payroll	103	58.59%
9	Working hour	91	61.48%
10	Purchase of material	84	64.15%
11	Registration	83	66.79%
12	Change of registration data	73	69.12%
13	Time of service	61	71.06%
14	Probationary period	58	72.90%
15	Hiring a professor	51	74.52%
16	Enrollment in courses and events	48	76.05%
17	Redistribution	41	77.35%
18	Retirement	39	78.59%
19	Service contracting	37	79.77%
20	Graduation ceremony	36	80.92%

Table 2 Most typical classes in the post-pandemic period

Position	Process group	Quantity	Pareto
1	Faculty Career Management	175	14.87%
2	Agreement	153	27.87%
3	Diploma	118	37.89%
4	TAE Career Management	80	44.69%
5	Purchase of material	67	50.38%
6	Graduation Ceremony	45	54.21%
7	Probationary period	41	57.69%
8	Payroll	36	60.75%
9	Enrollment in courses and events	32	63.47%
10	Official publication	31	66.10%
11	Redistribution	30	68.65%
12	Service contracting	25	70.77%
13	Absence	24	72.81%
14	Admission	23	74.77%
15	Reimbursement and general payment	19	76.38%
16	Time of service	19	77.99%
17	Official publication	16	79.35%
18	Registration	14	80.54%

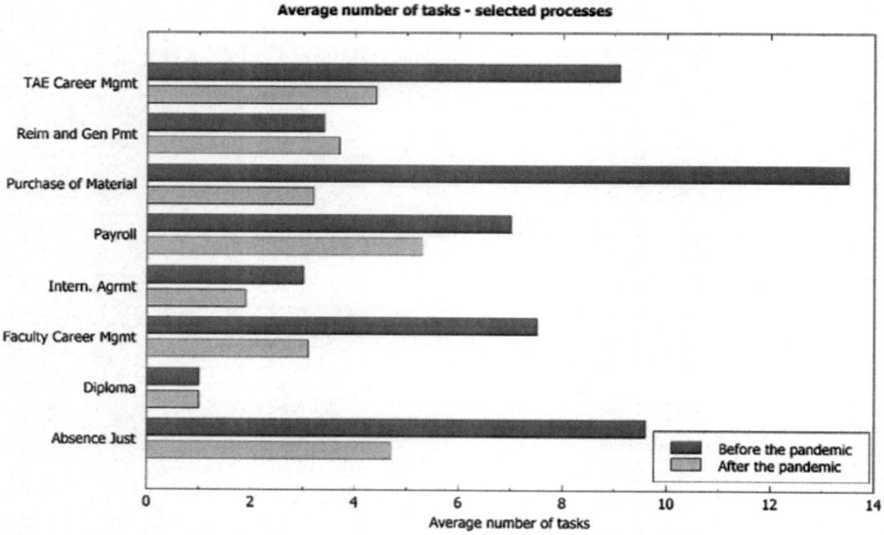

Fig. 4 The average of the number of tasks of the selected process

represent 80.54% of the total (1177 macro processes). Table 2 shows the list of these macro processes and their respective values.

We selected well-known procedures and plotted the average number of tasks in Fig. 4, which helps to discards the sub notification thesis since processes have more tasks after the home office policy than before. For example, Reimbursement and General Payment processes have more tasks after the pandemic than before.

Considering the duration, Fig. 5 shows the average duration of the same set of selected processes, and all average spans are smaller than before. We believe reductions related to moving times are insufficient to justify the considerable discrepancy observed, indicating structural changes in the processes.

Moving forward and analyzing a specific process, we show in Fig. 6 the Absence justification process, where a worker sends documents to human resources and health services to justify an absence. In this specific case, we discovered that an external federal system began to be used to register several steps of this process during the pandemic. Both duration and the number of tasks is smaller during the pandemic than before.

We also show the faculty career process in Fig. 7, and we observed the same reduction pattern. In this case, we discovered that a significant part of the process was officially not registered in the process records. Still, a couple of tasks were moved and reported to other activities.

Figure 8 shows the payroll process's duration and the number of tasks, and further investigation revealed that the organization digitized every step of this process. Then the signatures were substituted by public-key signatures since Brazil has implemented digital certificates under the common law.

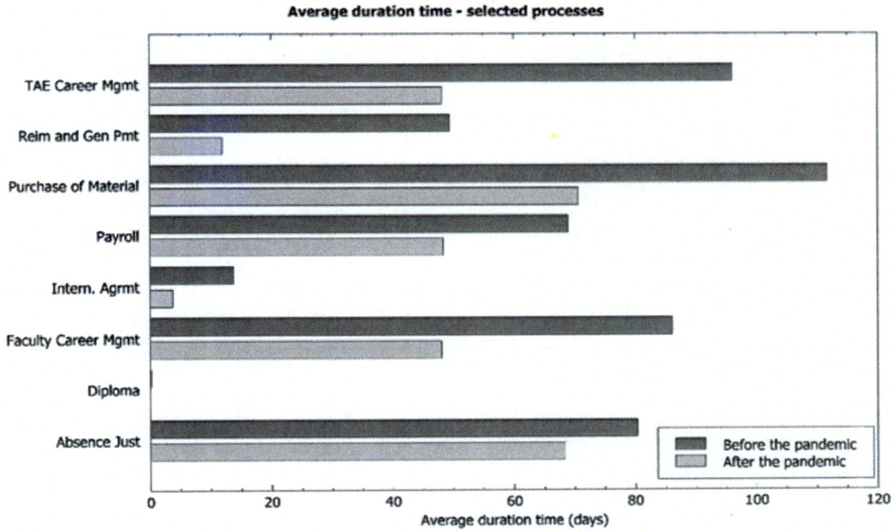

Fig. 5 The average duration of the selected process

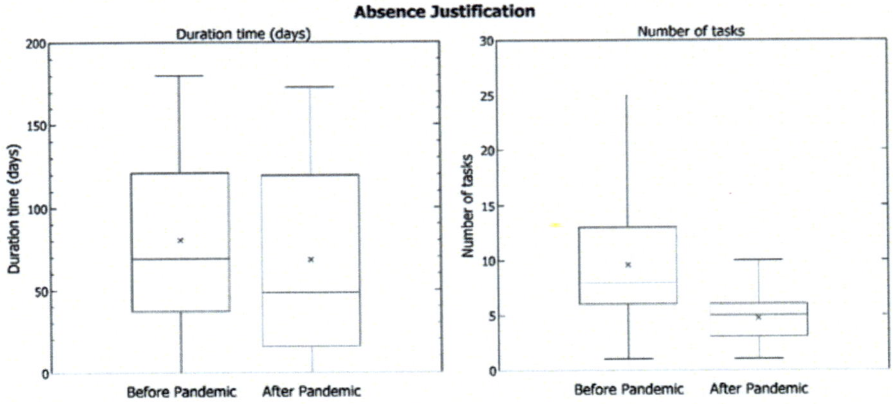

Fig. 6 Boxplot – Absence justification process

6 Conclusion

This paper aimed to study how the COVID-19 pandemic interfered in the administrative processes of a public organization dedicated to vocational and higher education using process mining techniques. There was a change in requests, execution time, and the number of stages relative to the processes between the pre- and post-pandemic periods.

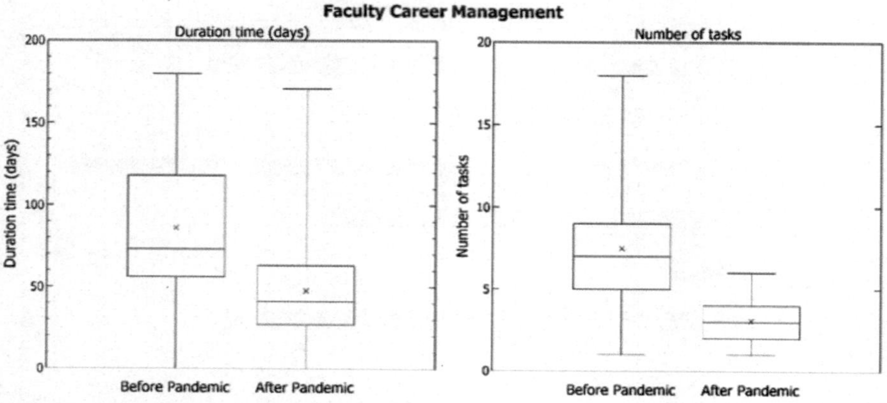

Fig. 7 Boxplot – Faculty Career Management

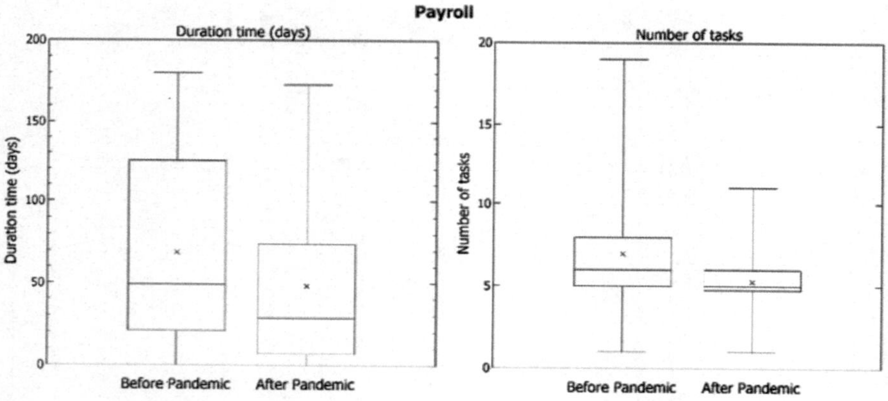

Fig. 8 Boxplot – Payroll

The demand pattern registered between the two analyzed periods did not change, while the needs of the institutional members changed a lot. It has happened because the demanded processes are the same, but they alternate between positions as they become less or more demanded. The work schedule has been frozen, and the rule changes to adapt to the contingencies of the virus have also caused some changes in process counting.

The duration time of the demands and the necessary steps for the realization of a request, on the other hand, change significantly between the two periods, indicating an acceleration in the digitalization of the execution of processes, which results in greater efficiency in execution.

This study shows through data that digitizing a company's processes brings efficiency and productivity gains. Therefore, society can benefit from this kind of change in administrative processes, which should become the new standard for services in general.

Future discussions in the medium and long term may indicate if the "new normal" imposed by the COVID-19 pandemic within this institution became the rule or were just mechanisms used to maintain the organization's functioning during the restrictive measures applied worldwide.

References

1. Pinzaru, F., Zbuchea, A., Anghel, L.: The impact of the COVID-19 pandemic on business expectations. Preliminary overview. In: Conference: Strategica 2020. Crisis and Risk Management. p. 721–730 (2020)
2. van der Aalst, W.M.P.: Process Mining: Discovery, Conformance and Enhancement of Business Processes. d.20. Springer, Heidelberg, Dordrecht, London, New York (2014)
3. Tiwari, A., Turner, C.J., Majeed, B.: A review of business process mining: state-of-the-art and future trends. Bus. Process. Manag. J. **14**(1), 5–22 (2008)
4. Savickas, T., Vasilecas, O.: Belief network discovery from event logs for business process analysis. Comput. Ind. **100**, 258–266 (2018)
5. van der Aalst, W., Adriansyah, A., van Dongen, B.: Replaying history on process models for conformance checking and performance analysis. Wiley Interdiscip. Rev. Data Min. Knowl. Disc. **2**, 182–192 (2012)
6. van der Aalst, W.M.P., Reijers, H.A., Song, M.: Discovering social networks from event logs. Comput. Supported Coop. Work. **14**(6), 549–593 (2005)
7. Maita, A., Martins, L., López Paz, C., Peres, S., Fantinato, M.: Process mining through artificial neural networks and support vector machines: a systematic literature review. Bus. Process. Manag. J. **21**(6), 391–1415 (2015)
8. Song, M., van der Aalst, W.M.P.: Towards comprehensive support for organizational mining. Decis. Support Syst. J. **46**(1), 300–317 (2008)
9. Appice, A., Di Pietro, M., Greco, C., Malerba, D.: Discovering and tracking organizational structures in event logs. Lect. Notes Comput. Sci. **9607**, 46–60 (2016)
10. Pika, A., Leyer, M., Wynn, M.T., Fidge, C., Hofstede, A., van der Aalst, W.M.P.: Mining resource profiles from event logs. ACM Trans. Manag. Inf. Syst. **8**(1), 1–29 (2017)
11. Barbastefano, R.G., Carvalho, D., Lippi, M.C.: Process mining classification with a weightless neural network. In: Thomé, A., Barbastefano, R., Reis, J. (eds.) Springer Proceedings in Mathematics & Statistics, vol. 337, pp. 349–356. Springer International Publishing (2020). https://doi.org/10.1007/978-3-030-56920-4_28
12. Chiavenato, I.: Administração Geral e Pública, 2nd edn. Elsevier, Rio de Janeiro (2008)
13. Thiede, M., Fuerstenau, D., Bezerra Barquet, A.P.: How is process mining technology used by organizations? Systemic literature review of empirical studies. Bus. Process. Manag. J. **24**(4), 900–922 (2018)

Circular Business Models: A Multiple Case Study in Manufacturing Companies in Northern Brazil

Marcia M. C. Bacovis and Miriam Borchardt

Abstract Research on Circular Business Models (CBM) supports companies in changing the business paradigm from linear to circular. The field of research on business models for CE is in a phase of maturation and consolidation, as several publications on the CBM topic have grown exponentially in the databases. However, little research has been carried out in Brazil to investigate how companies are applying EC principles in their business practices, what business models are adopted, and the value achieved. Thus, this study aims to contribute to knowledge about the adoption of CBM in Brazil, identifying the models adopted and which enabling elements are inducing and facilitating the transition. This research is exploratory, based on qualitative data, having as a research method the study of multiple cases carried out in five large manufacturing companies installed in an industrial park in northern Brazil, which are introducing the principles of CE in the form of doing business and achieving economic and environmental results. The survey results show that the companies surveyed are diversifying their business model towards models that slow down or close the production loop in a hybrid way (linear and circular), revealing that the creation of ecological value has been greater than the economic value. The role of several enabling elements was observed, but innovation is one of the main enablers for circularity and the creation of more sustainable organizations.

Keywords Circular economy · Circular business model · Enablers · Industrial Symbiose · Extending resource value

M. M. C. Bacovis (✉)
Federal Institute of Education, Science and Technology of Amazonas, Manaus, AM, Brazil
e-mail: mmbacovis@ifam.edu.br

M. Borchardt
Vale do Rio dos Sinos University (UNISINOS), São Leopoldo, RS, Brazil

V. M. López Sánchez et al. (eds.), *Industrial Engineering and Operations Management*, Springer Proceedings in Mathematics & Statistics 400,
https://doi.org/10.1007/978-3-031-14763-0_31

1 Introduction

Sustainable development has been a challenge for business organizations, which are called upon to balance their rate of exploitation of natural resources without compromising future generations [1]. In the context of sustainable development, the Circular Economy (CE) emerges as a strategy that aims to improve the efficiency of materials and energy use [2]. Circular Economy (CE) is a paradigm that suggests a redesign of the current linear economic system, largely based on linear flows of resources where a product is commonly used once and its embodied value is lost, to closed-loop resource flows. that can preserve the environmental and economic value embodied in the products over time [2–4]. In this context, the concept of a circular business model (CBM) emerges to help companies transition to a circular economy and adopt strategies such as reuse, repair and remanufacturing. In this way, commercial value creation is reconciled with the adoption of eco-efficiency strategies (eco-efficacy) [5].

Research on circular business models helps companies to change the business paradigm [6]. But the academic debate on the subject is still insufficient, as few studies investigate how companies can capture CE principles in their business practices [3, 7]. Manufacturers, who have been operating close to the linear industrial model, face difficulties in understanding how and why to change their business models; that is, what are the benefits of the circular model [3, 8]. Business models for CE have been widely studied and implemented in countries such as the Netherlands, China, Japan, and the USA [1, 9]. However, more studies are lacking in the context of emerging countries such as Brazil [10–12].

To contribute to the emerging knowledge about the adoption of circular business models in Brazil, to present propositions that guide other companies to make the transition to the circular economy, five case studies were carried out in large manufacturing companies that are introducing the EC principles and changing the way of doing business, achieving better economic, environmental and social results. This is an undeveloped region, in industrial terms, and which has a huge amount of available natural resources, which cannot be used in their entirety, without mechanisms that promote sustainable development, such as the principles of the circular economy., are applied correctly. In this type of region, economic success is almost always paid for with the depletion of natural resources, generating a loss of biodiversity and climate change, as well as impacts on the health of the local population [13].

Given this context, the research questions that guided this study were: What circular business models (CBM) are being implemented in manufacturing companies in emerging countries, such as Brazil? (Q1) What are the facilitators that promote the implementation of CBM in these companies? (Q2) What are the values created by implementing these CBMs?

The article is organized as follows: The first section contains the introduction and the main gaps. The second section presents the theoretical foundation on Business Models for CE and the facilitating elements. The third section describes the methodology used to carry out the case study research. The fourth section is the results and discussions. The fifth summarizes the contributions and limitations.

2 Theoretical Foundation

2.1 *Business Model*

A business model is a conceptual tool to help understand how a company does business and can be used for performance analysis, benchmarking and evaluation, management, communication, and innovation [14]. Four main characteristics of business models emerge from the literature, namely: (i) the value proposition; (ii) the value network, (iii) value capture; (iv) and value creation and delivery, referring to the key activities, resources, channels, technology, and standards that create value and the way value are then (re)distributed [4, 14].

The choice of business model guides the "business architecture" and expansion paths of the company. Once these paths are established, companies may find it difficult to change or adopt new business models [15]. The shift to a circular economy model (CBM) is a sea change that will require a new way of thinking and doing business [15, 16]. In a business model for EC, value creation is based on using the economic value retained in products after use, through the production of a new offering [5, 17, 18]. Thus, the creation, delivery, and capture of value happen by decreasing, closing, or narrowing the flows of energy and materials [15, 19, 20].

2.2 *Circular Business Model*

The field of research on the circular business model is in a phase of maturation and consolidation since several publications on the subject have appeared in the databases [4, 6, 15, 19, 20]. Thus, the literature has presented several taxonomies and archetypes of circular business models [4, 6, 15, 21–23], which will be detailed in Table 1, below.

According to Moreno et al. [22], the taxonomies of circular business models present a discourse based on economic terms of value generation, activities, and revenue sources. For Rizzos et al. [24], comprehensive knowledge of the design of CBM is necessary to stimulate and encourage the implementation of CE at a micro-level. Linking CE with business practice requires specific business models supported by a framework that captures CE principles [3]. For Lewandowski [6], the constituent elements of circular business models are derived from the CE principles. In the literature, such elements are understood and defined in a variety of ways, for example through the ReSOLVE framework [2], with new forms of circular value creation [15, 22, 25] and normative requirements for business models [6, 14].

Table 1 Circular business model present in the literature

References	Circular business archetypes
Bocken et al. [4]	1. Maximize Resource Efficiency 2. Create value from waste 3. Substitute for renewables and natural processes 4. Deliver functionality instead of ownership 5. Adopt a stewardship role 6. Encourage sufficiency 7. Re-purpose the business for society/environment 8. Develop scale-up solutions
Lacy and Rutqvist [21]	1. Circular supplies 2. Resource recovery 3. Extension of product life 4. Sharing platforms 5. Product service system 6. Virtualization.
Macarthur [2]	Framework ReSOLVE: 1. Regenerate 2. Share 3. Optimize 4. Close cycle 5. Virtualize 6. Exchange
Bocken et al. [15]	**Slowing** 1. Access and performance model 2. Classic long life 3. Extending product value 4. Encourage sufficiency **Closing** 1. Extending resource value 2. Industrial Symbiosis
Moreno et al. [22]	1. Sharing platform 2. Extending product value 3. Extending product life 4. Resource value 5. Circular supplies.

According to Sehnem et al. [11], CBM assumptions include circular supply – renewable energy, fuels, and bio-based products; waste as a resource from which useful resources and energy can be recovered; industrial symbiosis sharing platforms; get more benefits with the same volume of goods; the product as a service – dematerialization of products by providing "access to products", maintaining the producer's ownership.

Table 1 presents the most present business models in the researched literature.

Table 2 Enabling elements

Enablers	Description	References
Systemic approach	Integrate product and process design actions, business model and new forms of consumption, new ways of producing, consuming, and relating to customers.	[2, 10]
Collaboration/cooperation	The organization must collaborate internally and externally through formal and/or informal arrangements to create mutual value. Collaboration in supply chains is essential to closing the loop and converting waste into useful resources.	[10, 26]
Innovation	The organization must continually innovate to create value, allowing the sustainable management of resources through the design of processes, products/ services, and business models. Innovations can be technical innovations, organizational innovation, social innovation.	[4, 20, 28, 32]
Government incentives	Governments and government agencies should encourage companies that implement CE initiatives, allowing access to credit, and create legislation that stipulates rewards to companies that implement CE.	[4, 21, 33]
New technologies	Technological changes and new information technologies play an important role in the transition to the Circular Economy.	[24, 29–31]

2.3 *Enabling Elements for the Transition to the Circular Economy*

The successful implementation of an CE requires a series of actions and elements, such as collaboration/cooperation between supply chain companies and other parties [4, 26], innovation [4, 20, 27, 28], use of new technologies [24, 29–31], Systemic Approach [2, 32] and, government incentives and financial incentives [4, 21, 33].

The facilitators identified during the literature review are presented in Table 2, below.

3 Research Method

The literature review provided the theoretical basis for the multiple case study carried out here. Case studies are suitable for investigations of contemporary phenomena [33]. This research method was chosen due to the exploratory nature of the research, where the phenomenon studied is little studied or is not yet organized

Table 3 Multiple case study companies

	Case A	Case B	Case C	Case D	Case E
Activity	Electronics industry	Electronics industry, home appliances	Electronics industry	Chemical Industry (Vinyl Flooring)	Chemical Industry (Welding Technology)
Origin Country	USA	USA	South Korea	France	South Korea France USA
CBM	Product Service System; Extending product value; Extending resource value (electronic waste recycling)	Product Service System; Extending product value; Extend the value of resources (Package recycling)	Extending product value; Extending resource value (Remanufacturing, repair; Recycling)	Product Service System; Extending resource value; Reverse logistics of floors for recycling	Industrial Symbiosis; Extending resource value (Solder Sludge Recycling)

in an academically structured way. Content analysis was used to code and analyze the results of semi-structured interviews; for this, the NVIVO software version 11 was used.

After reviewing the literature, providing the main bibliographic reference, the next step was to choose the companies that would be part of the research. The main criteria for selecting companies were that they disclosed in their Sustainability Report, published on their website, that they were applying CE principles (energy efficiency, emission reduction, reuse of materials, redistribution, remanufacturing, recycling) and diversifying their business model for models of closing loops, delaying resource loops and reducing resource flows or regeneration and recovery processes [2, 15]. Thus, five manufacturing companies, subsidiaries of global corporations, were selected to explore business opportunities from a CE perspective. The companies are located in an industrial cluster in northern Brazil, with three companies operating in the electronics sector and two in the chemical sector. Despite the different sectors and business models, each company chosen for the case study represents a starting point to characterize elements and functions that contribute to the circular economy.

The five companies have had an Integrated Management System based on ISO 9001, ISO 14001, OHSAS 18001 standards for over 10 years. Companies in cases A, B, C, and E also meet the ROHS (Restriction of Certain Hazardous Substances) directive. Table 3 shows the companies participating in the multiple case study.

Data collection was carried out through semi-structured interviews with key informants (Environmental Analyst and Environment, Health and Safety Managers). The visits and interviews were carried out between January and May 2021, and the information was obtained on the companies' website and in documents provided by the interviewees; Phone contacts and interviews were also made through Zoom and Google Meet, due to the Covid-19 pandemic.

The results were compared through triangulation, involving data, perspectives, or theories, in response to the reliability and validity of the study with the theoretical findings.

4 Results and Discussion

The revised literature on the circular business model (Sect. 2.2) points out that organizations need a new approach to generate economic, social, and ecological value to the business in a more sustainable way [15, 19, 34].

Table 4 organizes the characteristics of the business models identified in the case studies, based on the proposal by Bocken et al. [15]. Then, Fig. 1 presents the circular business model strategies adopted by the five companies.

Table 4 and Fig. 1 show the consolidated results of the five cases are presented in Table 4. The companies in cases A, B, C, and D have CBM to decelerate the loop or slow the loop [15]. Companies in Cases A and B (electronics industry) are combining product life extension with Product as a Service (PSS model) to facilitate services that simultaneously increase product longevity (through reuse, maintenance, and remanufacturing), maintaining product ownership with the company. These strategies agree with studies by [4, 17] who claim that offering functionality rather than ownership through PSS ensures that the product returns and is remanufactured after use [15, 17, 35]. However, the product-as-a-service (PSS) model offered by companies in cases A, B, and D is still not well established. Respondents emphasized that they are testing this model and that they still face several problems.

The five companies in the case study Extend the value of resources through recycling [15, 20, 36, 37], from the "Closing the Loop" archetype. In this model, the value proposition is to explore the residual value of resources and convert them into new forms of value [15, 22]. Companies A, D, and E carry out recycling internally; companies B and C carry out this operation externally, through partners (recyclers, associations, or cooperatives). Extending the value of resources is the most common business model identified in the five cases [19, 35]; this result is in agreement with recent research carried out in manufacturing companies in Italy [38], demonstrating that many stakeholders still have the understanding that the EC is seen as waste management and recycling proposal. This result also confirms the claim that "CE implementation worldwide is in the early stages, focused on recycling rather than reuse [9, 35, 37].

As for the Industrial Symbiosis [15, 20], practiced by companies in cases A, D, and E. The company in case D has a partnership with intermediary companies to acquire PVC (pharmaceutical industry waste) while in case of E, the company carries out the entire process directly with its customers in the purchase of solder sludge and reverse logistics for the company's units. The literature emphasizes the

Table 4 Circular Business model archetypes in the companies of the study

Cases	Slowing the *Loop*	Closing the *Loop*
A	**Extending product life** – through the services of the Technical Assistance Network (maintenance); **Product as a Service** for notebooks. **Extending product value** – Factory remanufacturing of cell phones, printers, and notebooks.	**Extending resource value** – internal recycling, using the residual value of resources. **Industrial Symbiosis –** sends waste and by-products not used in the company to partners.
B	**Extending product life –** using quality inputs and through the services of the Technical Assistance Network; **Product as a Service (PSS)** – leasing of water purifiers; **Extending product value** – (Remanufacture of water purifiers)	**Extending resource value** – recycling of plastic parts, washing machines, and refrigerator motors – in partnership with ABREE (Brazilian Association for Recycling of Electronics and Appliances).
C	**Extending product life,** using quality inputs and through the services of the Technical Assistance Network; **Extending product value** (remanufacturing, repair) – Return program for old devices (cell phone) when you buy a new one.	**Extending resource value** – recycling electronic components, metal and plastic parts of televisions, smartphones, and other products; in partnership with Technical assistance network and with outsourced companies that carry out the reverse manufacture and recycling of components.
D	**Product as a Service** of vinyl floors for large events (sporadically).	**Extending resource value,** through internal recycling of floors and floors that return from post-installation. **Industrial Symbiosis** – acquisition of PVC inputs from partners; sending your waste to other partners (cement factory).
E		**Extending resource value** by recycling solder sludge; **Industrial Symbiosis –** acquisition of solder sludge generated in the customers' production process.

need for mutual competitive advantage between the two subjects participating in the Symbiosis [26, 38]. Ruggiere et al. [26] identified several factors (regulatory issues, fiscal and financial barriers, and consumer behavior) for organizational innovation and adoption of Industrial Symbiosis; providing value capture through potential savings using by-products and waste that would otherwise be wasted [20].

During the interviews, the interviewees confirmed several facilitating factors that leveraged the introduction of CE principles in the company, such as collaboration with partners in the value chain, use of new technologies in product and process

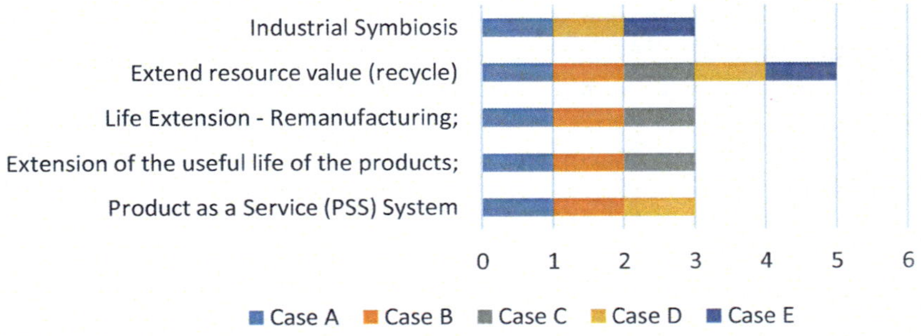

Fig. 1 Circular business model archetypes in the companies of the study

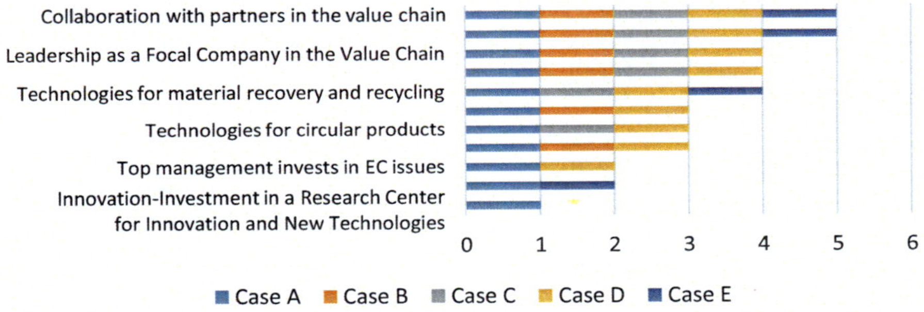

Fig. 2 Enabling Elements identified in the case study companies

development, use of information technologies to track products and for the recovery and recycling of materials, the commitment of top management to the adoption of EC principles, leadership as a focal company, the establishment of a culture of circularity and a change of mindset.

Figure 2 shows the enabling elements identified in each company.

Collaboration with partners in the value chain, present in the five cases, was identified as a key element for CBM development. Respondents highlighted that the company has implemented actions to strengthen the collaborative relationship with stakeholders and partners in the value chain, confirming the results of several studies [4, 26].

The adoption of new technologies is pointed out by the interviewees as the second factor in the transition process to CE. Companies have invested in technologies to make processes more eco-efficient (A, B, C, D, and E), technologies to develop circular products, using recycled inputs (case A, C, and D), technologies for recovery and recycling (cases A, C, D, and E). These results are in line with the works of [1, 29–31], which address the use of technologies, mainly information technologies (Big Data, IoT, mobile technology, data analysis), as an enabling element that can boost EC.

Innovation is another element identified as an enabler of CE. The main innovation pointed out by the interviewees is the question of "Development of products for circularity", present in the speech of the interviewees from A, B, C, and D. The interviewees from A and D have an understanding of the necessary change in the business model and those products need to be designed for many life cycles and use by multiple users, enhancing the benefits to the environment. The results in cases A and D are in line with works that addressed innovation as an enabler for CE [8, 27, 28, 39, 40].

The Circular Culture was pointed out as enabling element by the interviewees from A and D, and this had not been identified during the literature review. Respondents from A, B, and D reported that Senior Management invests in training for managers, engineers, and multidisciplinary teams (designers, chemical engineers, buyers, etc.) to understand CE, participation in workshops, and the dissemination of this content at meetings and internal training. In case A, an independent business unit was created, in addition to the Innovation and Technology Center to consolidate EC's actions.

The need for a systemic approach was pointed out by the interviewee of B, according to the excerpt "an internal reformulation of the productive processes from top to bottom is necessary; the move to CE needs to happen at all hierarchical levels". The managers and analysts interviewed realize that it will be necessary to rethink all the company's strategies, especially regarding the business model and society's awareness of this new paradigm, as the change is radical.

5 Final Considerations

This study expands the current state of knowledge about the adoption of circular economy principles and innovation in the business model for CE that are being adopted by multinational companies with factories installed in Brazil, supported by guidelines from the parent company. The companies surveyed are diversifying their business model towards models that slow down or close the production loop in a hybrid way, that is, in conjunction with the classic (linear) model that allows economic return and competitive advantage. However, research has revealed that ecological value creation has been greater than economic value. The study confirms that there is still no or little financial return; actions seeking to apply CE principles are still seen more as an investment.

The study confirmed the importance of several enabling elements, such as collaboration, use of new technologies for product and process design, focal company leadership in the ecosystem, and innovation as key drivers for circularity and value creation. The main findings and reflections are: (i) Collaboration between the various partners, including universities, will allow the creation of innovation ecosystems that seek sustainable solutions; (ii) new technologies will allow companies to gain a competitive advantage with EC initiatives/practices; (iii) investment in new technologies makes processes more eco-efficient; (iv) investment in innovations and technologies to develop circular products (circular design), using recycled inputs (circular supplies) facilitates remanufacturing, recycling with greater added value (upcycling); (v) circular business models seek to generate "circular value", not necessarily new products; (vi) the company's role as an orchestrator of the ecosystem to which it is a part is essential to manage the various interests and ensure alignment between the ecosystem partners.

About the creation and capture of circular value, the implementation of EC strategies and practices has generated savings through actions to close internal cycles, such as closing the water loop, using standardized packaging, reusing packaging, recycling of materials and products [2, 6]. The adoption of CBM can promote the achievement of several goals of the UN 2030 Agenda, the 17 Sustainable Development Goals (SDGs) [7, 10].

The interviewees recognized that innovations and diversification in the business model are the main actions for the transition to CE in companies, but that there are still many barriers to be overcome, such as double taxation of products and consumer acceptance, to acquire remanufactured products, or to become a user rather than an owner of a product. It is concluded that in Brazil, as in other emerging countries, CE still faces institutional voids and sustainability paradoxes [12], corroborating the study by Gusmerotti et al. [38] that the CE paradigm is still far from being institutionalized, even in large manufacturing companies.

Finally, the main limitation refers to the number of companies (five), the number of people interviewed per company (generally two), and the position/function, only in the sustainability area (EHS manager and environmental analysts), limiting the scope of discussions and opinions of the relevant aspects in the transition to the new production and commercialization model.

References

1. Agyemang, M., et al.: Drivers and barriers to circular economy implementation: an explorative study in Pakistan's automobile industry. Manag Decis. **57**, 971–994 (2019)
2. Ellen MacArthur Foundation, Zumwinkel, K., Stuchtey, M.R.: Growth Within a Circular Economy Vision for a Competitive Europe. Ellen MacArthur Foundation, Cowes (2015)
3. Lieder, M., Rashid, A.: Towards circular economy implementation: a comprehensive review in the context of the manufacturing industry. J. Clean. Prod. **115**, 36–51 (2016)
4. Bocken, N.M.P., et al.: A literature and practice review to develop sustainable business model archetypes. J. Clean. Prod. **65**, 42–56 (2014)
5. Linder, M., Williander, M.: Circular business model innovation: inherent uncertainties. Bus. Strateg. Environ. **26**(2), 182–196 (2017)
6. Lewandowski, M.: Designing the business models for circular economy—towards the conceptual framework. Sustainability. **8**(1), 43 (2016)
7. Merli, R., Preziosi, M., Acampora, A.: How do scholars approach the circular economy? A systematic literature review. J. Clean. Prod. **178**, 703–722 (2018)
8. Blomsma, F., Brennan, G.: The emergence of circular economy: a new framing around prolonging resource productivity. J. Ind. Econ. **21**(3), 603–614 (2017)
9. Stewart, R., Niero, M.: Circular economy in corporate sustainability strategies: a review of corporate sustainability reports in the fast-moving consumer goods sector. Bus. Strateg.

Environ. **27**(7), 1005–1022 (2018)
10. Silva, F.C., Shibao, F.Y., Kruglianskas, I., Barbieri, J.C., Sinisgalli, P.A.A.: Circular economy: analysis of the implementation of practices in the Brazilian network. Revista de Gestão. **26**(1), 39–60 (2019)
11. Sehnem, S., et al.: Circular business models: level of maturity. Manag. Decis. **57**, 1043–1066 (2019)
12. Jabbour, C.J.C., et al.: Stakeholders, innovative business models for the circular economy and sustainable performance of firms in an emerging economy facing institutional voids. J. Environ. Manag. **264**, 110416 (2020)
13. Bacovis, M.M.C., Borchardt, M.: Assessing the Influence of circular economy practices in companies that orchestrate an ecosystem of a Brazilian industrial cluster. In: Tavares Thomé, A.M., Barbastefano, R.G., Scavarda, L.F., Gonçalves dos Reis, J.C., Amorim, M.P.C. (eds.) Industrial Engineering and Operations Management. IJCIEOM 2021. Springer Proceedings in Mathematics & Statistics, vol. 367. Springer, Cham (2021)
14. Osterwalder, A., Pigneur, Y., Tucci, C.L.: Clarifying business models: Origins, present, and future of the concept. Commun. Assoc. Inf. Syst. **16**(1), 1 (2005)
15. Bocken, N.M.P., et al.: Product design and business model strategies for a circular economy. J. Ind. Prod. Eng. **33**(5), 308–320 (2016)
16. Frishammar, J., Parida, V.: Circular business model transformation: a roadmap for incumbent firms. Calif. Manag. Rev., [s. l.]. **61**(2), 5–29 (2019)
17. Mentink, B.A.S.: Circular business model innovation: a process framework and a tool for business model innovation in a circular economy (2014)
18. Baudet, Y.: Circular pioneers in a linear economy the transition in value creation in circular business models (2017)
19. Geissdoerfer, M., Morioka, S.N., de Carvalho, M.M., Evans, S.: Business models and supply chains for the circular economy. J. Clean. Prod. **190**, 712–721 (2018)
20. Pieroni, M.P.P., Mcaloone, T.C., Pigosso, D.C.A.: From theory to practice: systematising and testing business model archetypes for circular economy. Resour. Conserv. Recycl. **162**, 105029 (2020)
21. Lacy, P., Rutqvist, J.: Waste to wealth: creating advantage in a circular economy. Accenture Strategy. **293** (2015)
22. Moreno, M., et al.: A conceptual framework for circular design. Sustainability. **8**(9), 937 (2016)
23. Lüdeke-Freund, F., Gold, S., Bocken, N.M.P.: A review and typology of circular economy business model patterns. J. Ind. Ecol. [s.l.]. **23**(1), 36–61 (2019)
24. Rizzos, V., Behrens, A., Van Der Gaast, W., Hofman, E., Ioannou, A., Kafyeke, T., Flamos, A., Rinaldi, R., Papadelis, S., Hirschnitz-Garbers, M., Topi, C.: Implementation of circular economy business models by small and medium-sized enterprises (SMEs): barriers and enablers. Sustainability. **8**(11), 1212 (2016)
25. Van Renswoude, K., Ten Wolde, A., Joustra, D.J.: Circular business models—part 1: an introduction to IMSA's circular business model scan. IMSA, Amsterdam (2015)
26. Ruggieri, A., et al.: A meta-model of inter-organizational cooperation for the transition to a circular economy. Sustainability. **8**(11), 1153 (2016)
27. Prieto-Sandoval, V., Jaca, C., Ormazabal, M.: Towards a consensus on the circular economy. J. Clean. Prod. **179**, 605–615 (2018)
28. De Jesus, A., et al.: Eco-innovation in the transition to a circular economy: An analytical literature review. J. Clean. Prod. **172**, 2999–3018 (2018)
29. De Sousa Jabbour, A.B.L., et al.: Industry 4.0 and the circular economy: a proposed research agenda and original roadmap for sustainable operations. Ann. Oper. Res. [s. l.]. **270**(1), 273–286 (2018)
30. Nobre, G.C., Tavares, E.: Scientific literature analysis on big data and internet of things applications on circular economy: a bibliometric study. Scientometrics. **111**(1), 463–492 (2017)

31. Bressanelli, G., et al.: The role of digital technologies to overcome circular economy challenges in PSS Business Models: an exploratory case study. Procedia CIRP. **73**, 216–221 (2018)
32. British Standards Institution – BSI 8001 – Framework for implementing the principles of the circular economy in organizations. Guide (2017)
33. Govindan, K., Hasanagic, M.: A systematic review on drivers, barriers, and practices towards a circular economy: a supply chain perspective. Int. J. Protein Res. **56**(1–2), 278–311 (2018)
34. Yin, R.K.: Case Study Research and Applications: Design and methods. Sage Publications, Los Angeles (2017)
35. Ghisellini, P., Cialani, C., Ulgiati, S.: A review on circular economy: The expected transition to a balanced interplay of environmental and economic systems. J. Clean. Prod. **114**, 11–32 (2016)
36. Tukker, A.: Product services for a resource-efficient and circular economy–a review. J. Clean. Prod. **97**, 76–91 (2015)
37. Kirchherr, J., Reike, D., Hekkert, M.: Conceptualizing the circular economy: an analysis of 114 definitions. Resour. Conserv. Recycl. **127**, 221–232 (2017)
38. Gusmerotti, N.M., Testa, F., Corsini, F., Pretner, G., Iraldo, F.: Drivers and approaches to the circular economy in manufacturing firms. J. Clean. Prod. **230**, 314–327 (2019)
39. Chertow, M.R.: Industrial symbiosis: literature and taxonomy. Annu. Rev. Energy Environ. **25**(1), 313–337 (2000)
40. Konietzko, J., Bocken, N., Hultink, E.J.: Circular ecosystem innovation: an initial set of principles. J. Clean. Prod. **253**, 119942 (2020)

Analysis of Prospective Scenarios: A Study on Educational REITS in Brazil

Lucas Ramon dos Santos Hermogenes, Pablo Luiz Berriel Do Carmo Teófilo, Isaque David Pereira de Almeida, Carlos Francisco Simões Gomes, and Marcos dos Santos

Abstract Among the problems caused by the pandemic of the new coronavirus (SARS-COV-2), besides the irreparable loss of loved ones and the damage to global health caused by the disease, the restrictions imposed and the economic losses incurred by them stood out. Given the sudden changes imposed on the routine of society and companies, many businesses went bankrupt, while the other survivors needed to adapt quickly, resulting in a routine based on home office, e-commerce and distance learning. The educational sector was strongly affected by these restrictions, as well as the assets linked to it, as highlighted by the cumulative annual drop of 22% of the IFIX (index of Real Estate Investment Funds), witnessed by investors during the arrival and spread of the pandemic in Brazilian territory. Thus, the work was based on the prospective analysis of scenarios through the Momentum method, producing three possible future scenarios for the recovery of the educational REITs (Pessimistic, Optimistic, and Trend), with the help of three financial planning specialists. At the end of this study, it was possible to configure the scenarios: "The recovery of education REITs" as optimistic scenario, "Challenges of education REITs" as trend scenario, and "The crisis of education REITs" as Pessimistic scenario.

Keywords MOMENTUM · COVID-19 · Education REITs · Prospective scenarios

L. R. dos Santos Hermogenes (✉) · P. L. B. Do Carmo Teófilo · I. D. P. de Almeida · C. F. S. Gomes
Fluminense Federal University (UFF), Niterói, RJ, Brazil

M. dos Santos
Military Institute of Engineering (IME), Urca, RJ, Brazil

V. M. López Sánchez et al. (eds.), *Industrial Engineering and Operations Management*, Springer Proceedings in Mathematics & Statistics 400,
https://doi.org/10.1007/978-3-031-14763-0_32

1 Introduction

Nowadays, a new outbreak has taken the role of the most dangerous to peoples health [1], coming up as a new virus from the coronavirus family (SARS-COV-2), causative of COVID-19 [2], with the pandemic, popularly called the "Coronavirus", companies needed to adopt measures of social isolation and withdrawal, in order to slow down the disease's progress and prevent healthcare system collapse [3]. Due to the great concern for the people's protection, society's routine has been severely affected, paralyzing several usual activities worldwide, like shopping, studying, traveling, and many others. Furthermore [4] estimate that a scenario with no combating coronavirus' strategies established could result in about seven billion infections and 40 million deaths. Concerning business activities [5], point out the related risks to the economic interruption and its ripple effect towards supply chains col-lapse, companies' bankruptcy, and market uncertainties. According to [6], the damage caused to the global economy during the pandemic of COVID-19 approached to trillions of dollars, presenting a devaluation in the price of various assets as the spread of the new coronavirus evolved. For [7], Albeit many assets have seemed to be a great business opportunity among the devaluation presented, many are still the uncertainties related to their recovery, such as the scenario maintained for part of the Brazilian Real Estate Investment Trusts [8].

According to [9], during the height of fears regarding the arrival of the new coronavirus in Brazil in March 2020, the capital market witnessed a sharp fall of 15.85% for the IFIX, the Real Estate Investment Trusts index (Contemplating an annual fall of 22%). Given the challenges and the high complexity offered by the pandemic of COVID-19, robust techniques are required so as to produce feasible future visions about the problem through the available data, providing medium and long-term prospective scenarios in order to support strategic decision-making in the present.

The current research aims to provide three feasible prospective scenarios (Pessimistic, Optimistic, and Trend) for the Brazilian Education REITs during the COVID-19 pandemic, considering its challenges and consequences, by the Momentum method in the light of three financial planning specialists analysis.

The article is divided into five steps, which are: Introduction, Bibliometric Analysis, Theoretical Rationale, Case Study and Analysis of results and final considerations.

2 Bibliometric Analysis

This article brings an innovative approach, as it demonstrates the analysis of prospective scenarios in Educational Real Estate Investments Trusts, since there are no applications with this kind of approach, bringing optimistic, pessimistic, and trend scenarios, considering relevant variables for the study, where for each scenario it is pertinent that different decisions be taken. Seeking to understand the idea of Real Estate Investments Trusts, as well as the respective elements and restrictions of the system, a search was conducted in the existing literature for Real Estate Funds from the SCOPUS database, accessed through the Capes portal (www.capes.gov.br), in August 2021.

Initially, the theme was investigated using the following string of words: TITLE-ABS-KEY ("real estate funds" OR "reits" OR "reit" OR "real estate investment trusts" OR "real estate investment trust"), obtaining 1918 documents, being considered for the present study only the published articles, which corresponded to a total of 1610.

By analyzing the documents obtained, the strategy of limiting the articles to the following areas was adopted: Economics, Econometrics and Finance, Business, Administration and Accounting, and Decision Sciences, corresponding to the following string of words: TITLE-ABS-KEY ("real estate funds" OR "reits" OR "reit" OR "real estate investment trusts" OR "real estate investment trust") AND (LIMIT-TO (DOCTYPE, "ar")) AND (LIMIT-TO (SUBJAREA, "ECON") OR LIMIT-TO (SUBJAREA, "BUSI") OR LIMIT-TO (SUBJAREA, "DECI")) with 1306 items.

Aiming to grasp the REITs intellectual's structure distribution, starting from the 1306 selected documents, a distribution analysis was performed by journals, pointing to a total of 159 journals, with 15 journals standing out, which together corresponded to 68% of the publications, as shown. Journal Of Real Estate Finance and Economics – 228 papers, Real Estate Economics – 135 papers, Journal Of Real Estate Portfolio Management – 111 papers, Journal Of Property Investment and Finance – 108 papers, Journal Of Real Estate Research – 58 papers, Pacific Rim Property Research Journal – 36 papers, Journal Of European Real Estate Research – 31 papers, Real Estate Taxation – 31 papers, Journal Of Real Estate Literature – 24 papers, Journal Of Portfolio Management – 15 papers, Applied Financial Economics – 14 papers, Economic Modelling – 14 papers, Managerial Finance – 14 papers, International Journal of Strategic Property Management – 12 papers, Journal Of Property Investment Finance – 11 papers.

In an attempt of understanding the current research context, one more seeking was produced, focusing on the Prospective Scenarios' usage towards Real Estate Investment Trusts. Then, the following string of words was investigated: TITLE-ABS-KEY ("Prospective Scenarios" OR "Prospective Scenario") AND ("real estate funds" OR "Reits" OR "Reit" OR "real estate investment trusts" OR "real estate investment trust"), but no one documents were found. With it, this possible research gap may be explored with this current paper.

3 Theoretical Rationale

3.1 Prospective Scenarios

For [10] state that prospective scenarios development is not related to the prediction of future scenarios but rather to the production of admissible visions of future scenarios, understood within logical boundaries, obtained through studies of political-economic, socio-cultural and technological, environmental and legal factors. According to [11], the difference between prediction and prospection may be summarized following as displayed in Table 1.

For [12] states that prospective scenarios may be defined as a form of representation of possibilities that could happen, given the analysis of possible futures, which have the objective of guiding the decisions taken in the present, considering a set formed by future situations, as well as their central players and the estimated probability of occurrence for uncertain events, whereas [13] describes scenario forecasting as a methodology for imagining possible futures, aiming to obtain detailed sets of possibilities, leading decision makers to consider various changes that might be hidden during applications of other methods.

3.2 Momentum Method

According to [10, 14] Momentum can be defined with a unified strategic planning method, which starts from a hybrid approach that brings together the main resources held among the following methodologies in the literature: Morphological analysis [13]; a systematic technique that aims structures the feasible actions and combina-

Table 1 Prediction × prospection

Prediction	Prospection
Focuses on certainties and hides uncertainties	Focuses on the uncertainties legitimizing their acknowledgement
It originates a single image of the future	It originates vary but logical images of the future
Prioritizes continuity	Prioritizes the considerations of ruptures
Low influence of qualitative	Seeks to align the qualitative with the quantitative
A single prediction can hide risks	The plurality os "futures" points out the risks
May generate inertia	Promotes flexibility
It originates in simple models that become complex	It starts from complex futures and seeks simplification
May generate a sectorial approach	It adopts a global approach from the start

Source: Ref. [5]

tions contained in a given system, Scenario construction Methodology of Business environments [15]; Technique aimed at the business environment that focuses its efforts to the study of uncertainties and their causes, combining the existing variables during the creation of possible scenarios, always considering the evaluation of competitive advantages maintained in the market, Scenario Methodology [12, 16]; Technique directed towards the creation of scenarios utilizing the precise structuring of the problem, the consideration of resources such as the retrospective analysis of variables, the definition of the system's actors and the study of strategic objectives; and the Interax Method [16], which consists of the most probable distribution of events using the analysis of crossed-impacts.

The Momentum method consists in the execution of the following steps [10]:

(1) System overview; (2) Relevant actors' establishment; (3) Variables' study; (4) SWOT Analysis; (5) Uncertainties and causalities analysis; (6) Variables' selection; (7) Key indicators establishment; (8) Prospective scenarios production. Given its benefits, many studies have been carried out with the Momentum supporting, pointing out: [7] with the analysis of prospective scenarios for the feasibility of urban mining in Brazil, [17] providing prospective scenarios for a public university, [18] with the proposal of development of prospective scenarios and the multicriteria decision making aiding and [19] with investments strategies during the pandemic crisis mixing the resources from prospective scenarios (Momentum) and the Multicriteria Decision Analysis (SAPEVO-M-NC).

3.3 Education REITs

Real estate funds are investments based on resources that are raised through the sale of quotas and that can be used for the acquisition of urban, rural, built, under construction, commercial or residential real estate, as well as for the acquisition of securities linked to the real estate sector, such as quotas of other FIIs, Real Estate Letter of Credit (LCI), Certificate of Real Estate Receivables (CRI), shares of real estate companies and many others. Real estate funds are usually classified according to their investment strategies and performance [20]. According to [21], Brazil has invested in education a very low percentage in relation to its GDP - showing figures ranging from 3% to 5% over the last 30 years. Thus, a consensus was built among educators that Brazilian education should provide investments to education, including investments in better salaries for professionals, valuing those duly.

According to [22], encouraging students' interest in Science, Technology, Engineering and Mathematics is one, among the many challenges faced by the Brazilian educational system. Teachers have the mission of encouraging students to enter these academic areas. However, in order for them to fulfill their duties it is necessary that the government works to guarantee adequate salaries and working conditions.

According to [23], the quality of teaching and student performance in developing countries are factors sensitive to teacher remuneration.

With this, excellent approaches have been produced taking the complexity of education evaluation as a challenge, highlighting among them: Graduate program evaluation using a networked data envelopment analysis model [24]; International academic mobility: Student selection using a multicriteria approach [25]; Graduate engineering programs: An analysis of quality and productivity [26]; and Performance evaluation of calculus classes using operations research tools [27]. With support from [20, 28–30], the available data on educational REITs was collected and structured to generate an understanding of the current context for the assets, as shown below:

- **FAED11 – Real Estate Investment Fund – FII Anhanguera Educacional**

According to [28], FAED11 was established in August 2009 and has the purpose of obtain lands, commercial properties, or the rights thereto for expansion and operation through leasing to Anhanguera Educational Group. Its basic policy will be to make long-term real estate invest-ments, aiming, essentially, to acquire real estate that meets the object of FAED11.

- **FCFL11 – Real Estate Investment Fund – FII Campus Faria Lima**

The FCFL11 was established in February 2010 with the objective of investing in real estate projects, through the acquisition of commercial properties, or rights related to them, ready or under construction. These properties may be developed in commercial enterprises, with the purpose of sale, typical or atypical built to suit leasing. FCFL11's basic policy will be to make long-term real estate investments.

- **GESE11 – Real Estate Investment Fund – FII General Severiano**

According to [28], GESE11 was established in July 2016, counting with a 20 years deadline, having the goal of increase the value of its shares, exclusively through the acquisition and subsequent leasing and/or renting of the property located at General Severiano street, in the City and State of Rio de Janeiro. Thus, the fundamental objective basically boils down to obtaining income.

- **MINT11 – Real Estate Investment Fund – MINT Educacional**

According to [28], MINT11 was established in March 2020 and it has as target to obtain income and capital gains through the acquisition of real estate in the education sector.

- **RBED11 – Real Estate Investment Fund – Rio Bravo Renda Educacional – FII**

According to [28], RBED11 was established in September 2011 aiming to make long-term investments in REITs through both acquisition and subsequent management of commercial properties, also having the possibility of acquiring lands for building, notably related to educational activities, such as universities, colleges, technical schools, as well as negotiating quotas of other REITs.

4 Case Study

To apply the Momentum method, its eight steps were accomplished, as shown in detail in the following parts. Some of the Momentum steps accomplishment has counted with the Nord Research's mentoring program at some points, which had provided knowledgeable contributions and insights for some analysis based on financial specialists' expertise and experience [31]. For [32] in requirements, an analysis of the customer's needs is performed; in design, the technical specification of the solution is elaborated; in development, the proposal is constituted in software; and in the test stage, it is verified if the deliverable is in conformity with the specified requirements. In the scenario analysis, it is necessary to verify the variables and indicators. For [33] the best option to structure the problem was from the Soft Systems Methodology (SSM), for [34] the application of Value-Focused Thinking (VFT) resulted in a better result and for [35] the structuring of the problem only focused on the existing alternatives. However, after several researches, we decided to use, in this work, the momentum method to structure the problem, once the generated result is robust because it occurs after the application of some metrics.

- **System overview**

While the system is trying to his recover, several uncertainties are still presents, especially for Educational REITs, since the return to classes in person still presents many challenges in Brazilian territory, whereas, in parallel, distance learning increasingly shows itself as a viable alternative as new variants of the new coronavirus spread worldwide, renewing the same uncertainties and the risks already experienced in the most pandemic's alarming moments.

- **Relevant actors' establishment**

According to [10], the definition of relevant actors is entirely tied to identifying entities that can cause influence in a given system. The actors identified for the present study were described as presented in Table 2.

- **Variables' study**

Following with the variables' study, the P.E.S.T.E.L. analysis were performed, layering the system's variables into Policy, Economic, Social, Technology, Environmental and Legal, as shown in Table 3:

Table 2 Main relevant actors

Actors	Examples
Shareholders	Individual investors
	Corporate investors
	Other investment funds
Health Ministry	Advancing vaccination policies
	Efficient contagion reduction policies
Ministry of Economics	Financial institutions controlling and supervising
	Economic and financial negotiations with governments, multilateral organizations, and government agencies
Main economies basic interest rate	Exchange rate environment
	International interest rate policy
	Market trends
International economy	International agreements and cooperation
	International interest rate policy
	Monetary policy

Table 3 Variables' study

Uncertainty	No	Variable
Policy	V1	Political Stabilization
Economic	V2	Dollar quotation
Economic	V3	Basic interest rate
Economic	V4	Inflation
Economic	V5	GDP
Economic	V6	Taxes on REITs' incomings and dividends
Economic	V7	Country risk
Social	V8	Strengthened investor culture
Technology	V9	Distance learning innovations
Environmental	V10	Delta Variant spreading
Environmental	V11	Water shortage crisis
Legal	V12	Reits legislation

- **SWOT Analysis**

Following with the required steps, the SWOT analysis was performed, aiming to identify the strengths, weaknesses, threats, and opportunities contained in the system, as illustrated in Table 4.

Table 4 SWOT analysis

STRENGTHS (S)	WEAKNESSES (W)
Diversification of investments, mitigating the risks of economic instability.	Assets devaluation due to short-term investor generalized panic
Financial protections usage	Assets risks elevation
Atypical leasings for long-term	Dividends reduction, following the economic restrictions
OPPORTUNITIES (O)	**THREATS (T)**
Good assets with discounted prices	New health crisis due to Delta Variant spreading
Fusions and acquisitions	Taxes proposal for Reits dividends
High returns with pandemic ending	Water shortage crisis
Funds strategies changing (Residencial, retail, corporative ...)	Distance learning overcoming the traditional learning

Impact \ Dependence	V1	V2	V3	V4	V5	V6	V7	V8	V9	V10	V11	V12	∑ (Impact)
V1	x	7	3	3	5	0	7	3	0	0	0	0	28
V2	0	x	5	7	7	0	7	1	0	0	0	0	27
V3	0	3	x	5	7	0	5	3	0	0	0	0	23
V4	0	5	5	x	7	0	7	1	0	0	0	0	25
V5	0	5	7	7	x	0	7	0	0	0	0	0	26
V6	0	0	0	0	1	x	0	-7	0	0	0	0	-6
V7	5	7	1	1	7	0	x	0	0	0	0	0	21
V8	0	0	0	0	0	0	0	x	0	0	0	0	0
V9	0	0	0	0	0	0	0	0	x	0	0	0	0
V10	1	7	7	7	7	0	7	0	7	x	0	0	43
V11	1	7	7	7	7	0	7	0	0	0	x	0	36
V12	0	0	0	0	0	-7	0	0	0	0	0	x	-7
∑ (Dependence)	7	41	35	37	48	-7	47	1	7	0	0	0	216

Fig. 1 Cross-impact matrix

- **Uncertainties and causalities analysis**

In order to measure and understand both the dependency and impact relationship maintained between the identified variables, the Cross-Impact Matrix was produced (Fig. 1), scored according to the predefined gradations in Table 5. Which is used in the cross-impact matrix in order to measure the impact and dependency relationship between the variables.

Table 5 Score gradations

Description	Degree
Very negative impact	–7
Negative impact	–5
Medium negative impact	–3
Little negative impact	–1
No impact	0
Little positive impact	1
Medium positive impact	3
Positive impact	5
Very positive impact	7

Through the cross-impact matrix analysis, it became possible to measure the impact generated by a given variable on the other variables and also to measure how much it was impacted by the other variables. The total impact value was calculated through the algebraic sum of the row of each respective variable and represented by the word "Impact" while the total dependency value was calculated through the algebraic sum of the column of each respective variable and represented by the word "Dependency".

- **Variables' selection**

In order to clearly display the distribution of each variable, a scatter plot was created with two cut lines, one vertical (Mean Impact Value) and one horizontal (Mean Dependency Value), dividing the variables into four quadrants, as exemplified in Fig. 2.

Upper right quadrant – This quadrant represents the set with the most relevant variables for the present study, since they present high impact and dependency indices, both above the average. Lower right quadrant – This quad-rant represents the set with the variables of intermediate relevance to the study, since they present impact indexes above the average and low dependence, with indexes below the average. Lower left quadrant – This quadrant represents the set with the least relevant variables for the pre-sent study, since they present low impact and dependence indices, both below the average. Although the quadrant analysis recommended the exclusion of the less relevant variables for the study, the authors chose to keep the variable Taxes on REITs' incomings and dividends, given the uncertainties about its possible approval.

- **Key indicators establishment**

In order to confer greater consistency in the choice of variables related to the problem, after the quadrant analysis, a correlation study between the quantitative selected variables was proposed. At the end of the correlation study, all quantitative variables were kept, with the addition of the variable Taxes on REITs' incomings and dividends, thus making it possible to create the three possible scenarios: "The

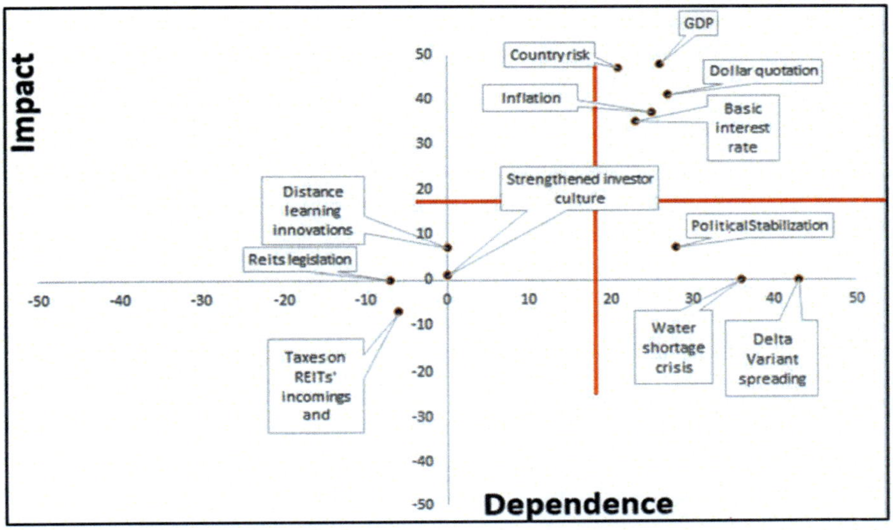

Fig. 2 Quadrant analysis

recovery of education REITs" (Optimistic scenario), "Challenges of education REITs" (Trend scenario), and "The crisis of education REITs" (Pessimistic scenario).

5 Analysis of Results and Final Considerations

During the construction of this work, literature searches were carried out that were able to denote the complexity and potential of education REITs, as well as their relevance in the national territory. The approach using the MOMENTUM method provided a meticulous structuring of the problem, with an analysis that contemplated both uncertainties and challenges present in the future trajectory of the recovery of education REITs, considering possible future visions guided by available economic data and also by the contributions of three financial market agents. The paper can contribute with the understanding of scenarios, and how the process of their evaluation can help in decision making, since when evaluating the indicators in each scenario, different decisions based on these scenarios must be taken. By evaluating the educational funds according to the scenario that may be inserted, the evaluator may analyze which fund is worth the investment and observe that the expansion of educational funds may reflect in a growth of poles and educational centers in Brazil. The work contributes significantly, since it is observed that there is a gap in the use of prospective scenarios in the analysis of real estate educational funds and the growth of investment in this type of fund reflects in the growth of more educational units in Brazil. As result three possible scenarios were proposed for the recovery of education REITs in the middle of the political-economic instability experienced in the country, these being: "The recovery of education REITs" as optimistic scenario; that considers a recovering of all key indicators, "Challenges of education REITs" as trend scenario, that considers the prevision stated by the financial organizations around the country, and "The crisis of education REITs" as Pessimistic scenario; that considers the worst values for the key indicators; thus reaching the objective of the present study. As a suggestion for future works, the use of multicriteria decision support tools would be indicated to order the most profitable funds for possible investments.

References

1. Huang, C., Wang, Y., Li, X., Ren, L., Zhao, J., Hu, Y., Zhang, L., Fan, G., Xu, J., Gu, X., Cheng, Z., Yu, T., Xia, J., Wei, Y., Wu, W., Xie, X., Yin, W., Li, H., Liu, M., et al.: Clinical features of patients infected with 2019 novel coronavirus in Wuhan, China. Lancet. **395**(10223), 497–506 (2020). https://doi.org/10.1016/S0140-6736(20)30183-5
2. Subramanian, B., Adolfo, B.P., Ponmalai, K.: Novel 2019 coronavirus structure, mechanism of action, antiviral drug promises and rule out against its treatment. J. Biomol. Struct. Dyn. (2020). https://doi.org/10.1080/07391102.2020.1758788
3. Hermogenes, L.R.S., Santos, M.D., Nascimento, P.F., Teixeira, L.F.H.S.B.: Importance of digital skills in times of crisis: some applications used during social isolation due to the COVID-19 pandemic. Rev. Augustus, Rio de Janeiro. **25**(51), 198–218 (2020)
4. Walker, P.G.T., Whittaker, C., Watson, O., Baguelin, M., Ainslie, K.E.C., Bhatia, S., Cuomo-Dannenburg, G.: Imperial-College-COVID19-Global-Impact-26-03-2020v2.pdf. (March), pp. 1–19 (2020)
5. Ahani, A., Nilashi, M.: Coronavirus outbreak and its impacts on global economy: the role of social network sites. J. Soft. Comput. Decis. Support Syst. **7**(2), 19–22 (2020)
6. Ozamiz-Etxebarria, N., Dosil-Santamaria, M., Picaza-Gorrochategui, M., Idoiaga-Mondragon, N.: Levels of stress, anxiety and depression in the first phase of the COVID-19 outbreak in a sample collected in northern Spain. Cadernos de Saúde Pública. **36**(4) (2020). https://doi.org/10.1590/0102-311x00054020
7. Teófilo, P.L.B.D.C., Dos Santos, M., Gomes, C.F.S.: Ranking Real Estate Funds during the economic crisis: an approach using value-focused thinking and multicriteria decision making support. In: XXIV Encontro Nacional de modelagem internacional e XII Encontro de Ciência e Tecnologia de materiais, Ilhéus, BA (2021)
8. Market and Consumption: The Shopping REITS falling, and why it may be an opportunity. Available in: https://mercadoeconsumo.com.br/2020/08/31/fundos-imobiliarios-de-shoppings-desabam-as-podem-ser-oportunidade (2020). Last accessed: 17/09/2021
9. B3 S.A.: B3 daily news. Available in: https://www.b3.com.br/pt_br/noticias. Last accessed: 10/09/2021
10. Gomes, L.F.A.M., Gomes, C.F.S.: Principles and Methods for Decision Making: A Multicriteria Approach, 6th edn. Atlas, São Paulo (2019)
11. European Commission: The practical guide to regional prospection in Portugal (2002)
12. Godet, M.A.: The art of scenarios and strategic planning: tools and pitfalls. Technol. Forecast. Soc. Chang. **65**(1), 3–22 (2000)
13. Schoemaker, P.J.H.: Scenario planning: a tool for strategic thinking. Sloan Manag. Rev. **36**(2), 25–40 (1995)

14. Gomes, C.F.S., Da Costa, H.G.: Proposal of the use of foresight in the multicriteria decision process. Prod Eng Res Rep. **13**(8), 94–114 (2013)
15. Porter, M.E.: Competitive Advantage: Creating and Sustaining Superior Performance. Editora Campus, Rio de Janeiro (1992)
16. Godet, M.A.: How to be Rigorous with Scenarios Planning. Futuribles, Paris (2000)
17. Teófilo, P.L.B.D.C., Dos Santos, M., Gomes, C.F.S.: Analysis of prospective scenarios: a study on the feasibility of urban mining in Brazil. In: Brazilian Symposium on Operational Research, 2021, João Pessoa, PB (2021)
18. Sanseverino, A.M., Gomes, C.F.S., Barcelos, M.R.S.: Prospective scenarios for a public university. Engevista. **20**(4), 601–622 (2018)
19. Vieira, J.A.M., Gomes, C.F.S., Braga, I.E.: Development of a scenario prospecting model with the use of multicriteria decision aiding: importance of environmental variables. Braz. J. Oper. Prod. Manag. **14**, 210–217 (2017)
20. Maêda, S.M.D.N., Costa, I.P.D.A.C., Dos Santos, M., Gomes, C.F.S., Teixeira, F.H.D.S.D.B.: Investments in times of pandemic – An approach by Momentum and SAPEVO-M-NC methods. In: XLI Encontro Nacional De Engenharia De Produção, Foz do Iguaçu, PR (2021)
21. Clube FII: Fiis Listing. Available in: https://www.clubefii.com.br/ (2021). Last accessed: 25/09/2021
22. Monlevade, J.A.: Public resources for education: irrevocable percent of GDP. Educ. Policy J. **8**(16) (2014)
23. Pontes, M.C.: Scientific education in Brazil. In: The Space Education Phenomenon at NASA, Brazil and Beyond. United Nations Industrial Development Organization Goodwill Ambassador, Brazil, pp. 240–248. Nova Science Publishers, Inc (2013)
24. Vegas, E.: Teacher labor markets in developing countries. Future Child. **17**(1), 219–232 (2007)
25. Meza, L.A., De Mello, J.C.C.B.S., Gomes Júnior, S.F., Moreno, P.: Evaluation of post-graduate programs using a network data envelopment analysis model. Dyna (Colombia). **85**(204), 83–90 (2018)
26. Torres, B.G., Naciff De Andrade, G., Soares De Mello, J.C.C.B.: International Academic Mobility: selection of students using a multicriteria approach. Espacios. **38**(35), 15 (2017)
27. Soares De Mello, J.C.C.B., Gomes, E.G., Angulo Meza, L., Soares De Mello, M.H.C., Soares De Mello, A.J.R.: Engineering post-graduate programmes: a quality and productivity analysis. Stud. Educ. Eval. **32**(2), 136–152 (2006)
28. Soares De Mello, J.C.C.B.S., Lins, M.P.E., De Mello, M.H.C.S., Gomes, E.G.: Evaluating the performance of calculus classes using operational research tools. Eur. J. Eng. Educ. **27**(2), 209–218 (2002)
29. Status Invest: Tudo sobre Fiis. Available in: https://statusinvest.com.br/fundos-imobiliarios. Last accessed: 20/09/2021
30. B3 S.A.: Fiis listed. Available in: https://www.b3.com.br/pt_br/produtos-e-servicos/negociacao/renda-variavel/fundos-de-investimentos/fii/fiis-listados. Last accessed: 18/09/2021
31. Funds Explorer: Ranking de Fiis. Available in: https://www.fundsexplorer.com.br/ranking. Last accessed: 20/09/2021
32. Jardim, R.R.J., Santos, M., Neto, E., da Silva, E., de Barros, F.: Integration of the waterfall model with ISO/IEC/IEEE 29148:2018 for the development of military defense system. IEEE Lat. Am. Trans. **18**, 2096–2103 (2020). https://doi.org/10.1109/TLA.2020.9400437
33. Almeida, I.D.P., et al.: Study of the location of a second fleet for the Brazilian navy: structuring and mathematical modeling using SAPEVO-M and VIKOR methods. In: International Conference of Production Research - ICPR (2020)
34. Hermogenes, L.R.S., Costa, I.P.A., Santos, M., Gomes, C.F.S.: Acquisition of a CNC router for a joinery in Brazil: an approach from VFT, SAPEVO-M and WASPAS methods. In: International Conference on Pervasive Computing and Social Networking ICPCSN (2022)
35. Santos, M., Costa, I.P.A., Gomes, C.F.S.: Multicriteria decision-making in the selection of warships: a new approach to the AHP method. Int J Analyt. Hierarchy Process. (2021). https://doi.org/10.13033/ijahp.v13i1.833

An Analysis of Different Variable Selection Methods in the Context of a Banking Institution

Jaime Velázquez Juárez and Ana Elisa Lage Ramírez

Abstract Variable selection has become important in some research areas such as data science and analytics, in which datasets with hundreds or even thousands of variables are what is available for model development. Among the objectives of variable selection, also called feature selection, is to improve the predictive capacity of input variables, as well as to provide faster and cost-effective predictors. A greater number of variables theoretically allow more information to be stored, but in practice this rarely happens due to the greater possibility of noise and redundancy in real-world data. Similarly, working with large data sets implies greater computational capacity and the principle of parsimony is no longer fulfilled. In this article, we use different methods for selecting variables, such as Gini, R^2, LASSO, and LARS to select variables in the context of predicting the probability of default of a mortgage loan database. A systematic way for variable selection is used taking advantage of the characteristics of the different selection methods.

Keywords Variable selection · Feature selection · Gini · R^2 · LASSO · LARS

1 Introduction

As data science and analytics aim to tackle more complex tasks, the problem of discerning what is the relevant information within a large amount of data has become increasingly important. For example, extracting a person's bank details often involves dealing with many features and variables. At the same time, the information provided by the credit bureau and the non-traditional data causes the number of variables to grow exponentially.

In this sense, we can find several advantages in the selection of variables, from facilitating the visualization and understanding of the information, to the reduction

J. V. Juárez (✉) · A. E. L. Ramírez
Universidad Anáhuac Norte México, Naucalpan, Mexico
e-mail: jaime.velazquezju@anahuac.mx

V. M. López Sánchez et al. (eds.), *Industrial Engineering and Operations Management*, Springer Proceedings in Mathematics & Statistics 400,
https://doi.org/10.1007/978-3-031-14763-0_33

of information processing and storage time; besides not having to face the curse of dimensionality. Some feature selection methods place more emphasis on one aspect than the other. Many times, the variable selection process does not necessarily provide an optimal solution for the construction of a predictive model, especially when working with redundant variables. On the other hand, a subset of variables that eliminates redundancies could leave out variables relevant to model development [1].

The purpose of this study was to test different variable selection methods that are currently used in analytics within the context of the possible non-payment of a mortgage loan in the financial sector. This context is relevant because, on the one hand, we have many variables, and on the other, there are strong redundancies in the data. In what follows, we first describe the different variable selection algorithms that we used, and we then discuss the results and implications of considering each method for variable selection. We conclude with a decision-making process based on the combination of results from the methods that were utilized.

2 Variable Selection Algorithms

Today there are several methods of variable selection that can help to discern from a group of variables which subset should be used for statistic modeling and which subset should be discarded. It is important to perform variable selection because typically databases contain some variables that provide useful information while others may provide unwanted bias. For this reason, it is necessary to carefully select only the variables that best describe the phenomenon to be modeled. By using a variable selection criterion, the dimensionality of the data can be reduced without losing useful information, and at the same time the information composed of bias can be minimized. In short, to make sure that the results obtained are good, it is necessary to carefully select the variables that will be used [2].

Given that the context of the study was related to credit risk, we decided to use some of the methods that are most used in this area. All the methods were implemented with SAS Enterprise Miner. The methods are:

1. Gini Statistic
2. R^2 (R square)
3. Clustering de variables
4. LASSO regression
5. LARS regression
6. Chi square

According to some authors [3] feature selection methods can be divided according to whether the target variable is taken into consideration in the selection method or not. If the target variable is considered, then the method is considered supervised, whereas, if the target variable is not considered for selection, then the method is unsupervised.

In our study, we used both approaches. For example, Gini statistic and clustering of variables are considered unsupervised, whereas R^2, LASSO, LARS and Chi square are considered supervised. Furthermore, within the supervised methods, there are also three types that can be considered: filter, wrapper, and intrinsic method [4]. In the filter method, features are selected based on how they correlate to the output. A wrapper feature selection method creates models using different subsets of input variables, selecting those subsets that provide better results according to some specified metric. The intrinsic method combines the qualities of both the filter and wrapper method. The model will train and check the accuracy of different subsets and select the best among them. Based on this way of classifying the supervised selection methods, we see that the R^2 can be considered a wrapper method, the LASSO and LARS can be considered intrinsic, and finally the chi-squared can be classified as a filter method.

In what follows we briefly describe each of the methods that were used in this study.

2.1 *Gini Statistic*

Since the Gini index is often used as a partitioning criterion in classification trees, the corresponding impurity is often referred to as Gini importance [5]. It is known that the importance of impurity is biased in favor of variables with many possible dividing points. This typically occurs when a categorical variable has many categories, with continuous variables [6], as well as variables with high category frequencies [7].

The Gini index is nothing more than the Gini coefficient but expressed as a percentage. This means that it varies from 0 to 100. The Gini index constitutes a measure of the uniformity or non-uniformity of a distribution. For lower values of the Gini index, we can assume that the variable is evenly distributed. On the contrary, higher values of the Gini index will point to non-uniformity of the distribution. In the context of the scorecard, the Gini index is used to measure the equality of event rates in the attributes of a descriptive variable [8]:

- The algorithm first sorts attributes in descending order of their event rates. For example, if a variable has *m* attributes, ordered from 1 to *m*, then, attribute 1 will have the highest event rate, attribute 2 will have the second highest event rate, and so on.
- Then, for each of the ordered attributes *i*, the number of events (n_i^{events}) and non-events ($n_i^{no\ events}$) is counted.
- Next, the total number of events (N^{event}) and the total number of non-events ($N^{no\ event}$) in the data are counted. We can observe in Fig. 1 the formula for calculating the Gini Statistic.

Once the formula has been calculated, a cut-off value must be determined. When the value of the Gini index for a feature is less than the cut-off value, then the feature is

$$Gini\ Statistic = \left(1 - \frac{2 * \sum_{i=2}^{m}(n_i^{no\ event} * \sum_{j=1}^{i-1} n_j^{no\ event}) + \sum_{k=1}^{m}(n_k^{event} * n_k^{no\ event})}{N^{event} * N^{no\ event}}\right)$$

Fig. 1 Gini statistic formula

rejected and not considered for model development. Otherwise, the feature will be considered for modeling. The most used cut-off value is .20.

2.2 *R-square (R^2)*

R^2 selection method uses a step-forward least squares regression that maximizes the R^2 value of the model. This criterion is used to identify those variables that could be more useful in predicting the target variable based on linear models [9]. The following steps apply when using R^2 for a model with a binary objective variable:

1. Calculate the square correlations: We first calculate the square correlation coefficient (simple R^2), also known as coefficient of determination. This coefficient refers to the proportion of the target variation explained by a single input variable. It ranges from 0 (no linear relationship between the input and the target variable) and 1 (the input variable explains all the variability of the output variable) [9]. The R^2 is first compared to a default minimum of 0.005. If the squared correlation coefficient is less than the cut-off criterion, then the feature is rejected, otherwise it is considered in the following step [10].
2. Step-forward regression: After step number 1, the remaining significant variables will be evaluated by a step-by-step R^2 regression. This process selects the input variable that explains the greatest amount of variability of the output variable. In each step, an additional variable will be chosen that provides the greatest increment for the R^2 model. The process will stop when there is no additional variable that can be added to the model that improves the R-square criterion [9].
3. Logistic regression for binary targets: if the target is a binary variable (0 and 1), then a logistic regression analysis is executed using the predicted values which are the output of the step-by-step selection process as the independent variable. Because there is one input variable, only the intercept and the slope are estimated. The range of predicted values is divided into a series of equidistant intervals, on which the logistic function is interpolated [9].

This method will help to reduce the number of input variables by rejecting those that do not help explain the output variable, given that they do not achieve a certain desired value of R^2.

2.3 Variable Clustering

When developing a predictive model, we would like to be able to choose those variables that are considered most relevant, that is, those variables that best explain the phenomenon of interest [11]. Many times, we can form clusters of variables that share certain characteristics or that provide information in common. When we have groups of variables with these characteristics, we would like to keep those variables that are representative of each cluster and that can help us explain the variability in the data but considering a smaller subset of input variables [12].

Grouping and selecting variables as described above, can help eliminate collinearity, decrease variable redundancy, and may help reveal the underlying structure of the input variables in the dataset [8]. In this sense, we can still separate the data and explain variability by considering the least number of variables possible.

Variable grouping is used with numeric variables. It separates the variables into hierarchical groups. Each resulting group can be represented by the first principal component of the cluster, which is called the cluster component and refers to a linear combination of the variables in the cluster. The first principal component is a weighted average of the variables of the cluster that explains as much variance as possible [8].

Since cluster components are constructed for each set of variables, the cluster components are not necessarily orthogonal as in Principal Component Analysis (PCA). However, it is guaranteed that the cluster component for each group provides the maximum variance for the group of variables in the cluster.

This algorithm can be used to replace a large set of variables with the cluster components with little loss of information [6]. On the other hand, if there is still a need to maintain the original variables for the model, it is easier to analyze by group which are the variables in each cluster that contribute with higher weights to each cluster component.

This algorithm is divisive and iterative. This means that all variables start in one cluster and in each step, clusters will be divided until the maximum number of clusters is reached or a stopping criterion is met.

Then, the algorithm repeats the following steps:

1. A cluster is chosen for splitting. There are two main criteria that can be followed for cluster selection. The first option is to select the cluster that has the smallest percentage of variation explained by its cluster component. The second option is to select the cluster that has the largest eigenvalue associated with the second principal component [6].
2. The selected cluster is divided into two groups by finding the first two principal components. Each variable will be assigned to the cluster with which they have a higher correlation with the principal component [8].
3. The process continues trying to maximize the variance explained by the cluster components [8].

In summary, variable grouping finds a group of variables that are as correlated as possible with each other within one cluster and as least correlated as possible with variables in other clusters.

Considering the aspects that were explained about the clustering algorithm, it is important to clarify that the database that was used for this study contained more than 100 variables. This grouping algorithm represents a computationally intensive process, so it was decided the perform the grouping in two stages to reduce the calculation times. In what follows, we describe the steps of the algorithm as described in [8]:

1. The variables are identified and then correlations are calculated for all pairs of variables in the dataset.
2. The correlation matrix is used to group variables into initial disjointed groups called global groups. Figure 2 shows the formula.

An example of the calculation of global clusters can be found in Fig. 3. Starting with 350 variables, in the first step of the algorithm the variables are separated into 5 global clusters.

After step 2, as can be seen in Fig. 4, the grouping of variables is carried out in each of the global clusters.

Fig. 2 Formula for the number of global clusters

$$\text{Number of clusters} = \text{INT}\left(\frac{\text{number of variables}}{100} + 2\right)$$

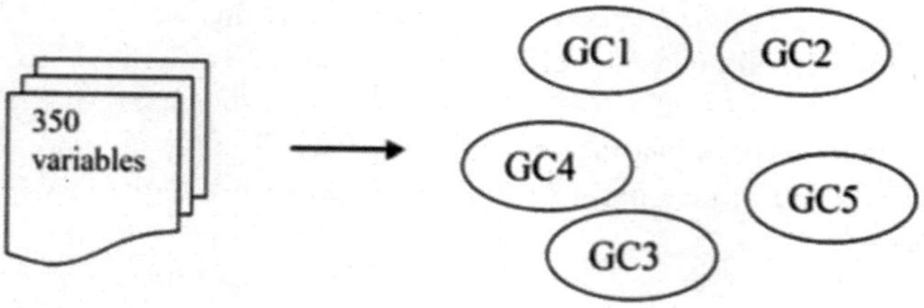

Fig. 3 calculation of global clusters [8]

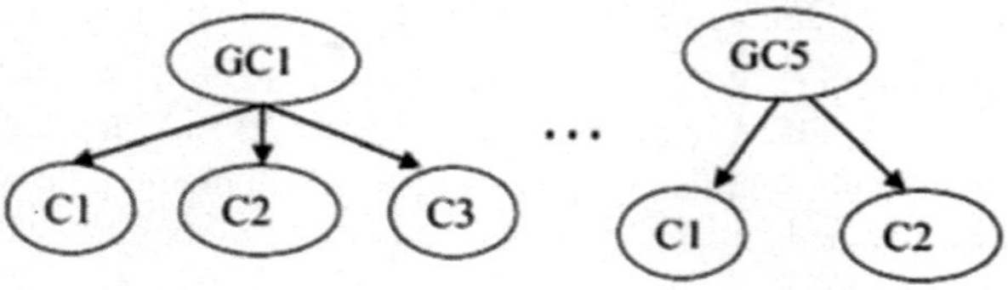

Fig. 4 Cluster construction after global cluster construction [8]

2.4 LASSO (Least Absolute Shrinkage and Selection Operator)

Since classical variable selection techniques perform a discrete process of exploring the modelling space (each variable is selected or discarded), these often suffer from high variability, which in turn can impair their performance in terms of prediction. Instead, regularization techniques are useful in avoiding overfitting on one hand, and on the other, they tend to be more continuous and less variable procedures, which present them as interesting alternatives [13]. Furthermore, regularization makes it possible to train complex models with relatively small data sets, decreasing the risk of overfitting by controlling the complexity of the model.

LASSO was introduced in 1996 [14] and it refers to a regression analysis method that performs both variable selection and regularization in order to enhance prediction results as well as interpretability of the resulting model. It was first introduced as a modeling technique and later used as a variable selection algorithm.

LASSO regularization penalizes the sum of the absolute value of the regression coefficients $\left(||\beta||_1 = \sum_{j=1}^{p} |\beta_j|\right)$. This penalty is known as *l1* and has the effect of forcing the coefficients of the predictors to tend to zero. Since a predictor with zero regression coefficient does not influence the model, LASSO manages to exclude the least relevant predictors. The degree of penalty is controlled by the hyperparameter λ. When $\lambda = 0$, the result is equivalent to that of a linear model by ordinary least squares. As λ increases, the greater the penalty and the more predictors are excluded. Therefore, LASSO performs some variable selections, retaining only the variables for which the estimated coefficients are different from zero. As a result, the models generated by LASSO are generally much easier to interpret, given that it produces sparse models—that only involve a subset of the original variables.

2.5 LAR (Least Angle Regression)

Minimum angle regression (LAR) provides a very quick way to make a forward selection of variables [15]. Like the direct selection method used with regression models, this algorithm also produces a sequence of regression models. In each step a parameter is added to the model and the sequence will stop when all parameters have been added to the model [8].

The algorithm starts by centering and scaling all the variables. Then, the variable that is most correlated with the residual is identified. The regression line will move in this direction. The length of the step will determine the coefficient of this input variable. The step length will be selected so that another input and the predicted response have the same correlation with the current residual. The predicted response will then move in the direction that is equiangular between the two input variables, until a third input variable has the same correlation with the current residual as the two input variables that are already included in the model. The process will continue until all the input variables are considered [8].

2.6 Chi-squared

The Chi-squared test is used to determine whether two variables are independent (from a statistical point of view) or not. This means that if the probability of occurrence of the different possible values of one variable depends on which category of the variable occurs, then we say that the two variables are dependent on each other [16].

The test statistic is constructed by comparing for each cell (combination of categories of both variables) the observed values against the expected values under the hypothesis of independence. The larger the differences between the observed and expected values, the greater the propensity to reject the null hypothesis which would imply that there is no association between the two variables.

3 Analysis and Results

The database that was used for the analysis contains 650,000 records and 344 variables, including categorical, quantitative variables and including a binary target variable that represents whether an individual will fail to cover the corresponding payment of her loan the following month.

The different methods described in the previous section were used with the aim of selecting the most predictive variables that could later be used for model development.

Each algorithm returned a different number of variables in the selection process:

- The Gini statistic method selected 102 variables using as a reference a Gini coefficient of 0.20 and greater.
- The R^2 method selected 10 variables.
- For the clustering variable method 54 variables were selected.
- LASSO yields 46 predictive variables.
- LAR selected 30 variables.
- Finally, Chi-squared, the method that yields 11 variables.

Because each method returned a very different number of variables, we decided to create a table containing those variables that were selected by a greater number of methods. Table 1 shows the 10 variables that were selected in at least 3 different methods.

The Gini statistic was the method that selected the most variables, whereas the R^2 and Chi-squared were the more rigorous methods in terms of variable selection. In the case of the Chi-squared method, even though the target is a binary variable, since many of the variables are continuous, it is possible that the contingency tables resulting for the analyses are numerous and with few values in each of them making it harder to reject the null hypothesis of independence.

Table 1 Selected in at least 3 algorithms

Variable	Gini	R^2	Clustering	LASSO	LARS	Chi square	All methods	Only in 5	Only in 4	Only in 3	Only in 2
Var 1	✓	✓	✓	✓	✓	✓	✓	✓	✓	✓	✓
Var 2	✓	✓	✓	✓	✓	✓	✓	✓	✓	✓	✓
Var 3	✓	✗	✓	✓	✓	✗	✗	✗	✓	✓	✓
Var 4	✓	✓	✗	✓	✓	✗	✗	✗	✓	✓	✓
Var 5	✓	✓	✗	✓	✗	✓	✗	✗	✓	✓	✓
Var 6	✓	✗	✓	✗	✗	✓	✗	✗	✗	✓	✓
Var 7	✓	✗	✗	✓	✓	✗	✗	✗	✗	✓	✓
Var 8	✗	✓	✓	✓	✗	✗	✗	✗	✗	✓	✓
Var 9	✓	✗	✗	✓	✓	✓	✗	✗	✓	✓	✓
Var 10	✓	✗	✗	✓	✓	✓	✗	✗	✓	✓	✓

Not all the selection methods take into consideration the target variable. For example, the Gini algorithm and the variable selection method, both consider the variability of the data itself and not the relationship with the target variable. Given that both methods were the ones that returned a greater number of variables, it is probable that the group of variables still include some redundancies within the data and resulting in more complex modeling that does not necessarily yields to better results.

Finally, the LASSO and LARS selection methods were the more consistent methods in terms of the variables that were selected by each of them. However, it was considered that 46 and 30 variables were still a large number and that it might probably lead to an overfitting of the model.

The two variables that were selected by all the methods were the most relevant in terms of the context, given that both variables were related to late payments. The other relevant variables had to do with the delay the client has had outside the financial institution as well. In this sense, the selected variables help to describe customer behavior regarding their loans which is what is sought to predict from the business point of view.

4 Conclusions

For this study, different methods were used to select variables for a modeling problem in the context of a financial institution that grants loans to its clients and that seeks to predict the non-payment of its clients from a database that has 650,000 registers and 344 variables. Variable selection methods are usually of great interest in this context because there is usually a large number of variables that can present redundancies, collinearity and that can lead to overfitting the models.

Six different methods were used for the selection of variables. The feature selection included supervised and unsupervised methods as well as filter, wrapper, and intrinsic methods. Instead of selecting a particular method, the selection was made with each one of them and in the end those variables that appeared in a greater number of methods were considered as the selected variables. This approach was decided since there are methods that are more restrictive than others and therefore, it was sought to take advantage of the analysis perspective that each one of them offers.

Finally, the selected variables were analyzed in terms of the business context, and it was concluded that the final variables had predictive meaning, so they were used for modeling.

References

1. Avrim, B., Pat, L.: Selection of relevant features and examples in machine learning. Artif. Intell. **97**, 245–271 (1997)
2. Gualdrón Guerrero, O.E.: Artist, Desarrollo de diferentes métodos de selección de variables para sistemas multisensoriales. [Art]. Universitat Rovira i Virgili (2006)
3. Kuhn, M., Johnson, K.: Applied Predictive Modeling, vol. 26, p. 13. Springer, New York (2013)
4. Cai, J., Luo, J., Wang, S., Yang, S.: Feature selection in machine learning: a new perspective. Neurocomputing. **300**, 70–79 (2018)
5. Nembrini, S., König, I.R., Wright, M.N.: The revival of the Gini importance? Bioinformatics. **34**(21), 3711–3718 (2018)
6. Strobl, C., Boulesteix, A.: Bias in random forest variable importance measures: illustrations, sources, and a solution. BMC Bioinf. **8**, 25 (2007)
7. Nicodemus, K.: On the stability and ranking of predictors from random forest variable importance measures. Brief. Bioinform. **12**(4), 369–373 (2011)
8. SAS, SAS Enterprise Miner™ 14.3: Reference Help, SAS Documentation, 30 8 2017. [En línea]. Available: https://documentation.sas.com/doc/en/emref/14.3/p1qzwz7onopjqcn11uc04i18urg7.htm#n1dfqqxdcj2ulzn16l5vt1bum8o1. Último acceso: 7 2 2022
9. SAS, SAS Enterprise Miner 15.1: Ayuda de referencia, SAS Documentation, 18 3 2020. [En línea]. Available: https://documentation.sas.com/doc/en/emref/15.1/n1m7rvh6yyb3mmn0zavezsher4ml.htm#p0bf1cmcy9dg36n141yf71alb1lk. Último acceso: 7 2 2022
10. Kunhn, M., Johnson, K.: Feature Engineering and Selection: A Practical Approach for Predictive Models. CRC Press, Boca Raton (2019)
11. Fraiman, R., Justel, A., Svarc, M.: Selection of variables for cluster analysis and classification rules. J. Am. Stat. Assoc. **103**(483), 1294–1303 (2008)
12. Fowlkes, E., Gnanadesikan, R., Kettenring, J.: Variable selection clustering. J. Classif. **5**, 205–228 (1988)
13. Hastie, T., Friedman, J.H., Tibshirani, R.: The Elements of Statistical Learning, Segunda ed. Springer, New York (2009)
14. Tibshirani, R.: Regression shrinkage and selection via the Lasso. J. R. Stat. Soc. **58**(1), 267–288 (1996)
15. Efron, B., Hastie, T., Johnstone, I., Tibshirani, R.: Least angle regression. Ann. Stat. **32**(2), 407–451 (2004)
16. Wubetie, H.: Application of variable selection and dimension reduction on predictors of MSE's development. J. Big Data. **6**, 6–17 (2019)